Lignocellulose Biodegradation

ACS SYMPOSIUM SERIES **889**

Lignocellulose Biodegradation

Badal C. Saha, Editor
Agricultural Research Service, U.S. Department of Agriculture

Kyoshi Hayashi, Editor
National Food Research Institute

Sponsored by the
ACS Division of Cellulose and Renewable Materials

American Chemical Society, Washington, DC

Library of Congress Cataloging-in-Publication Data

Lignocellulose Biodegradation / Badal C. Saha, editor, Kyoshi Hayashi, editor; sponsored by the ACS Division of Cellulose and Renewable Materials.

p. cm.—(ACS symposium series ; 889)

"Developed from a symposium titled Advances in biodegradation and biotransformation of lignocellulosics, presented at the 225th National Meeting of the American Chemical Society (ACS) in New Orleans, Louisiana, March 23–27, 2003 ..."—Pref.

Includes bibliographical references and index.

ISBN 0–8412–3876–6 (alk. paper)

1. Lignocellulose—Biodegradation.

I. Saha, Badal C., 1949- II. Hayashi, Kyoshi, 1952- III. American Chemical Society. Division of Cellulose and Renewable Materials. IV. American Chemical Society. Meeting (225th : 2003 : New Orleans, La.) V. Series.

TP248.65.L54L535 2004
661′.802—dc22 2004049677

The paper used in this publication meets the minimum requirements of American National Standard for Information Sciences—Permanence of Paper for Printed Library Materials, ANSI Z39.48–1984.

Distributed by Oxford University Press

PRINTED IN THE UNITED STATES OF AMERICA

Foreword

The ACS Symposium Series was first published in 1974 to provide a mechanism for publishing symposia quickly in book form. The purpose of the series is to publish timely, comprehensive books developed from ACS sponsored symposia based on current scientific research. Occasionally, books are developed from symposia sponsored by other organizations when the topic is of keen interest to the chemistry audience.

Before agreeing to publish a book, the proposed table of contents is reviewed for appropriate and comprehensive coverage and for interest to the audience. Some papers may be excluded to better focus the book; others may be added to provide comprehensiveness. When appropriate, overview or introductory chapters are added. Drafts of chapters are peer-reviewed prior to final acceptance or rejection, and manuscripts are prepared in camera-ready format.

As a rule, only original research papers and original review papers are included in the volumes. Verbatim reproductions of previously published papers are not accepted.

ACS Books Department

Contents

Enzyme Characterization

Applications

Indexes

Preface

Advances have been made in lignocellulose biodegradation and applications in the production of fuels and chemicals. It is timely to provide a book that can assist practicing scientists, engineers, and graduate students with effective tools for tackling the longstanding challenges in lignocellulose degradation.

This book was developed from a symposium titled *Advances in Biodegradation and Biotransformation of Lignocellulosics*, presented at the 225th National Meeting of the American Chemical Society (ACS) in New Orleans, Louisiana, March 23–27, 2003 and sponsored by the ACS Division of Cellulose and Renewable Materials. It presents a compilation of twelve symposium manuscripts and nine solicited manuscripts representing recent advances in lignocellulose biodegradation research. The chapters in the book have been organized in four sections: pretreatment, biodegradation, enzyme characterization, and applications. An overview chapter on lignocellulose biodegradation and applications in biotechnology has been included.

We are fortunate to have contributions from world-class researchers in the field of biomass conversion. We are taking this opportunity to express our sincere appreciations to the contributing authors, the reviewers who provided excellent comments to the editors, the ACS Division of Cellulose and Renewable Materials, and the ACS Books Department for making possible the publication of this book.

We hope that this book will actively serve as a valuable multidisciplinary (chemistry, biochemistry, microbiology, molecular biology, and biochemical engineering) contribution to the continually expanding field of liognocellulose biodegradation.

Badal C. Saha
Fermentation Biotechnology Research Unit
National Center for Agricultural Utilization Research
Agricultural Research Service
U.S. Department of Agriculture
1815 North University Street
Peoria, IL 61604
(309) 681–6276 (telephone)
(309) 681–6427 (fax)
sahabc@ncaur.usda.gov (email)

Kiyoshi Hayashi
Planning Section
National Food Research Institute
2–1–12, Kannondai
Tsukuba, Ibaraki 305–8642
Japan
+81–298–38–7991 (telephone)
+81–298–38–8005 (fax)
khayashi@affrc.go.jp (email)

Lignocellulose Biodegradation

Overview

Chapter 1

Lignocellulose Biodegradation and Applications in Biotechnology

Badal C. Saha

Fermentation Biotechnology Research Unit, National Center for Agricultural Utilization Research, Agricultural Research Service, U.S. Department of Agriculture, 1815 North University Street, Peoria, IL 61604

Lignocellulosic biomass such as agricultural and forestry residues and herbaceous energy crops can serve as low cost feedstocks for production of fuel ethanol and other value-added commodity chemicals. However, development of efficient pretreatment and cost-effective enzymatic conversion of any lignocellulosic biomass to fermentable sugars is a key issue. In this overview chapter, various pretreatment options (dilute acid, steam explosion, alkaline peroxide) and enzymes (mainly cellulases and hemicellulases) involved in lignocellulose degradation are presented. Mixed sugars generated by lignocellulose biodegradation are fermented to fuel ethanol, xylitol, 2-3-butanediol and other value-added products. Recent advances in the developments on lignocellulose biodegradation and applications in biotechnology are reviewed.

In 2003, about 2.81 billion gallons of ethanol are produced annually in the United States, with approximately 95% derived from fermentation of corn starch. With increased attention to clean air and oxygenates for fuels, opportunities exist for rapid expansion of the fuel ethanol industry. Various lignocellulosic biomass such as agricultural residues, wood, municipal solid wastes and wastes from pulp and paper industry can serve as low cost and abundant feedstocks for production of fuel ethanol or value-added chemicals. It is estimated that approximately 50 billion gallons of ethanol could be produced from current biomass wastes with the potential to produce up to 350 billion gallons from dedicated energy farms in the USA (*1*). At present, the degradation of lignocellulosic biomass to fermentable sugars represents significant technical and economic challenges, and its success depends largely on the development of highly efficient and cost-effective enzymes for conversion of pretreated lignocellulosic substrates to fermentable sugars. In this overview chapter, the author reviews the current knowledge on lignocellulose biodegradation and use of lignocellulosic hydrolyzates as feedstocks for developing bio-based products and processes.

Structure and Composition of Lignocellulosic Biomass

Lignocellulosic biomass includes various agricultural residues (straws, hulls, stems, stalks), deciduous and coniferous woods, municipal solid wastes (MSW, paper, cardboard, yard trash, wood products), waste from pulp and paper industry and herbaceous energy crops (switchgrass, barmudagrass). The compositions of these materials vary. The major component is cellulose (35-50%), followed by hemicellulose (20-35%) and lignin (10-25%). Proteins, oils and ash make up the remaining fraction of lignocellulosic biomass (*1*). The structures of these materials are complex with recalcitrant and heterogeneous characteristics and native lignocellulose is resistant to an enzymatic hydrolysis. In the current model of the structure of lignocellulose, cellulose fibers are embedded in a lignin-polysaccharide matrix. Xylan may play a significant role in the structural integrity of cell walls by both covalent and non-covalent associations (*2*).

Cellulose is a linear polymer of D-glucose units linked by 1,4-ß-D-glucosidic bonds. Hemicelluloses are heterogeneous polymers of pentoses (xylose, arabinose), hexoses (mannose, glucose, galactose), and sugar acids. Unlike cellulose, hemicelluloses are not chemically homogeneous. Hardwood hemicelluloses contain mostly xylans, whereas softwood hemicelluloses contain mostly glucomannans (*3*). Xylans of many plant materials are heteropolysaccharides with homopolymeric backbone chains of 1,4-linked β-D-xylopyranose units. Besides xylose, xylans may contain arabinose, glucuronic acid or its 4-*O*-methyl ether, and acetic, ferulic and *p*-coumaric acids. The frequency and composition of branches are dependent on the source of xylan (*4*). The backbone consists of *O*-acetyl, α-L-arabinofuranosyl, α-1,2-linked glucuronic or 4-*O*-methylglucuronic acid substituents. However, unsubstituted linear xylans have also been isolated from guar seed husk, esparto grass and tobacco stalks (*5*).

Xylans can thus be categorized as linear homoxylan, arabinoxylan, glucuronoxylan and glucuronoarabinoxylan.

Xylans from different sources, such as grasses, cereals, softwood and hardwood, differ in composition. Birch wood (Roth) xylan contains 89.3 % xylose, 1% arabinose, 1.4% glucose and 8.3% anhydrouronic acid (*6*). Rice bran neutral xylan contains 46% xylose, 44.9% arabinose, 6.1% galactose, 1.9% glucose and 1.1% anhydrouronic acid (*7*). Wheat arabinoxylan contains 65.8% xylose, 33.5% arabinose, 0.1% mannose, 0.1% galactose and 0.3% glucose (*8*). Corn fiber xylan is one of the complex heteroxylans containing β-(1,4)-linked xylose residues (*9*). It contains 48-54% xylose, 33-35% arabinose, 5-11% galactose and 3-6% glucuronic acid (*10*). About 80% of the xylan backbone is highly substituted with monomeric side-chains of arabinose or glucuronic acid linked to *O*-2 and/or *O*-3 of xylose residues and also by oligomeric side chains containing arabinose, xylose and sometimes galactose residues (*11*). The heteroxylans, which are highly cross-linked by diferulic bridges, constitute a network in which the cellulose microfibrils may be imbedded (*12*). Structural wall proteins might be cross-linked together by isodityrosine bridges and with feruloylated heteroxylans, thus forming an insoluble network (*13*). Ferulic acid is covalently cross-linked to polysaccharides by ester bonds and to components of lignin mainly by ether bonds (*14*). In softwood heteroxylans, arabinofuranosyl residues are esterified with *p*-coumaric acids and ferulic acids (*15*). In hardwood xylans, 60-70% of the xylose residues are acetylated (*16*). The degree of polymerization of hardwood xylans (150-200) is higher than that of softwoods (70-130).

Pretreatment of Lignocellulosic Biomass

The pretreatment of any lignocellulosic biomass is crucial before enzymatic hydrolysis. The objective of pretreatment is to decrease the crystallinity of cellulose which enhances the hydrolysis of cellulose by cellulases (*17*). Various pretreatment options are available to fractionate, solubilize, hydrolyze and separate cellulose, hemicellulose and lignin components (*1, 18-20*). These include concentrated acid (*21*), dilute acid (*22*), SO_2 (*23*), alkali (*24, 25*), hydrogen peroxide (*26*), wet-oxidation (*27*), steam explosion (autohydrolysis) (*28*), ammonia fiber explosion (AFEX) (29), CO_2 explosion (*30*), liquid hot water (*31*) and organic solvent treatments (*32*). In each option, the biomass is reduced in size and its physical structure is opened. Some methods of pretreatment of Lignocellulose is given in Table I.

The effectiveness of dilute acids to catalyze the hydrolysis of hemicellulose to its sugar components is well known . Two categories of dilute acid pretreatment are used: High temperature (> 160°C) continuous-flow for low solids loading (5-10%, w/w) and low temperature (< 160°C) batch process for high solids loading (10-40%, w/w) (*33*). Dilute acid pretreatment at high temperature usually hydrolyzes hemicellulose to its sugars (xylose, arabinose and other sugars) that are water

Table I. Methods for pretreatment of lignocellulosic biomass

Method	*Example*
Autohydrolysis	Liquid hot water, steam pressure, steam explosion, supercritical CO_2explosion
Acid treatment	Dilute acid (H_2SO_4), Concentrated acid (H_2SO_4)
Alkali treatment	Sodium hydroxide, lime, ammonia, alkaline hydrogen peroxide
Organic solvent with water	Methanol, ethanol, butanol, phenol

soluble (*18*). The residue contains cellulose and often much of the lignin. The lignin can be extracted with solvents such as ethanol, butanol, or formic acid. Alternatively, hydrolysis of cellulose with lignin present produces water-soluble sugars and the insoluble residues that are lignin plus unreacted materials. Torget et al. (*34*) achieved both high xylan recovery and high simultaneous saccahrification and fermentation (SSF) conversion while applying extremely dilute H_2SO_4 (0.07 wt %) in a counter-current flowthrough configuration. A major problem associated with the dilute acid hydrolysis of lignocellulosic biomass is the poor fermentability of the hydrolyzates. A drawback of the concentrated acid process is the costly recovery of the acid.

Steam explosion provides effective fractionation of lignocellulosic components at relatively low costs (*35*). Optimal solubilization and degradation of hemicellulose are generally achieved by either high temperature and short residence time (270°C, 1 min) or lower temperature and longer residence time (190°C, 10 min) steam explosion (*36*). The use of SO_2 as a catalyst during steam pretreatment results in the enzymatic accessibility of cellulose and enhanced recovery of the hemicellulose derived sugars (*37*). Steam pretreatment at 200-210°C with the addition of 1% SO_2 (w/w) was superior to other forms of pretreatment of willow (*38*). A glucose yield of 95%, based on the glycan available in the raw material, was achieved. Steam explosion can induce hemicellulose degradation to furfural and its derivatives and modification of the lignin-related chemicals under high severity treatment (> 200°C, 3-5 min, 2-3% SO_2) (*39*). Boussaid et al. (*40*) recovered around 87% of the original hemicellulose component in the water-soluble stream by steam explosion of Douglas fir softwood under low severity conditions (175°C, 7.5 min, 4.5% SO_2). More than 80% of the recovered hemicellulose was in monomeric form. Enzymatic digestibility of the steam-exploded Douglas-fir wood chips (105°C, 4.5 min, 4.5% SO_2) was significantly improved using an optimized alkaline peroxide treatment (1% H_2O_2, pH 11.5 and 80°C, 45 min) (*41*). About 90% of the lignin in the original wood was solubilized by this procedure, leaving a cellulose-rich residue that was completely hydrolyzed within 48 h, using an enzyme (cellulase) loading of 10FPU/g cellulose. Saccharification of 100 g sugarcane bagasse with enzymes after steam explosion with 1% H_2SO_4 at 220°C for 30 sec at water to solid ratio of 2:1 yielded 65.1g sugar (*42*).

A pretreatment method involves steeping of the lignocellulosic biomass (using corn cob as a model feedstock) in dilute NH_4OH at ambient temperature to remove lignin, acetate and extractives (*43*). This is followed by dilute acid treatment that readily hydrolyzes the hemicellulose fraction to simple sugars, primarily xylose. The residual cellulose fraction of biomass can then be enzymatically hydrolyzed to glucose. Sugarcane bagasse, corn husk and switchgrass were pretreated with ammonia water to enhance enzymatic hydrolysis (*44*). Garrote et al. (*45*) treated *Eucalyptus* wood substrates with water under selected operational conditions (autohydrolysis reaction) to obtain a liquid phase containing hemicellulose decomposition products (mainly acetylated xylooligosaccharides, xylose and acetic acid). In a further acid catalyzed step (posthydrolysis reaction), xylooligosaccharides were converted into xylose. Wet oxidation method can be used for fractionation of lignocellulosics into solubilized hemicellulose fraction and a solid cellulose fraction susceptible to enzymatic saccharification. Bjerre et al. (*46*) found that combination of alkali and wet oxidation did not generate furfural and 5-hydroxymethyl furfural (HMF). Klinke et al. (*47*) characterized the degradation products from alkaline wet oxidation (water, sodium carbonate, oxygen, high temperature and pressure) of wheat straw. Apart from CO_2 and water, carboxylic acids were the main degradation products from hemicellulose and lignin. Aromatic aldehyde formation was minimized by the addition of alkali and temperature control. Oxygen delignification of kraft pulp removed up to 67% of the lignin from softwood pulp and improved the rate and yield from enzymatic hydrolysis by up to 111% and 174%, respectively (*48*). Palm and Zacchi (*49*) extracted 12.5 g of hemicellulose oligosaccharides from 100 g of dry spruce using a microwave oven at 200°C for 5 min.

Supercritical CO_2 explosion was found to be effective for pretreatment of cellulosic materials before enzymatic hydrolysis (*50, 51*). Zheng et al. (*52*) compared CO_2 explosion with steam and ammonia explosion for pretreatment of sugarcane bagasse and found that CO_2 explosion was more cost-effective than ammonia explosion and did not cause the formation of inhibitory compounds that could occur in steam explosion.

Phenolic compounds from lignin degradation, furan derivatives (furfural and HMF) from sugar degradation and aliphalic acids (acetic acid, formic acid and levulinic acid) are considered to be fermentation inhibitors generated from pretreated lignocellulosic biomass (*53*). The formation of these inhibitors depends on the process conditions and the lignocellulosic feedstocks (*54*). Various methods for detoxification of the hydrolyzates have been developed (*55*). These include treatment with ion-exchange resins, charcoal or ligninolytic enzyme laccase, pre-fermentation with the filamentous fungus *Trichoderma reesei*, removal of non-volatile compounds, extraction with ether or ethyl acetate and treatment with alkali (lime) or sulfite. Treatment with alkali (overliming) has been widely used for detoxification of lignocellulosic hydrolyzates prior to alcohol fermentation. However, overliming is a costly method which also produces low-value byproducts such as gypsum (*56*). Softwood hydrolyzate, when overlimed with wood ash, improved its fermentability to ethanol which is due to the reduction of the inhibitors such as furan and phenolic compounds and to nutrient effects of some inorganic components from the wood ash on the fermentation (57). Persson et al. (*58*) employed countercurrent flow supercritical fluid extraction to detoxify a dilute acid hydrolyzate of spruce prior to ethanol fermentation with baker's yeast. Weil et al. (*59*) developed a method for the removal of furfural from biomass hydrolyzate by using a polymeric adsorbent, XAD-4, and desorption of the furfural to regenerate the adsorbent using ethanol. Bjorklund et al. (*60*) explored the possibility of using lignin residue left after acid hydrolysis of lignocellulosic material for detoxification of spruce dilute acid hydrolyzates prior to fermentation with *Saccharomyces cerevisiae*. Treatment with the lignin residue removed up to 53% of the phenolic compounds and up to 68% of the furan aldehydes in a spruce dilute acid hydrolyzate. Up to 84% of the lignin-derived compounds can be extracted with organic solvents (ethyl acetate and diethyl ether) from *Eucalyptus* wood acid hydrolyzate (*61*). The phenolic compounds extracted by solvents showed antioxidant activity.

Each pretreatment method offers distinct advantages and disadvantages. The pretreatment of lignocellulosic biomass is an expensive procedure with respect to cost and energy.

Cellulose Biodegradation

Effective hydrolysis of cellulose to glucose requires the cooperative action of three enzymes: endo-1, 4-ß-glucanase (EC 3.2.1.4), exo-1, 4-ß-glucanase (EC 3.2.1.91) and ß-glucosidase (EC 3.2.1.21). Cellulolytic enzymes with ß-glucosidase act sequentially and cooperatively to degrade crystalline cellulose to glucose. Endoglucanase acts in a random fashion on the regions of low crystallinity of the cellulosic fiber whereas exoglucanase removes cellobiose (ß-1, 4 glucose dimer) units from the non-reducing ends of cellulose chains. Synergism between these two enzymes is attributed to the *endo-exo* form of cooperativity and has been studied extensively between cellulases in *T. reesei* in the degradation of cellulose (*62*). Besides synergism, the adsorption of the cellulases on the insoluble substrates is a necessary step prior to hydrolysis. Cellobiohydrolase appears to be the key enzyme for the degradation of native cellulose (*63*). The catalytic site of the enzyme is covered by long loops, resulting in tunnel morphology (*64*). The loops can undergo large movements, leading to the opening or closing of the tunnel roof (*65*). An endo type attack of the polymeric substrates becomes possible when the roof is open and once entrapped inside the catalytic tunnel, a cellulose chain is threaded through the tunnel and sequentially hydrolyzed one cellobiosyl unit at a time. Kleywegt et al. (*66*) revealed the presence of shorter loops that create a groove rather than a tunnel in the structure of the enzyme EGI from *T. reesei*. In most organisms, cellulases are modular enzymes that consist of a catalytic core connected to a cellulose-binding domain (CBD) through a flexible and heavily glycosylated linker region (*67*). The CBD is responsible for bringing the catalytic domain in an appropriate position for the breakdown of cellulose. Binding of cellulases and the formation of cellulose-cellulase complexes are considered critical steps in the hydrolysis of insoluble cellulose (*68*). ß-Glucosidase hydrolyzes cellobiose and in some cases cellooligosaccharides to glucose. The enzyme is generally responsible for the regulation of the whole cellulolytic process and is a rate limiting factor during enzymatic hydrolysis of cellulose as both endoglucanase and cellobiohydrolase activities are often inhibited by cellobiose (*69-71*). Thus, β-glucosidase not only produces glucose from cellobiose but also reduces cellobiose inhibition, allowing the cellulolytic enzymes to function more efficiently. However, like ß-glucanases, most ß-glucosidases are subject to end-product (glucose) inhibition (*72*). *C. peltata* produces a highly glucose tolerant β-glucosidase with a K_i value of 1.4 M (252 mg/ml) for glucose (*73*). The kinetics of the enzymatic hydrolysis of cellulose including adsorption, inactivation and inhibition of enzymes have been studied extensively (*74*). For a complete hydrolysis of cellulose to glucose, the enzyme system must contain the three enzymes in right proportions. *T. reesei* (initially called *T. viride*) produces at least five endoglucanases (EGI, EGII, EGIII, EGIV and EGV), two exoglucanases (CBHI and CBHII) and two β-glucosidases (BGLI and BGLII) (*75*). An exo-exo synergism between the two cellobiohydrolases was also observed (*76*). The fungus produces up to 0.33 g protein per g of utilizable carbohydrate (*77*).

Product inhibition, thermal inactivation, substrate inhibition, low product yield and high cost of cellulase are some barriers to commercial development of the enzymatic hydrolysis of cellulose. Many microorganisms are cellulolytic. However, only two microorganisms (*Trichoderma* and *Aspergillus*) have been studied extensively for cellulase. A newly isolated *Mucor circinelloides* strain produces a complete cellulase enzyme system (*78*). The endoglucanase from this strain was found to have a wide pH stability and activity. There is an increasing demand for the development of thermostable, environmentally compatible, product and substrate tolerant cellulases with increased specificity and activity for application in the conversion of cellulose to glucose in the fuel ethanol industry. Thermostable cellulases offer certain advantages such as higher reaction rate, increased product formation, less microbial contamination, longer shelf-life, easier purification and better yield.

The cellulose hydrolysis step is a significant component of the total production cost of ethanol from wood (*79*). Achieving a high glucose yield is necessary (>85% theoretical) at high substrate loading (>10% w/v) over short residence times (<4 days). Simultaneous saccharification (hydrolysis) of cellulose to glucose and fermentation of glucose to ethanol (SSF) improve the kinetics and economics of biomass conversion by reducing accumulation of hydrolysis products that are inhibitory to cellulase and β-glucosidase, reducing the contamination risk because of the presence of ethanol, and reducing the capital equipment requirements (*80*). An important drawback of SSF is that the reaction has to operate at a compromised temperature of around 30°C instead of enzyme optimum temperature of 45-50°C. Enzyme recycling, by ultrafiltration of the hydrolyzate, can reduce the net enzyme requirement and thus lower costs (*81*). A preliminary estimate of the cost of ethanol production for SSF technology based on wood-to-ethanol process is $1.22/gal of which the wood cost is $0.459/gal (*82*). A separate fungal enzyme hydrolysis and fermentation process for converting lignocellulose to ethanol were also evaluated (*83*). The cellulase enzyme was produced by the fungal mutant *Trichoderma* Rut C-30 (the first mutant with greatly increased β-glucosidase activity) in a fed batch production system that is the single most expensive operation in the process. The conversion of lignocellulosic biomass to fermentable sugars requires the addition of complex enzyme mixtures tailored for the process and parallel reuse and recycle the enzymes until the cost of enzymes comes down. Enzyme recycling may increase the rates and yields of hydrolysis, reduce the net enzyme requirements and thus lower costs (*84*). As mentioned earlier, the first step in cellulose hydrolysis is considered as the adsorption of cellulase onto cellulosic substrate. As the cellulose hydrolysis proceeds, the adsorbed enzymes (endo- and exo-glucanase components) are gradually released in the reaction mixture. The β-glucosidase does not adsorb onto the substrate. These enzymes can be recovered and reused by contacting the hydrolyzate with the fresh substrate. However, the amount of enzyme recovered is limited because some enzymes remain attached to the residual substrate, and some enzymes are thermally inactivated during hydrolysis. Poor recovery of cellulase was achieved in the case of substrates containing a high proportion of lignin (*85*). Addition of surfactant to enzymatic hydrolysis of lignocellulose increases the conversion of cellulose to soluble sugars. Castanon and Wilke (*86*) reported a 14% increase in glucose yield and more than twice as much recovered enzyme from newspaper saccharification when Tween 80 was added. Karr and Holtzapple (*87*) studied the effect of Tween on the enzymatic hydrolysis of lime pretreated corn stover and concluded that Tween improves corn stover hydrolysis through three effects: enzyme stabilizer, lignocellulose disrupter and enzyme effector. The enhancement is due to reduction of the unproductive enzyme adsorption to the lignin part of the substrate as a result of hydrophobic interaction of surfactant with lignin on the lignocellulose surface, which releases nonspecifically bound enzyme (*88*).

Cellolignin is an industrial residue obtained during the production of furfural from wood and corn cobs when pretreated by dilute H_2SO_4 at elevated temperature. It was completely converted to glucose by cellulase from *T. viride* and *A. foetidus* (*89*). The concentration of glucose in the hydrolyzate reached 4-5.5% with about 80% cellulose conversion. Kinetic analysis of cellolignin hydrolysis, using a mathematical model of the process, has shown that, with product inhibition, nonspecific adsorption of cellulase onto lignin and substrate induced inactivation seem to affect negatively the hydrolysis efficiency. Borchert and Buchholz (*90*) investigated the enzymatic hydrolysis of different cellulosic materials (straw, potato pulp, sugar beet pulp) with respect to reactor design. The kinetics were studied including enzyme adsorption, inhibition and inactivation. The results suggest the use of reactors with plug flow characteristics to achieve high substrate and product concentrations and to avoid back-mixing to limit the effect of product inhibition. For efficient use of cellulases, a reactor with semipermeable hollow fiber or an ultrafilter membrane was used, and this allowed cellulases to escape end-product inhibition (*91-94*). A totally integrated biotechnology of rice straw conversion into ethanol was reported (*95*). It dealt with (a) ethanol refining of rice straw to segregate cellulose from pentose sugars and lignin, (b) preparation of highly active mixed cellulase enzymes, (c) a novel reactor system allowing rapid product formation involving enzymatic hydrolysis of cellulose to sugars followed by microbial conversion of the later into ethanol and its simultaneous flash separation employing a programmed recompression of ethanol vapors and condensation and (d) concentration of ethanol via alternative approaches. Use of cellulase enzymes improves ink detachment from old newspapers giving similar or better results in place of classical chemicals (*96*).

Hemicellulose Biodegradation

Hemicellulases are either glycosyl hydrolases or carbohydrate esterases. The total biodegradation of xylan requires endo-β-1,4-xylanase (EC 3.2.1.8), β-xylosidase (EC 3.2.1.37) and several accessory enzymes, such as α-L-arabinofuranosidase (EC 3.2.1.55), α-glucuronidase (EC 3.2.1.131), acetylxylan esterase (EC 3.1.1.72), ferulic acid esterase (EC 3.1.1.73) and *p*-coumaric acid esterase, which are necessary for hydrolyzing various substituted xylans (*97*). The endo-xylanase attacks the main chains of xylans and β-xylosidase hydrolyzes xylooligosaccharides to xylose. The α-arabinofuranosidase and α-glucuronidase remove the arabinose and 4-*O*-methyl glucuronic acid substituents, respectively, from the xylan backbone. The esterases hydrolyze the ester linkages between xylose units of the xylan and acetic acid (acetylxylan esterase) or between arabinose side chain residues and phenolic acids, such as ferulic acid (ferulic acid esterase) and *p*-coumaric acid (*p*-coumaric acid esterase). It is stated that hindrance of lignocellulose biodegradation is associated with phenolic compounds (*98*).The phenolic acids are produced via the phenylpropanoid biosynthetic pathway (*99*). They act as a cross-linking agent between lignin and carbohydrates or between carbohydrates. β-Mannanase (EC 3.2.1.78) hydrolyzes mannan-based hemicellulases and liberate β-1,4-manno-oligomers, which can be further degraded to mannose by β-mannosidase (EC 3.2.1.25).

Many microorganisms, such as *Penicillium capsulatum* and *Talaromyces emersonii*, possess complete xylan degrading enzyme systems (*100*). Significant synergistic interactions were observed among endo-xylanase, β-xylosidase, α-arabinofuranosidase and acetylxylan esterase of the thermophilic actinomycete *Thermomonospora fusca* (*101*). Synergistic action between depolymerizing and side-group cleaving enzymes has been verified using acetylated xylan as a substrate (*102*). Many xylanases do not cleave glycosidic bonds between xylose units which are substituted. The side chains must be cleaved before the xylan backbone can be completely hydrolyzed (*103*). On the other hand, several accessory enzymes only remove side chains from xylooligosaccharides. These enzymes require a partial hydrolysis of xylan before the side chains can be cleaved (*104*). Although the structure of xylan is more complex than cellulose and requires several different enzymes with different specificities for complete hydrolysis, the polysaccharide does not form tightly packed crystalline structures like cellulose and is, thus, more accessible to enzymatic hydrolysis (*105*).

Corn fiber, a byproduct of corn wet milling facility, contains about 20% starch in addition to 15% cellulose and 35% hemicellulose (*106*). The xylan from corn fiber is highly resistant to enzymatic degradation by commercially available hemicellulases (*22*). Dilute acid (1% H_2SO_4 v/v, 15% solids) pretreatment at a relatively low temperature (120°C, 1 h) to minimize the formation of inhibitory compounds, followed by enzymatic saccharification of the cellulosic portion, is an excellent workable process for generating fermentable sugars (85-100% yield) from corn fiber (*22*). A partial saccharification of corn fiber was achieved using a crude enzyme preparation from *Aureobasidium* sp. (*107*). Christov et al. (*108*) showed that crude enzyme preparation from *A. pullulans* was only partially effective in the removal of xylan from dissolving pulp. Two newly isolated fungal cultures (*Fusarium proliferatum* NRRL 26517, *F. verticillioides* NRRL Y-26518) have the capability to utilize corn fiber xylan as growth substrate (*109-112*). The crude enzyme preparations from these fungi were able to degrade corn fiber xylan well but the purified endo-xylanases could not degrade corn fiber xylan. The purified β-xylosidase released xylose from xylobiose and other short-chain xylooligosaccharides. For effective hydrolysis of xylan substrates, a proper mix of endo-xylanase with several accessory enzymes is essential. *A. pullulans* produces a highly thermostable novel extracellular α-L-arabinofuranosidase that has the ability to rapidly hydrolyze arabinan and debranched arabinan and release arabinose from various arabinoxylans (*113*). Arabinose-rich lignocellulosic hydrolyzates can be used for production of the enzyme (*114*). Spagnuolo et al. (*115*) reported that incubation of beet pulp with α-L-arabinofuranosidase and end-arabinase produced a hydrolyzate consisting mainly arabinose.

Ferulic acid esterase breaks the ester linkage between ferulic acid and the attached sugar and release ferulic acid from complex cell walls such as wheat bran, sugar beet pulp, barley spent grain and oat hull (*116-119*). The ability of *Thermomyces lanuginosus* to produce high levels of cellulase-free thermostable xylanase has made the fungus an attractive source of the enzyme with potential as a bleach-boosting agent in the pulp and paper industry and as an additive in the baking industry (*120*).

Lignin Biodegradation

Lignin is a long chain heterogeneous aromatic polymer (average molecular weight 8,000-11,000) composed largely of phenylpropane units most commonly linked by ether bonds. It effectively protects the woody plants against microbial attack and only a few organisms including rot-fungi and some bacteria can degrade it (*121*). The conversion of cellulose and hemicellulose to fuels and chemicals will generate lignin as a by-product that can be burned to provide heat and electricity, converted to low-molecular weight chemicals and used in the manufacture of various polymeric materials. As lignin makes up 15-25% in some lignocellulosic biomass, the selling price of lignin has a very large impact on ethanol price (*35*).

Efficient removal of lignin from lignin-carbohydrate complex (LCC) is important in pulp and paper industry. The lignin barrier can be disrupted by a variety of pretreatment rendering the cellulose and hemicellulose more susceptible to enzymatic attack (*122*). There are many papers about microbial breakdowns of lignin, the enzymes and the pathways (*123-126*). Several white rot fungi have the ability to delignify kraft pulp. Their lignin-degrading capacity is attributed to extracellular oxidative enzymes that function together with low molecular weight cofactors (*127*). The degradation of lignin by the white rot fungus (WRF) *Phanerochaete chrysosporium* is catalyzed by extracellular peroxidases (lignin peroxidase, LiP, EC 1.11.1.14 and manganese peroxidase, MnP, EC 1.11.1.13) in a H_2O_2-dependent process (*128, 129*). LiP seems to use veratryl cation radical as mediator (*130*). The WRF *Ceriporiopsis subvermispora* produces laccase (EC 1.10.3.2) and MnP isozymes, as well as hemicellulases and a poor complex of cellulases lacking in cellobiohydrolase activity (*131*). Laccase is a family of 'blue-copper' oxidases containing four copper ions. It oxidizes the phenolic but not the non-phenolic subunits of lignin. Redox mediators drive laccase towards the oxidation of non-phenolic subunits, particularly the benzyl alcohol groups. Each laccase may have a preferred low molecular mass mediator substrate, which may represent a major secreted metabolite (*130*). A laccase and mediators with NO, NOH or HRNOH groups can be combined in a laccase-mediated system (Lignozyme process) that are effective in delignifying wood in a pilot pulp and paper process. A pre-oxidation of the α-hydroxy-β-arylether subunits in wood pulp by the laccase/violuric acid system appears to be promising for weakening the network of lignin, thereby activating it towards subsequent oxydelignification treatments (*132*). It was demonstrated that the WRF *Pycnoporus cinnabarinus* degrades lignin in the absence of both Lip and MnP (*133*). The fungus was found to produce predominantly laccase, and neither LiP nor MnP was produced. The factors involved in lignin biodegradation process are not yet fully understood (*134*).

The biodegradation of lignocellulose by *Rigidoporus lignosus* was greatly stimulated in solid state cultivation when compared with liquid culture (*135*). This WRF produces laccase and MnP. The edible mushroom *Pleurotus ostreatus* degrades lignin efficiently and selectively (*136*). It can, therefore, serve to upgrade lignocellulosic wastes. Lignin degradation by *Agaricus bisporus* accounts for a 30% increase in bioavailable holocellulose during cultivation on compost (*137*). Table II lists some of the enzymes involved in lignocellulose degradation.

Direct Microbial Conversion

In direct microbial conversion of lignocellulosic biomass into ethanol that could simplify the ethanol production process from these materials and reduce ethanol production costs, *Clostridium thermocellum,* a thermoanaerobe was used for enzyme production, hydrolysis and glucose fermentation (*138*). Cofermentation with *C. thermosaccharolyticum* simultaneously converted the hemicellulosic sugars to ethanol. However, the formations of by-products such as acetic acid and low ethanol tolerance are some drawbacks of the process. *Neurospora crassa* produces extracellular cellulase and xylanase and has the ability to ferment cellulose to ethanol (*139*).

In nature, cellulosic materials are degraded with the cooperation of many microorganisms. A mixed culture of one cellulolytic bacterium together with another non-cellulolytic bacterium was found to be effective for cellulose degradation (*140, 141*). Recently, Haruta et al. (*142*) obtained a microbial community from rice straw compost that had the capability of degrading 60% of rice straw within 4 days at 50°C. The community structure consisting of both aerobic and anaerobic bacteria remained constant after multiple subcultures exceeding 2 years.

Table II. Enzymes involved in lignocellulose degradation

Enzyme	Systematic name	EC number	Mode of action
Endo-1,4-β-glucanase	1,4-β-D-Glucan-4-glucanohydrolase	3.2.1.4	Endo-hydrolysis of 1,4-β-D-glucosidic linkages
Exo-1,4-β-glucanase	1,4-β-D-Glucan cellobiohydrolase	3.2.1.91	Hydrolysis of 1,4-β-D-glucosidic linkages releasing cellobiose
β-Glucosidase	β-D-Glucoside glucohydrolase	3.2.1.21	Hydrolyzes cellobiose and short chain cello-oligosaccharides to glucose
Endo-1,4-β- xylanase	1,4-β-D-Xylan xylanohydrolase	3.2.1.8	Hydrolyzes mainly interior β-1,4-xylose linkages of the xylan backbone
α-L-Arabinofuranosidase	α-L-Arabinofuranoside arabinofurano-hydrolase	3.2.1.55	Hydrolyzes terminal nonreducing α-arabinofuranose fromarabinoxylans
α-Glucuronidase	α-Glucuronoside glucanohydrolase	3.2.1.31	Releases glucuronic acid from glucuronoxylans
Acetylxylan esterase	Acetyl-ester acetylhydrolase	3.1.1.6	Hydrolyzes acetylester bonds in acetyl xylans
Ferulic acid esterase	Carboxylic ester hydrolase	3.1.1.1	Hydrolyzes feruloylester bonds in xylans
Lignin peroxidase		1.11.1.7	Oxidation of benzilic alcohols, cleavage of C-C bonds, cleavage of C-O bonds.
Manganese peroxidase		1.11.1.7	Catalytically dependent on H_2O_2 and Mn^{2+} ions
Laccase	Donar: hydrogen peroxide oxidoreductase	1.10.3.2	Oxidizes phenolic subunits of lignin

Applications in Biotechnology

Production of Fuel Ethanol

Lignocellulosic biomass can serve as low-cost feedstocks for production of fuel ethanol. It generates a mixture of sugars upon pretreatment itself or in combination with enzymatic hydrolysis. The sugar mixture may contain any combination of xylose, arabinose, glucose, galactose, mannose, fucose and rhamnose depending on the source. Although traditional *S. cerevisiae* and *Zymomonas mobilis* ferment glucose to ethanol rapidly and efficiently, they cannot ferment other sugars such as xylose and arabinose to ethanol. Some yeasts (*Pachysolen tannophilus*, *Pichia stipitis*, *Candida shehatae)* have the capability to ferment xylose to ethanol (*143, 144*). These yeasts have low ethanol tolerance and slow rates of fermentation and cannot be used in industrial application (*145, 146*). Xylose can be converted to xylulose using the enzyme xylose isomerase and traditional yeasts can ferment xylulose to ethanol *(147, 148*). However, the process is not cost-effective. Only a few yeast strains can hardly ferment arabinose to ethanol (*149, 150*). Thus, no naturally occurring yeast and bacterium can ferment mixed sugars to ethanol.

Some bacteria such as *Escherichia coli*, *Klebsiella* , *Erwinia*, *Lactobacillus*, *Bacillus* and *Clostridia* can utilize mixed sugars but produce no or limited quantity of ethanol. These bacteria generally produce mixed acids (acetate, lactate, propionate, succinate) and solvents (acetone, butanol, 2,3-butanediol). Several microorganisms have been genetically engineered to produce ethanol from mixed sugar substrates by using two different approaches: (a) divert carbon flow from native fermentation products to ethanol in efficient mixed sugar utilizers such as *Escherichia*, *Erwinia* and *Klebsiella* and (b) introduce the pentose utilizing capability in the efficient ethanol producers such as *Saccharomyces* and *Zymomobilis* (*151-154*). Recombinant *E. coli* K011, *E. coli* SL40, *E. coli* FBR3, *Zymomonas* CP4 (pZB5) and *Saccharomyces* 1400 (pLNH32) strains fermented corn fiber hydrolyzates to ethanol (21-34 g/L) with yields of 0.41-0.50 g of ethanol per gram of sugar consumed (*155-156)*. Increasing gene expression through the replacement of promoters and the use of a higher gene dosage (plasmids) substantially eliminated the apparent requirement for large amounts of complex nutrients of ethanologenic recombinant *E. coli* strain (*157*). Ethanol tolerant mutants of recombinant *E. coli* have been developed that can produce up to 6% ethanol (*158*). The increased ethanol tolerance in the *E. coli* mutant LY01 appears to result from increased glycine metabolism, increased production of the osmolyte betaine from choline, loss of FNR function, increased production of *mar* drug resistance proteins, increased metabolism of serine and pyruvate and decreased production of organic acids (*159*). The recombinant *Z. mobilis* in which four genes from *E. coli*, *xylA* (xylose isomerase), *xylB* (xylulokinase), *tal* (transaldolase) and *tktA* (transketolase) were inserted, grew on xylose as the sole carbon source and produced ethanol at 86% of the theoretical yield (Figure 1) (*153*). It was demonstrated that phosphorylation is a vital step for metabolism of xylose through the pentose phosphate pathway (*160*). The gene *XKS1* (encoding xylulokinase) from *S. cerevisiae* and the heterologous genes from *XYL1* and *XYL2* (from *P. stipitis*) were inserted into a hybrid host, obtained by classical breeding of *S. uvarum* and S. *diastaticus*, which resulted in *Saccharomyces* strain pLNH32, capable of growing on xylose alone. Chromosomal integration of a single copy of the *XYL1-XYL2-XYLS1* cassettee in *S. Cerevisiae* resulted in strain TMB3001 (*161*). This strain attained specific uptake rates (g/g.h) of 0.47 and 0.21 for glucose and xylose, respectively, in continuous culture using a minimal medium. Martin et al. (*162*) studied ethanol production from steam exploded (205 and 215°C, 10 min) enzymatic (cellulase, β-glucosidase) hydrolyzates of sugar cane bagasse using the recombinant *S. cerevisiae* strain TMB 3001. The hydrolyzates were detoxified by treatment with laccase and also by overliming. The ethanol yield was 0.32-0.35 g/g of total sugar from the detoxified hydrolyzates. Partial xylose utilization with low xylitol formation was observed.

Sedlak and Ho (*163*) expressed genes [*arab* (L-ribulokinase), *araA* (L-arabinose isomerase) and *araD* (L-ribulose-5-phosphate)] from the *araBAD* operon encoding the arabinose metabolizing genes from *E. coli* in *S. cerevisiae* but the transformed strain was not able to produce any detectable amount of ethanol from arabinose. Zhang et al. (*164*) constructed one strain of *Z. mobilis* (PZB301) with seven plasmid borne genes encoding xylose- and arabinose metabolizing genes and pentose phosphate pathway (PPP) genes. This recombinant strain was capable of fermenting both xylose and arabinose in a mixture of sugars with 82-84% theoretical yield in 80-100 h at 30°C. Richard et al. (*165*) reported that overexpression of all five enzymes (aldose reductase, L-arabinitol 4-dehydrogenase, L-xylulose reductase, xylitol dehydrogenase and xylulokinase) of the arabinose catabolic pathway in *S. cerevisiae* led to growth of *S. cerevisiae* on arabinose (Figure 2). The recombinant *S. cerevisiae* produced ethanol from arabinose at a very slow rate.

Microorganisms metabolically engineered with improved inhibitor tolerance could reduce the need for detoxification process. Larsson et al. (*166*) developed a *S. cerevisiae* strain with enhanced resistance to phenolic fermentation inhibitors in lignocellulose hydrolyzates by heterologous expression of laccase.

Three technology options (concentrated acid process, CHAP; SO_2/dilute acid process, CASH; enzymatic hydrolysis process) were evaluated for ethanol production from pine feedstock using a uniform platform (*167*). Even though each process suffers from certain different disadvantages, none of the three processes can be eliminated as less economical than the other two with an ethanol production price of US $1.89-2.04/gallon. A feasibility study of using softwood forest thinning as a biomass source for ethanol production in California was performed (*168*). A two-stage dilute acid (190°C, 0.7% H_2SO_4, 3 min; 220°C, 1.6% H_2SO_4, 3 min)

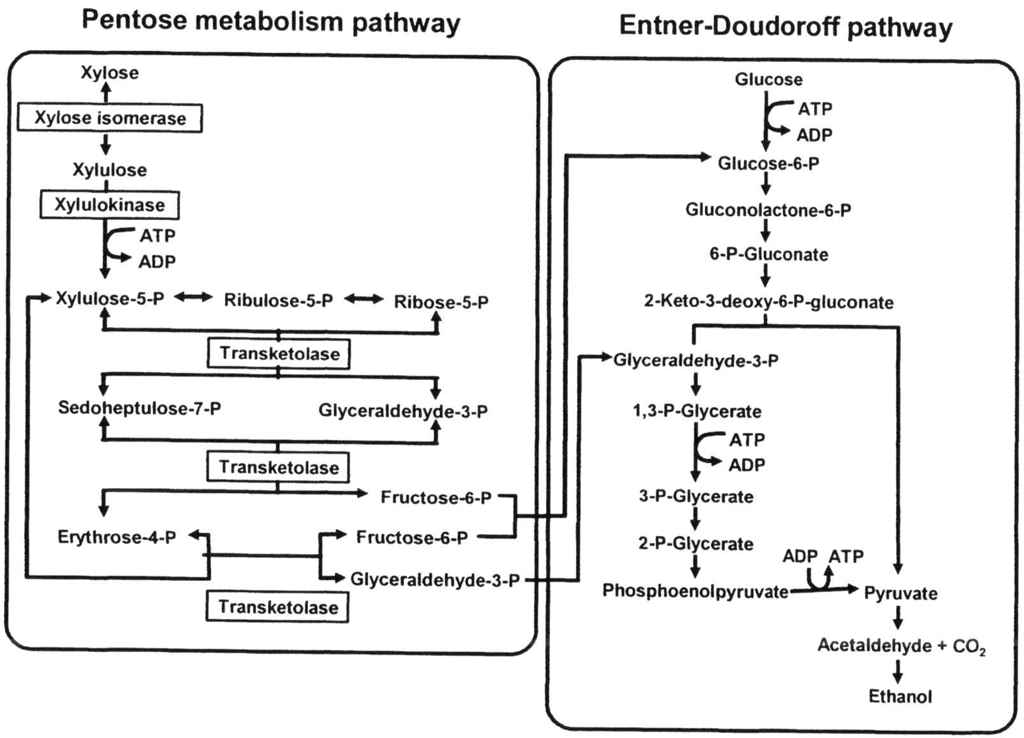

Figure 1. Proposed pentose metabolism and Entner-Doudoroff pathways in engineered Zymomonas mobilis. (Reproduced with permission from reference 153. Copyright 1995.)

hydrolysis process is used for the production of ethanol from softwoods using a recombinant xylose-fermenting yeast, and the residual lignin is used to generate steam and electricity. It is concluded that such a biomass to ethanol plant seems to be an appealing proposition for California, if ethanol replaces MTBE, which is slated for a phase out. With increased research efforts for developing a stable, ethanol tolerant and robust recombinant ethanologenic organism capable of tolerating common fermentation inhibitors generated during pretreatment, competitive and economical production of fuel ethanol from lignocellulose holds strong promise.

Production of Xylitol

Xylitol, a sugar alcohol, has potential use as a natural food sweetener, a dental caries reducer and a sugar substitute for diabetics. It is produced by chemical reduction in alkaline conditions of the xylose derived mainly from wood hydrolyzate (*169*). The recovery of xylitol from the xylan fraction is about 50-60% or 8-15% of the raw material employed. Drawbacks of the chemical process are the requirements of high pressure (up to 50 atm) and temperature (80-140°C), use of an expensive catalyst (Raney-Nickel) and use of extensive separation and purification steps to remove the by-products that are mainly derived from the hemicellulose hydrolyzate (*170*). The bulk of xylitol produced is consumed in various food products such as chewing gum, candy, soft drinks and ice cream. It gives a pleasant cool and fresh sensation due to its high negative heat of solution.

Many yeasts and mycelial fungi possess NADPH dependent xylose reductase (EC 1.1.1.21), which catalyzes the reduction of xylose to xylitol as a first step in xylose metabolism (*171*). Xylitol can be subsequently oxidized to xylulose by the action of xylitol dehydrogenase, which preferentially uses NAD as an acceptor (*172*). In xylose fermenting yeasts, the initial reactions of xylose metabolism appear to be rate-limiting (*173*). This results in accumulation of xylitol in the culture medium, the degree varying with the culture conditions and the yeast strain used (*174*). A surplus of NADH during transient oxygen limitation inhibiting the activity of xylitol dehydrogenase results in xylitol accumulation (*175*). The pathway for xylose utilization in microorganisms is shown in Figure 3. Some of the natural xylose-fermenting yeasts that are known to produce xylitol are: *C. boidini*, *C. entomaea*, *C. guillermondii*, *C. peltata* , *C. tropicalis*, *C. parapsilosis* and *Debaryomyces hansenii* (*176-179*). Oxygen plays an important role in xylose uptake by yeasts. Phosphate limitation stress induces xylitol overproduction by *D. hansenii* (*180*). The highest xylitol concentration attained in microbial processes using xylose as substrate have been in the range of 200 to 220 g/L (*181-184*). Nakano et al. (*185*) reported very high xylitol (356 g/L) production by *C. magnoliae* in fed-batch culture under a microaerobic condition maintained by

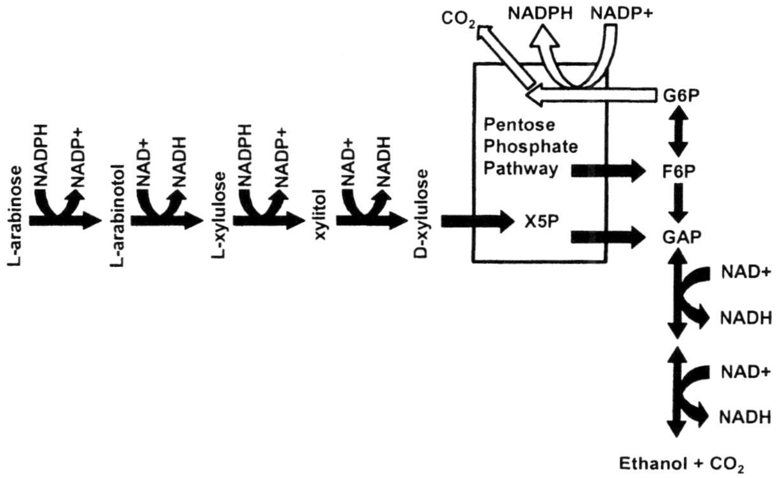

Figure 2. Redox cofactor requirement in L-arabinose catabolism. L-Arabinose conversion to equimolar amounts of CO_2 and ethanol is redox neutral, i.e. anaerobic fermentation to ethanol should be possible. However, the conversion of L-arabinose to D-xylulose requires NADPH and NAD^+ and produces NADH and $NADP^+$. NADPH is mainly regenerated in the oxidative part of the pentose phosphate pathway, where the reduction of $NADP^+$ is coupled to CO2 production. The abbreviations are: G6p, glucose-6-phosphate; F6P, fructose 6-phosphate; X5P, D-Xylulose 5-phosphate; GAP, D-glyceraldehyde 3-phosphate. (Reproduced from Ref. *165* with permission from Elsevier Science)

simple fuzzy control with a yield of 0.75, which corresponded to 82% of the theoretical yield..

The fermentation of sugarcane bagasse hemicellulose hydrolyzate to xylitol by a hydrolyzate-acclimatized yeast strain *Candida* sp. B-22 was studied (*186*). A final xylitol concentration of 94.74 g/L was obtained from 105.35 g/L xylose in hemicellulose hydrolyzate after 96 h of incubation. *C. guilliermondii* FTI 20037 was able to ferment a sugar cane bagasse hydrolyzate producing 18.4 g/L xylitol from 29.5 g/L of xylose, at a production rate of 0.38 g/L.h (*187*). This lower value, compared to that (0.66 g/L.h) of the synthetic medium, may be attributed to the various toxic substances that interfere with microbial metabolism (e.g., acetic acid). Dominguez et al. (*188*) studied different treatments (neutralization, activated charcoal and neutralization, cation-exchange resins and neutralization) of sugar cane bagasse hemicellulose hydrolyzate to overcome the inhibitory effect on xylitol production by *Candida* sp. 11-2. The highest xylitol productivity (0.205 g/L.h), corresponding to 10.54 g/L, was obtained from hydrolyzates treated with activated charcoal (initial xylose, 42.96 g/L). To obtain higher xylitol productivity, treated hydrolyzates were concentrated by vacuum evaporation in rotavator to provide higher initial xylose concentration. The rate of xylitol production increased with increasing initial xylose concentration from 30 to 50 g/L, reaching a maximum of 28.9 g/L after 48 h fermentation. The decrease in xylitol production was dramatic with further increases in the initial xylose concentration. Parajo et al. (*189*) reported a xylitol production of 39-41 g/L from concentrated *Eucalyptus globulus* wood acid hydrolyzate containing 58-78 g xylose/L by *D. hansenii* NRRL Y-7426 using an initial cell concentration of 50-80 g/L. Recently, Rivas et al (*190*) achieved a xylitol concentration of 71 g/L and volumetric productivity of 1.5 g/L.h when *D. hansenii* NRRL Y-7426 (12 g dry mass/L) was grown semiaerobically using detoxified corn cob hydrolyzate produced by autohydrolysis-posthydrolysis at starting xylose concentration of 100 g/L.

Hydrolyzed hemicellulosic fractions of sugar cane bagasse and rice straw were tested for xylitol production in batch fermentation by *C. guilliermondii* under semiaerobic condition and compared these with synthetic medium containing xylose (*191*, *192*). Simultaneous utilization of hemicellulosic sugars (glucose and xylose) was observed, and the highest substrate uptake rate was attained in sugar cane bagasse medium. Increased xylitol concentration (40 g/L) was achieved in synthetic and rice straw media, although the highest xylitol production rate was obtained in sugar cane bagasse hydrolyzate. Both hydrolyzates can be converted into xylitol with satisfactory yields and productivities. Xylitol production by *C. guilliermondii* was evaluated using rice straw hemicellulose hydrolyzate under different conditions of initial pH, nitrogen sources and inoculum level (*193,194*). The xylitol yields were 0.68 g/g for the medium containing ammonium sulfate at pH 5.3 and 0.66 g/g with urea at pH 4.5. Under appropriate inoculum conditions, rice straw hemicellulose hydrolyzate was converted into xylitol by the yeast with 77% yield. Mayerhoff et al. (*195*) evaluated 30 different yeast strains belonging to 4 different genera (*Candida*, *Debaryomyces*, *Hansenula* and *Pichia*) for xylitol production from rice straw hemicellulose hydrolyzate. The best performer was *C. mogii* NRRL Y-17032, which yielded 0.65 g xylitol/g at 0.40 g/L.h over 75 h. Preziosi-Belloy et al. (*196*) investigated the production of xylitol from aspenwood hemicellulose hydrolyzate by *C. guilliermondii*. The hydrolyzate was supplemented by yeast extract, and the maximum xylitol yield (0.8 g/g) and productivity (0.6 g/L.h) were reached by controlling oxygen input. A two-stage sequential fermentation scheme for production of xylitol and arabitol from a mixture of sugars by *C. guilliermondii* was developed (*197*). Following glucose consumption, cells

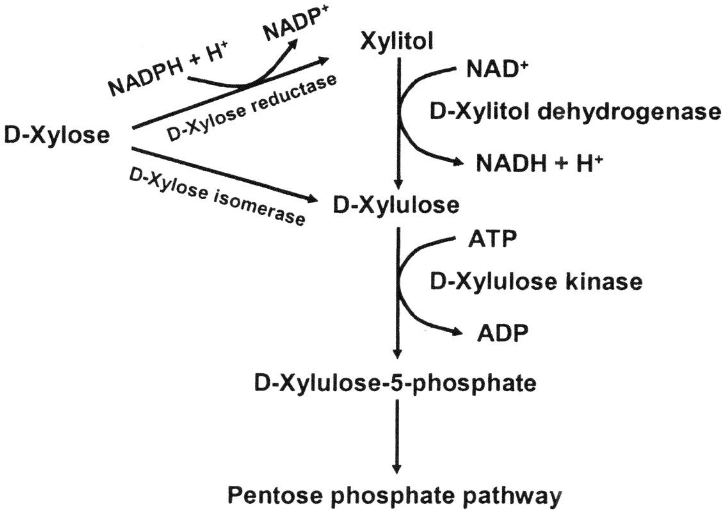

Figure 3. Pathway for xylose utilization in microorganisms. (Reproduced from reference 176. Copyright 1997 American Chemical Society.)

were removed from mixed sugar cultures and replaced with cells from cultures grown on xylose alone. In the second fermentation stage, xylose and arabinose were successfully fermented to xylitol and arabitol. Dilute acid hydrolyzate of corn fiber after treatment with a mixed-bed deionization resin was suitable for the two-stage fermentation process. Xylitol production was studied from barley bran hydrolyzates by continuous fermentation with *D. hansenii* (*198*). The optimum xylitol productivity (2.53 g/L.h) was reached at a dilution of rate of 0.284/h with cell recycle after membrane separation. Xylitol was produced in a two-substrate (xylose, glucose) batch fermentation by *C. tropicalis* with cell recycling (*199*). The optimized cell recycle fermentation resulted in xylitol yield of 0.823 g/g xylose with a productivity of 4.94 g/L.h and a final xylitol concentration of 189 g/L. Xylitol productivity up to 1.14 g xylitol/L.h with 86% conversion efficiency was achieved with a strain of *C. guilliermondii* in continuous culture using a membrane bioreactor at a dilution rate of 0.03/h (*200*).

Carvalho et al. (*201*) achieved maximum xylitol concentration of 20.6 g/L with a volumetric productivity of 0.43 g/L.h and yield of 0.47 g/g after 48 h fermentation during batch xylitol production from concentrated sugarcane bagasse hydrolyzate by *C. guilliermondii* cells, immobilized in calcium-alginate beads. Aeration rate strongly influences xylitol production from sugarcane bagasse hydrolyzate by immobilized cells of *C. guilliermondii* in a fluidized bed reactor (*202*).

E. coli JM 109 was used as a host for xylitol production by expressing xylose reductase gene (*xyrA*) of *C. tropicalis* IFO 0618 (*203*). When xylose (50 g/L) and glucose (5 g/L) were added to IPTG-induced cells, 13.3 g/L of xylitol was produced during 20 h of cultivation. A number of recombinant *S. cerevisiae* strains have been created by expressing the xylose reductase gene (*XYL1*) from *P. stipitis* and *C. shehate*, and production of xylitol from xylose by these recombinant strains in batch and fed batch fermentations have been investigated (*204-206*). Lee et al. (*207*) studied the fermentation characteristics of recombinant *S. cereseviae* containing a xylose reductase gene from *P. stipitis*. Xylitol was produced with a maximum yield of 0.95 g/g xylose consumed in the presence of glucose used as co-substrate for co-factor regeneration. However, addition of glucose caused inhibition of xylose transport and accumulation of ethanol. By adopting a glucose-limited fed batch fermentation where a high ratio of xylose to glucose was maintained, a xylitol concentration of 105.2 g/L was achieved with a 1.69 g/l.h productivity with this recombinant yeast. Kim et al. (*208*) reported that *S. cerevisiae* containing multiple XYL1 genes of *P. stipitis* on the chromosome is much more efficient for xylitol production in the long term non-selective culture than S. *cerevisiae* harboring the episomal plasmid containing the XYL1 gene. Such an improvement in the integrated recombinant strain was supported by the fact that the mitotic stability of the XR gene along with its high expression level worked in a cooperative manner (*209*).

Continuous enzymatic production of xylitol with simultaneous coenzyme regeneration in a charged membrane reactor was studied (*210*). An NADH dependent xylose reductase from *C. tenuis* catalyzed the reduction of xylose. This was coupled to enzymatic oxidation of glucose by glucose dehydrogenase from *Bacillus cereus* to make achievable an up to 10,000-fold regeneration of NADH per cycle of discontinuous conversion. Under suitable conditions, 300 g/L of substrate could be converted in yields above 96% in one single batch reaction.

The recovery of xylitol from fermented sugarcane bagasse hydrolyzate was studied (*211*). The best clarifying treatment was found by adding 20 g activated carbon to 100 ml fermented broth at 80°C for 1 h at pH 6.0. The clarified medium was treated with ion exchange resins after which xylitol crystallization was attempted. The ion exchange resins were not efficient but the crystallization technique showed good performance, although the crystals were involved in a viscous and colored solution. Recently, Faveri et al. (*212*) reported xylitol recovery by crystallization from synthetic solutions and fermented hemicellulose hydrolyzates. The method involves evaporation of dilute solution up to super saturation, cooling of super saturation solution, separation of crystals by centrifugation and final filtration. Using two sets of tests on xylitol-xylose synthetic solutions and one set on fermented hardwood hemicellulose hydrolyzate, the best results in terms of either crystallization (0.56) yield or purity degree (1.00) were obtained with quite concentrated solutions of 730 g/L at relatively high temperature

(-5°C). They concluded that xylitol separation by crystallization from fermented hemicellulose hydrolyzate is feasible.

Thus, xylose rich hemicellulosic materials can serve as abundant and cheap feedstocks for production of xylitol by fermentation. It is possible to introduce the pathway for conversion of arabinose to xylitol in the xylitol producing yeast. In that case, xylitol can be produced from both xylose and arabinose.

Production of 2,3-butanediol

2,3-Butanediol, otherwise known as 2,3-butylene glycol (2,3-BD), is a valuable chemical feedstock because of its application as a solvent, liquid fuel and as a precursor of many synthetic polymers and resins. With a heating value of 27,200 J/g, 2,3-BD compares favorably with ethanol (29,100 J/g) and methanol (22,100 J/g) for use as a liquid fuel and fuel additive (*213*). Dehydration of 2,3-BD yieldsthe industrial solvent methyl ethyl ketone which is much more suited as a fuel because of its much lower boiling point. Further dehydration yields 1,3-butanediene, which is the starting material for synthetic rubber and is also an important monomer in the polymer industry (*214*). During World War II, it was needed for conversion to 1,3-butanediene. Methyl ethyl ketone can be hydrogenated to yield high octane isomers suitable for high quality aviation fuels. Diacetyl, formed by catalytic dehydrogenation of the diol, is a highly valued food additive (*215*). A wide variety of chemicals can also be easily prepared from 2,3-BD (*216, 217*). There is an interest in industrial scale production of 2,3-BD from various agricultural residues as well as logging, pulp and paper, and food industry wastes (*215*).

2,3-BD can occur in two enantiomeric forms: D- (-) and L-(+) as well as an optically inactive meso-form. *B. polymyxa* produces D-(-) 2,3-BD whereas *Klebsiella pneumoniae* (*Aerobacter aerogenes*) produce meso-form and also some of the L-(+) form. *B. subtilis*, *Seratia marcescens* and *A. hydrophia* produce mixtures of different forms (*218*). A newly isolated *Enterobacter cloacae* NRRL B-23289 produces meso-2,3-butanediol (0.35-0.43 g/g sugar) from a variety of sugar substrates including corn fiber hydrolyzates (*219*). The typical 2,3-BD yield was 0.30-0.45 g/g sugar (*214, 220*). Ui et al. (*221*) cloned a gene fragment including genes coding for three enzymes (α-acetolactate synthase, α-acetolactate decarboxylase and meso2,3-BD dehydrogenase (D-acetoin forming) involved in the formation of meso-BD of *K. pneumoniae* IAM 1063 in *E. coli* JM 109 after its insertion into pUC118. The resulting *E. coli* JM109/pBDO118 produced 17.7 g of meso-BD from 100 g of glucose per L.

Butanediol is produced during oxygen limited growth by a fermentative pathway known as mixed acid-butanediol pathway (Figure 4) (*222*). The 2,3-BD pathway and the relative proportions of acetoin and butanediol serve to maintain the intracellular NAD/NADH balance in changing culture conditions. The theoretical maximum yield of 2,3-BD from monosaccharides is 0.5 g/g (*223*). The efficient biological conversion of all the available sugars in agricultural biomass residues to fuels and chemicals is crucial to the efficiency of any process intended to compete economically with petrochemical products (*217*).

The high boiling point of 2,3-BD, its high affinity for water, and the dissolved and solid substances of the fermentation broth make it difficult for 2,3-BD to be purified and recovered from fermentation slurry (*224*). Various methods such as solvent extraction, liquid-liquid extraction and salting out have been used to recover butanediol. Another feasible method to recover butanediol is by countercurrent stream stripping (*225*).

Production of Vanillin

Ferulic acid is the major cinnamic acid found in a variety of plant cell walls. Corn fiber contains about 3% ferulic acid. Wheat bran is another source of ferulic acid (0.5-1%). Faulds et al. (*226*) developed a laboratory scale procedure to produce free ferulic acid (5.7 g) from wheat bran (1 kg) by using a *Trichoderma* xylanase preparation and *A. niger* ferulic acid esterase. Using filamentous fungi, a two-stage process for vanillin formation was developed in which a strain of *A. niger* was first used to convert ferulic acid to vanillic acid, which was then reduced to vanillin by a laccase-deficient strain of *Pycnoporus cinnabarinus* (*227*).

Concluding Remarks

Lignocellulose biodegradation and its conversion to a wide variety of commodity chemicals holds enormous potential. At present, the conversion of lignocellulosic biomass to fermentable sugars is not cost-effective. Some of the emerging pretreatment methods such as alkaline peroxide and AFEX generate solubilized and partially degraded hemicellulosic biomass that need to be treated further with enzymes or other means to produce fermentable sugars from them. With the development of a suitable pretreatment method minimizing the formation of inhibitory compounds for fermentative organisms and use of proper mixture of cellulases and hemicellulases (enzyme cocktail) tailored for each biomass conversion, this vast renewable resource can be utilized for production of fuels and chemicals by fermentation or enzymatic means. Research emphasis should be to develop efficient and cost-effective pretreatment method, enzymes for use in cellulose and hemicellulose conversion in an industrial scale, robust efficient microorganism to ferment lignocellulosic hydrolyzates in a cost-competitive way and method for cost-effective recovery of fermentation products. Finally, integration of various process steps such as biomass pretreatment, enzymatic

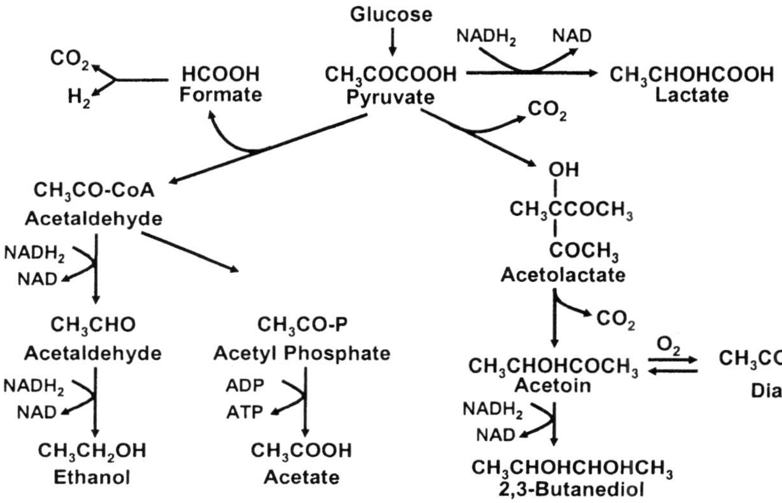

Figure 4. Metabolic pathway of 2,3-butanediol production from glucose.

saccharification, detoxification, fermentation of the hydrolyzates and recovery of products will greatly aid in reducing the overall cost of using lignocellulose for practical purposes.

References

1. Wyman, C. E. *Bioresour. Technol.* **1994**, *50*, 3-16.
2. Thomson, J. A. *FEMS Microbiol. Rev.* **1993**, *104*, 65-82.
3. McMillan, J. D. In: Himmel, M. E.; Baker, J. O.; Overend, R. P., Eds.; *Enzymatic conversion of Biomass for Fuel Production*. American Chemical Society, Washington, D.C., **1993**. pp 292-323.
4. Aspinall, G. O. In: Preiss, J., Ed.; *The Biochemistry of Plants (A Comprehensive Treatise), Vol. 3. Carbohydrates: Structure and Function*. Academic Press, New York, NY, **1980,** pp 473-500.
5. Eda, S.; Ohnishi, A.; Kato, K. *Agric. Biol. Chem.* **1976**, *40*, 359-364.
6. Kormelink, F. J. M.; Voragen, A. G. J. *Appl. Microbiol. Biotechnol.* **1993**, *38*, 688-695.
7. Shibuya, N.; Iwasaki, T. *Phytochemistry* **1985**, *24*, 285-289.
8. Gruppen, H.; Hamer, R. J.; Voragen, A. G. J. *J. Cereal Sci.* **1992**, *16*, 53-67.
9. Saha, B. C. *Biotechnol. Adv.* **2000**, *18*, pp 403-423.

10. Doner, L. W.; Hicks, K. B. *Cereal Chem.* **1997**, *74*, 176-181.
11. Saulnier, L.; Marot, C.; Chanliaud, E.; Thibault, J. F. *Carbohydr. Polymers* **1995**, *26*, 279-287.
12. Saulnier, L.; Thibault, J. F. *J. Sci. Food Agric.* **1999**, 79, 396-402.
13. Hood, E. E.; Hood, K. R.; Fritz, S. E. *Plant Sci.* **1991**, *79*, 13-22.
14. Scalbert, A.; Monties, B.; Lallemand, J. Y.; Guittet, E.; Rolando, C. *Phytochemistry* **1985**, *24*, 1359-1362.
15. Mueller-Hartley, I.; Hartley, R. D.; Harris, P. J.; Curzon, E. H. *Carbohydr. Res.* **1986**, *148*, 71-85.
16. Timell , T. E. *Wood Sci. Technol.* **1967**, *1*, 45-70.
17. Focher, B.; Marzetti, A.; Cattaneo, M..; Beltrame, P. I.; Carniti, P. *J. Appl. Polym. Sci.* **1981**, *26*, 1989-1999.
18. Bungay, H. *Enzyme Microb. Technol.* **1992**, *14*, 501-507.
19. Dale, B. E.; Moreira, M. J. *Biotechnol. Bioeng. Symp.* **1982**, *12*, 31-43.
20. Weil, J.; Westgate, P.; Kohlmann, K.; Ladisch, M. R. *Enzyme Microb. Technol.* **1994**, *16*, 1002-1004.
21. Goldstein, I. S.; Easter, J. M. *Tappi J.* **1992**, *75*, 135-140.
22. Saha, B. C.; Bothast, R. J. *Appl. Biochem. Biotechnol.* **1999**, *76*, 65-77.
23. Clark, D. P.; Mackie, K. L. *J. Wood Chem. Technol.* **1987**, *7*, 373-403.
24. Koullas, D. P.; Christakopoulos, P. E.; Kekos, D.; Koukios, E. G.; Macris, B. J. *Biomass Bioenergy* **1993**, *4*, 9-13.
25. Kaar, W. E.; Holtzaple, M. T. *Biomass Bioenergy* **2002**, *18*, 189-199.
26. Gould, J. M. *Biotechnol. Bioeng.* **1984**, *26*, 46-52.
27. Schmidt, A. S.; Thomsen, A. B. *Bioresour. Technol.* **1998**, *64*, 139-151.
28. Fernandez-Bolanos, J.; Felizon, B.; Heredia, A.; Rodriguez, R.; Guillen, R.; Jimenez, A. *Bioresour. Technol.* **2001**, *79*, 53-61.
29. Dale, B. E.; Leong, C. K.; Pham, P. K.; Esquivel, V. M.; Rios, L.; Latimer, V. M. *Bioresour. Technol.* **1996**, *56*, 111-116.
30. Dale, B. E.; Moreira, M. J. *Biotechnol. Bioeng. Symp.* **1982**, *12*, 31-43.
31. Laser, M.; Schulman, D.; Allen, A. G.; Lichwa, J.; Antal, M. A. Jr.; Lynd, L. R. *Bioresour. Technol.* **2002**, *81*, 33-44.
32. Chum, H. L.; Johnsoon, D. K.; Black, S. *Biotechnol. Bioeng.* **1988**, *31*, 643-649
33. Sun, Y.; Cheng, J. *Bioresour. Technol.* **2002**, *83*, 1-11.
34. Torget, R. W.; Kadam, K. L.; Hsu, T. -A.; Philippidis, G. P.; Wyman, C. E. *US Patent* **1998**, 5, 705, 369.
35. Nguyen, Q. A.; Saddler, J. N. *Bioresour. Technol.* **1991**, *35*, 275-282.
36. Duff, S. J. B.; Murray, W. D. *Bioresour. Technol.* **1996**, *55*, 1-33.
37. Brownell, H. H.; Saddler, J. N. *Biotechnol. Bioeng. Symp.* **1984**, *14*, 55-68.
38. Eklund, R.; Zacchi, G. *Enzyme Microb. Technol.* **1995**, *17*, 255-259.
39. Ando, S.; Arai, I.; Kiyoto, K.; Hanai, S. *J. Ferment. Technol.* **1986**, *64*, 567-570.

40. Bosssaid, A.; Robinson, J.; Cal, Y.-J.; Gregg, D. J.; Saddler, J. N. *Biotechnol. Bioeng.* **1999**, *64*, 284-289.
41. Yang, B.; Boussaid, A.; Mansfield, S. D.; Gregg, D. J.; Saddler, J. N. *Biotechnol. Bioeng.* **2002**, *77*, 678-684.
42. Morjanoff, P. J.; Gray, P. P. *Biotechnol. Bioeng.* **1987**, *29*, 733-741.
43. Cao, N. J.; Krishnan, M. S.; Du, J. X.; Gong, C. S.; Ho, N. W. Y.; Chen, Z. D.; Tsao, G. T. *Biotechnol. Lett.* **1996**, *18*, 1013-1018.
44. Kurakake, M.; Kisaka, W.; Ouchi, K.; Komaki, T. *Appl. Biochem. Biotechnol.* **2001**, *90*, 251-259.
45. Garrote, G.; Dominguez, H.; Parajo, J. C. *Bioresour. Technol.* **2001**, *79*, 155-164.
46. Bjerre, A. B.; Olesen, A. B.; Fernqvist, T.; Ploger, A.; Schmidt, A. S. *Bioresour. Technol.* **1996**, *49*, 568-577.
47. Klinke, H. B.; Ahring, B. K.; Schmidt, A. S.; Thomson, A. B. *Bioresour. Technol.* **2002**, *82*, 15-26.
48. Draude, K.; Kurniawan, C. B.; Duff , S. J. B. *Bioresour. Technol.* **2001**, *79*, 113-120.
49. Palm, M.; Zacchi, G. *Biomacromolecules* **2003**, *4*, 617-623.
50. Kim, K. H.; Hong, J. *Bioresour. Technol.* **2001**, *77*, 139-144.
51. Zheng, Y.; Lin, H. M..; Wen, J.; Cao, N.; Yu, X.; Tsao, G. T. *Biotechnol. Lett.* **1995**, *17*, 845-850.
52. Zheng, Y. Z.; Lin, H. M.; Tsao, G. T. *Biotechnol. Prog.* **1998**, *14*, 890-896.
53. Palmqvist, E.; Hahn-Hagerdal, B. *Bioresour. Technol.* **2000**, *74*, 25-33.
54. Taherzadeh, M. J.; Eklund, R.; Gustafsson, L.; Niklasson, C.; Linden, G. *Ind. Eng. Chem. Res.* **1997**, *36*, 4659-4665.
55. Palmqvist, E.; Hahn-Hagerdal, B. *Bioresour. Technol.* **2000**, *74*, 117-124.
56. Von Silver, M.; Zucchi, G.; Olsson, I.; Hahn-Hagerdal, B. *Biotechnol. Prog.* **1994**, 10, 555-560.
57. Miyafuji, H.; Danner, H.; Neureiter, M.; Thomasser, C.; Braun, R. *Biotechnol. Bioeng.* **2003**, *84*, 390-393.
58. Persson, P.; Larsson, S.; Jonsson, L. J.; Nilvebrant, N. O.; Sivik, B.; Munteanu, F.; Thorneby, L.; Gorton, L. *Biotechnol. Bioeng.* **2002**, *79*, 694-700.
59. Weil, J. R.; Dien, B.; Bothast, R.; Hendrickson, R.; Mosier, N. S.; Ladisch, M. R. *Ind. Eng. Chem. Res.* **2002**, *41*, 6132-6138.
60. Bjorklund, L.; Larsson, S.; Jonsson, L. J.; Reimann, A.; Nilvebrant, N.-O. *Appl. Biochem. Biotechnol.* **2002**, *98-10*, 563-575.
61. Cruz, J. M.; Dominguez, J. M.; Dominguez, H.; Parajo, J. C. *Food Chem.* **1999**, *67*, 147-153.
62. Henrissat, B.; Driguez, H.; Viet, C.; Schulein, M. *Bio/Technol.* **1985**, *3*, 722-726.

63. Divne, C.; Stahlberg, J.; Ruohonen, L.; Pettersson, G.; Knowles, J. K. C.; Teeri, T. T.; Jones, T. A. *Science* **1994**, *265*, 524-528.
64. Divne, C.; Stahlberg, J.; Teeri, T. T.; Jones, T. A. *J. Mol. Biol.* **1998**, *275*, 309-325.
65. Zou, J.-Y.; Kleywege, G. J.; Stalberrg, J.; Driguez, H.; Nerinckx, W.; Clayssens, M.; Koivula, A.; Teeri, T,; Jones, T. A. Crystalllographic evidence for ring distortion and protein conformational changes during catalysis in Cellobiohydrolase Cel6A from *Trichoderma reesei.* Structure 7; 1035-1042.
66. Kleywegt, G. J.; Zou, J. Y.; Divne, C.; Davies, G. J.; Sinning, I.; Stahlberg, J.; Reinikainen, T.; Sridodsuk, M.; Teeri, T. T.; Jones, T. A. *J. Mol. Biol.* **1997**, *272*, 383-397.
67. Gilkes, N. R.; Henrissat, B.; Kilburn, D. C.; Miller, R. C. Jr.; Warren, R. A. *Microbiol. Rev.* **1991**, *55*, 303-315.
68. Grethlein, H. E. *Bio/Technol.* **1985,** *3*, 155-60.
69. Woodward, J.; Wiseman. A. *Enzyme Microb. Technol.* **1982**, 2, 73-79.
70. Coughlan, M. P. *Biotechnol. Genet. Eng. Rev.* **1985**, 3, 39-109.
71. Kadam, S. K.; Demain, A. L. *Biochem. Biophys. Res. Commun.* **1989**, 161, 706-711.
72. Saha, B. C.; Freer, S. N.; Bothast, R. J. In: Saddler, J. N.; Penner, M. H. Eds.; *Enzymatic Degradation of Insoluble Carbohydrates*, American Chemical Society, Washington, D. C. **1995**, pp 197-207.
73. Saha, B. C.; Bothast, R. J. *Appl. Environ. Microbiol.* **1996**, *62*, 3165-3170.
74. Ladisch, M. R.; Lin, K. W.; Voloch, M.; Tsao, G. T. *Enzyme Microb. Technol.* **1983**, 5, 82-102.
75. Kubicek, C. P.; Penttila, M. E. In: Harman, G. E.; Kubicek, C. P. Ed.; *Trichoderma* and *Gliocladium*, Vol. 2, Taylor & Francis, Ltd, London, UK, **1998**, p 49-72.
76. Fagerstam, L. G.; Pettersson, G. *FEBS Lett.* **1980**, *119*, 97-100.
77. Esterbauer, H.; Steiner, W.; Laudova, I.; Hermann, A.; Hayn, M. *Bioresour. Technol.* **1991**, *36*, 51-65.
78. Saha, B. C. *Process Biochem.* **2004**, (published online Dec. 4, 2003).
79. Nguyen, Q. A.; Saddler, J. N. *Bioresour. Technol.* **1991**, 35, 275-282.
80. Philippidis, G. P.; Smith, T. K.; Wyman, C. E. *Biotechnol. Bioeng.* **1993**, 41, 846-853.
81. Tan, L. U. L.; Yu, E. K. C.; Mayers, P.; Saddler, J. N. *Appl. Microbiol. Biotechnol.* **1987**, 26, 21-27.
82. Hinman, N. D.; Schell, D. J.; Rieley, C. J.; Bergeron, P. W.; Walter, P. J. *Appl. Biochem. Biotechnol.* **1992**, 34/35, 639-649.
83. Wright, J. D.; Power, A. J.; Douglas, L. J. *Biotechnol. Bioeng. Symp.* **1986**, 17, 285-302.
84. Lee, D.; Yu, A. H. C.; Saddler, J. N. *Biotechnol. Bioeng.* **1995**, 45, 328-336.
85. Tanaka, M.; Fukui, M.; Matsuno, R. *Biotechnol. Bioeng.* **1988**, 32, 897-902.

86. Castanon, M.; Wilke, C. *Biotechnol. Bioeng.* **1981**, *23*, 1365-1372.
87. Kaar, W. E.; Holtzapple, M. T. *Biotechnol. Bioeng.* **1998**, *59*, 419-426.
88. Eriksson, T.; Borjesson, J.; Tjerneld, F. *Enzyme Microb. Technol.* **2002**, *31*, 353-364.
89. Gusakov, A. V.; Sinitsyn, A. P.; Manenkova, J. A.; Protas, O. V. *Appl. Biochem. Biotechnol.* **1992**, 34/35, 625-637.
90. Borchert, A.; Buchholz, K. *Process Biochem.* **1987**, 22, 173-180.
91. Henley, R. G.; Young, R. Y. K.; Greenfield, P. F. *Enzyme Microb. Technol.* **1980**, 2, 206-208.
92. Klei, H. E.; Sundstrom, D. W.; Coughlin, K. W.; Ziolkowski, K. *Biotechnol. Bioeng. Symp.* **1981**, *11*, 593-601.
93. Ohlson, I.; Trayardh, G.; Hahn-Hagerdal, B. *Biotechnol. Bioeng.* **1984**, 26, 647-653.
94. Kinoshita, S.; Chua, J. W.; Kato, N.; Yoshida, T.; Taguchi, T. *Enzyme Microb. Technol.* **1986**, *8*, 691-695.
95. Ghose, T. K.; Ghosh, P. In: Ladisch, M. R.; Bose, A. Eds.; *Harnessing Biotechnology for the 21st Century*, American Chemical Society, Washington, D. C., **1992**, pp 505-506.
96. Pelach, M. A.; Pastor, F. J.; Puig, J.; Vilasecca, F.; Mutje, P. *Process Biochem.* **2003**, *38*, 1063-1067.
97. Saha, B. C.; Bothast, R. J. In: Imam, S. H.; Greene, R. V.; Zaidi, B. R. Eds.; *Biopolymers: Utilizing Nature's Advanced Materials*, American Chemical Society, Washington, D. C., **1999**, pp 167-194.
98. Hartley, R. D.; Ford, C. W. In: Lewis, L. G.; Paice, M. G.; Eds.; *Plant Cell Wall Polymers, Biogenesis and Biodegradation*, American Chemical Society, Washington, D. C., **1989**, pp 135-145.
99. Faulds, C. B.; Williamson, G. *J. Gen. Microbiol.* **1991**, *137*, 2339-2345.
100. Filho, E. X. F.; Touhy, M. G.; Puls, J.; Coughlan, M. P. *Biochem. Soc. Trans.* **1991**, *19*, 25S.
101. Bachmann, S. L.; McCarthy, A. J. *Appl. Environ. Microbiol.* **1991**, *57*, 2121-2130.
102. Poutanen, K.; Puls, J. In: Lewis, G.; Paice, M. Eds.; *Biogenesis and Biodegradation of Plant Cell Wall Polymers*, American Chemical Society, Washington, D. C., **1989**, pp 630-640.
103. Lee, S. F.; Forsberg, C. W. *Can. J. Microbiol.* **1987**, *33*, 1011-1016.
104. Poutanen , K.; Tenkanen, M.; Korte, H.; Puls, J. In: Leatham , G. F.; Himmel, M. Eds.; *Enzymes in Biomass Conversion*, American Chemical Society, Washington, DC, **1991**, pp 426-436.
105. Gilbert, H. J.; Hazlewood. G. P. *J. Gen. Microbiol.* **1993**, *139*, 187-194.
106. Saha, B. C.; Dien, B. S.; Bothast, R. J. *Appl. Biochem. Biotechnol.* **1998**, *70-72*, 115-125.
107. Leathers, T. D.; Gupta, S. C. *Appl. Microbiol. Biotechnol* **1997**, *59*, 337-347.

108. Christov, L. P.; Myburgh, J.; van Tonder, A.; Prior, B. A. *J. Biotechnol.* **1997**, *55*, 21-29.
109. Saha, B. C. *Appl. Microbiol. Biotechnol.* **2001**, *56*, 762-766.
110. Saha, B. C. *Process Biochem.* **2002**, *37*, 1279-1284.
111. Saha, B. C. *J. Ind. Microbiol. Biotechnol.* **2001**, *27*, 241-145.
112. Saha, B. C. *Bioresour. Technol.* **2003**, *90*, 33-38.
113. Saha B. C.; Bothast, R. J. *Appl. Environ. Microbiol.* **1998**, *64*, 216-220.
114. Saha B. C.; Bothast, R. J. *Curr. Microbiol.* **1998**, *37*, 337-340.
115. Spagnuolo, M.; Crecchio, C.; Pizzigallo, M. D. R.; Ruggiero, P. *Biotechnol. Bioeng.* **1999**, *64*, 685-691.
116. Faulds, C. B.; Williamson, G. *Appl. Microbiol. Biotechnol.* **1995**, *43*, 1082-1087.
117. Kroon, P. A.; Faulds, C. B.; Williamson, G. *Biotechnol. Appl. Biochem.* **1996**, *23*, 255-262.
118. Bartolome, B.; Faulds, C. B.; Williamson, G. *J. Cereal Sci.* **1997**, *25*, 285-288.
119. Yu, P.; Maenz, D. D.; McKinnon, J. J.; Racz, V. J.; Christensen, D. A. *J. Agric. Food Chem.* **2002**, *50*, 1625-1630.
120. Singh, S.; Madlala, A. M.; Prior, B. *FEMS Microbiol. Rev.* **2003**, *27*, 3-16.
121. Polvinen, K.; Lehtonen, P.; Leisola, M.; Visuri, K. In: Leatham, G. F.; Himmel, M. E. Eds.; *Enzymes in Biomass Conversion*, American Chemical Society, Washington, D. C., **1991**, pp 225-235.
122. Chang, M. M.; Chou, T. Y. C.; Tsao, G. T. *Adv. Biochem. Eng.* **1981**, 20, 15-42.
123. Tien, M.; Kirk, T. K. *Science* **1983**, 221, 661-663.
124. Tien, M. *CRC Crit. Rev. Microbiol.* **1987**, 15, 141-168.
125. Umezawa, T.; Higuchi, T. In: Leatham, G. F.; Himmel, M. E. Eds.; *Enzymes in Biomass Conversion*, American Chemical Society, Washington, D. C., **1991**, pp 236-246.
126. Fiechter, A. *J. Biotechnol.* **1993**, 30, 49-55.
127. De Jong, E.; Field, J. A.; de Bont, J. A. M. *FEMS Microbiol. Rev.* **1994**, *13*, 153-187.
128. Shoemaker, H. E.; Leisola, M. S. A. *J. Biotechnol.* **1990**, 13, 101-109.
129. Zimmermann, W. *J. Biotechnol.* **1990**, *13*, 119-130.
130. Call, H. P.; Mucke, I. *J. Biotechnol.* **1997**, *53*, 163-202.
131. Ferraz, A.; Cordova, A. M.; Machuca, A. *Enzyme Microb. Technol.* **2003**, *32*, 59-65.
132. Barreca, A. M.; Fabbrini, M.; Galli, C.; Gentili, P.; Ljunggren, S. *J. Mol. Catal. B: Enzym.* **2003**, *26*, 105-110.
133. Galliano, H.; Gas, G.; Boudet, A. M. *FEMS Microbiol. Lett.* **1990**, *67*, 295-300.
134. Hadar, Y.; Kerem, Z.; Gorodecki, B. *J. Biotechnol.* **1993**, *30*, 133-139.

135. Vicuna, R. *Enzyme Microb. Technol.* **1988**, 10, 646-655.
136. Eggert, C.; Temp, U.; Eriksson, K. E. *FEBS Lett.* **1997**, *407*, 89-92.
137. ten Have, R.; Wijngaard, H.; Aries-Kronenburg, N. A. E.; Straatsma, G.; Schaap, P. J. *J. Agric. Food Chem.* **2003**, *51*, 2242-2245.
138. Lynd, L. In: Fiechter, A., Ed.; *Adv. Biochem. Eng. Biotechnol.* **1989**, 38, pp 1-52.
139. Despande, V.; Keskar, S.; Mishra, C.; Rao, M. *Enzyme Microb. Technol.* **1986**, *8*, 149-152.
140. Odom, J. M.; Wall, J. D. *Appl. Environ. Microbiol.* **1983**, *45*, 1300-1305.
141. Lewis, S. M.; Montgomery, L.; Garleb, K. A.; Barger, L. L.; Fahey, G. C. Jr. *Appl. Environ. Microbiol.* **1988**, *54*, 1163-1169.
142. Haruta, S.; Cui, Z.; Huang, Z.; Li, M.; Ishi, M.; Igarashi, Y. *Appl. Microbiol. Biotechnol.* **2002**, *59,* 529-534.
143. Schneider, H.; Wang, P. Y.; Chan, Y. K.; Maleszka, R. *Biotechnol. Lett.* **1981**, *3*, 89-92.
144. Bothast , R. J.; Saha, B. C. *Adv. Appl. Microbiol.* **1997**, *44*, 261-286.
145. Du Preez, J. C. *Enzyme Microb. Technol.* **1994**, *16*, 944-956.
146. Hahn-Hagerdal, B.; Jeppsson, H.; Skoog, K.; Prior, B. A. *Enzyme Microb. Technol.* **1994**, *16*, 933-943.
147. Gong, C. S.; Chen, L. F.; Flickinger, M. C.; Chiang, L. C.; Tsao, G. T. *Appl. Environ. Microbiol.* **1981**, *41*, 430-436.
148. Hahn-Hagerdal, B.; Berner, S.; Skoog, K. *Appl. Microbiol. Biotechnol.* **1986**, *24*, 287-293.
149. Saha, B. C.; Bothast, R. J. *Appl. Microbiol. Biotechnol.* **1996**, *45*, 299-306.
150. Dien, B. S.; Kurtzman, C. P.; Saha, B. C.; Bothast, R. J. *Appl. Biochem. Biotechnol.* **1996**, *57/58*, 233-242.
151. Ingram, L. O.; Alterthum, F.; Ohta, K.; Beall, D. S. In: Pierce, G. E. Ed,; *Developments in Industrial Microbiology*, Vol. 31, **1990**, pp 21-30.
152. Ho, N. W. Y.; Chen, Z.; Brainard. A. P. *Appl. Environ. Microbiol.* **1998**, *64*, 1852-1856.
153. Zhang, M.; Eddy, C.; Deanda, K.; Finkelstein, M.; Picataggio, M. *Science* **1995**, *267*, 240-243.
154. Hahn-Hagerdal, B.; Wahlborm, C. F.; Gardonyi, M.; van Zyl, W. H.; Cordero Otero, R. R.; Jonsson, L. J. *Adv. Biochem. Eng. Biotechnol.* **2001**, *73*, 53-84.
155. Bothast, R. J.; Nichols, N. N.; Dien, B. S. *Biotechnol. Prog.* **1999**, *15*, 867-875.
156. Ingram, L. O.; Aldrich, H. C.; Borges, A. C. C.; Causey, T. B.; Martinez, A.; Morales, A.; Saleh, A.; Underwood, S. A.; Yomano, Y. P.; York, S. W.; Zaldivar, J.; Zhou, S. *Biotechnol. Prog.* **1999**, *15*, 855-866.
157. Martinez, A.; York, S. W., Yomano, Y. P.; Pineda, V. L.; Davis, F. C.; Shelton, J. C.; Ingram, L. O. *Biotechnol. Prog.* **1999**, *15*, 891-897.

158. Yomano, L. P.; York, S. W.; Ingram, L. O. *J. Ind. Microbiol.* **1998**, *20*, 132-138.
159. Gonzalez, R.; Tao, H.; Purvis, J. E.; York, S. W.; Shanmugam, K. T.; Ingram, L. O. *Biotechnol. Prog.* **2003**, *19*, 612-623.
160. Deng, X. X.; Ho, N. W. Y. *Appl. Biochem. Biotechnol.* **1990**, *24/25*, 193-199.
161. Eliasson, A.; Christensson, C.; Wahborn, C. F.; Hahn-Hagerdahl , B. *Appl. Environ. Microbiol.* **2000**, *66*, 3381-3386.
162. Martin, C.; Galbe, M.; Wahlborn, C. F.; Hahn-Hagerdal, B.; Jonsson, L. J. *Enzyme Microb. Technol.* **2002**, *31*, 274-282.
163. Sedlak, M.; Ho, N. W. Y. *Enzyme Microb. Technol.* **2001**, *28,* 16-24.
164. Zhang , M.; Chou, Y.; Picataggio, S.; Finkelstein, M. *US Patent* **1998**, *5*, 843, 760.
165. Richard , P.; Verho, R.; Putkonen, M.; Londesborough, J.; Penttila, M. *FEMS Yeast Res.* **2003**, *3*, 185-189.
166. Larsson, S.; Cassland, P.; Jonsson, L. J. *Appl. Environ. Microbiol.* **2001**, *67*, 1163-1170.
167. Von Sivers, M.; Zacchi, M. *Bioresour. Technol.* **1995**, *51*, 43-52.
168. Kadam, K. L.; Wooley, R. J.; Aden, A.; Nguyen, Q. A.; Yancey, M. A.; Ferraro, F. M. *Biotechnol. Prog.* **2000**, *16*, 947-957.
169. Malaja, A.; Hamalainen, L. *US Patent* **1977**, 4,008,285.
170. Winkelhausen, E.; Kuzmanova, S. *J. Ferment. Bioeng.* **1998**, *86*, 1-14.
171. Chiang, C.; Knight, K. G. *Nature (London)* **1960**, *188*, 79-81.
172. Hofer, M.; Betz, A.; Kotyk, A. *Biochim. Biophys. Acta* **1971**, *252*, 1-12.
173. Nolleau, V.; Preziosi-Belloy, L.; Navarro, J. M. *Biotechnol. Lett.* **1995**, *17*, 417-422.
174. van Dijken, J.P.; Scheffer, W. A. *FEMS Microbiol. Rev.* **1986**, *32*, 199-224.
175. Granstorm, T.; Ojamo, H.; Leisola, M. *Appl. Microbiol. Biotechnol.* **2001**, *55*, 36-42.
176. Saha, B. C.; Bothast, R. J. In: Saha, B. C.; Woodward, J. Eds.; *Fuels and Chemicals from Biomass*, American Chemical Society, Washington, D. C., **1997**, pp 307-319.
177. Silva, S. S.; Felipe, M. G. A.; Mancilha, I. M. *Appl. Biochem. Biotechnol.* **1998**, *70-72*, 331-339.
178. Saha, B. C.; Bothast, R. J. *J. Ind. Microbiol. Biotechnol.* **1999**, *22*, 633-636.
179. Saha, B. C.; Bothast, R. J. *Appl. Microbiol. Biotechnol.* **1996**, *45*, 299-306.
180. Tavares, J. M.; Duarte, L. C.; Amaral-Collaco, M. T.; Girio, F. M. *FEMS Microbiol. Lett.* 1999, *171*, 115-120.
181. Silva, S. S.; Afschar, A. S. *Bioprocess Eng.* **1994**, *11*, 129-134.
182. Kim, S. Y.; Kim, J. H.; Oh, D. K. *J. Ferment. Technol.* **1997**, *83*, 267-270.
183. Meyrial, V.; Delgenes, J. P.; Moletta, R.; Navarro, J. M. *Biotechnol. Lett.* **1991**, *13*, 281-286.

184. Cao, N.-J.; Tang, R.; Gong, C. S.; Chen, L. F. *Appl. Biochem. Biotechnol.* **1994**, *45/46*, 515-519.
185. Nakano, K.; Katsu, R.; Tada, K.; Matsumura, M. *J. Biosci. Bioeng.* **2000**, *89*, 372-376.
186. Chen, L. F.; Gong, C. S. *J. Food Sci.* **1985**, *50*, 226-228.
187. Pfeifer, M. I.; Silva, S. S.; Felipe, M. G. A.; Roberto, I. C.; Mancilha, I. M. *Appl. Biochem. Biotechnol.* **1996**, *57/58*, 423-430.
188. Dominguez , J. M.; Gong, C. S.; Tsao, G. T. *Appl. Biochem. Biotechnol.* **1996**, *57/58*, 49-56.
189. Parajo, J. C.; Dominguez, H.; Dominguez, J. M. *Enzyme Microb. Technol.* **1997**, *21*, 18-24.
190. Rivas, B.; Torre, P.; Dominguez, J. M.; Perego, P.; Converti, A.; Parajo, J. C. *Biotechnol. Prog.* **2003**, *19*, 706-713.
191. Roberto, I. C.; Mancilha, I. M.; Souza, C. A. D.; Felipe, M. G. A.; Sato, S.; Castro, H. F. D. *Biotechnol. Lett.* **1994**, *16*, 1211-1216.
192. Roberto, I. C.; Felipe, M. G. A.; Mancilha, I. M.; Vitola, M.; Sato. S.; Silva, S. S. *Bioresour. Technol.* **1995**, *51*, 255-257.
193. Roberto, I. C.; Sato, S.; Mancilha, I. M. *J. Ind. Microbiol.* **1996**, *16*, 348-350.
194. Roberto, I. C.; Silva, S. S.; Felipe, M. G. A.; de Mancilha, I. M.; Sato, S. *Appl. Biochem. Biotechnol.* **1996**, *57/58*, 339-347.
195. Mayerhoff, Z. D. V. L.; Roberto, I. C.; Silva, S. S. *Biotechnol. Lett.* **1997**, *5*, 407-409.
196. Preziosi-Belloy, L.; Nolleau, V.; Navarro, J. M.. *Biotechnol. Lett.* **2000**, *22*, 239-243.
197. Leathers, T. D.; Dien, B. S. *Process Biochem.* **2000**, *35*, 765-769.
198. Cruz, J. M.; Dominguez, J. M.; Parajo, J. C. *Biotechnol. Lett.* **2000**, *22*, 1895-1898.
199. Choi, J. H.; Moon, K. H.; Ryu, Y. W.; Seo, J. H. *Biotechnol. Lett.* **2000**, *22*, 1625-1628.
200. Faria, L. F. F.; Pereira, N. Jr.; Nobrega, R. *Desalination*, **2002**, *249*, 231-236.
201. Carvalho, W.; Silva, S. S.; Converti, A.; Vitolo, M. *Biotechnol. Bioeng.* **2002**, *79*, 165-169.
202. Santos, J. C.; Carvalho, W.; Silva, S. S.; Converti, A. *Biotechnol. Prog.* **2003**, *19*, 1210-1215.
203. Suzuki, T.; Yokoyama, S.; Kinoshita, Y.; Yamada, H.; Hatswu, M.; Takamizawa, K.; Kawai, K. *J. Biosci. Bioeng.* **1999**, *87*, 280-284.
204. Hallborn, J.; Walfridsson, M.; Airaksine, U.; Ojamo, H.; Hahn-Hagerdal, B. *Bio/Technol.* **1991**, *9*, 1090-1095.
205. Hallborn, J.; Gorwa, M. F.; Meinander, N.; Penttila, M.; Keranen, S.; Hahn-Hagerdal, B. *Appl. Microbiol. Biotechnol.* **1994**, *42*, 326-333.
206. Govinden, R.; Pillay, B.; van Zyl, W. H.; Pillay, D. *Appl. Microbiol. Biotechnol.* **2001**, *55*, 76-80.

207. Lee, W. J.; Ryu, Y. W.; Seo, J. H. *Process Biochem.* **2000**, *35*, 1199-1203.
208. Chung, Y. S.; Kim, M. D.; Lee, W. J.; Ryu, Y. W.; Kim, J. H.; Seo, J. H. *Enzyme Microb. Technol.* **2002**, *30*, 809-816.
209. Kim, Y. S.; Lim, S. Y.; Kim, J. H.; Kim, S. C. *J. Biotechnol.* **1999**, *67*, 159-171.
210. Nidetzky, B.; Neuhauser, W.; Haltrich, D.; Kulbe, K. D. *Biotechnol. Bioeng.* **1996**, *52*, 387-396.
211. Gurgel, P. W.; Mancilha, I. M.; Pecanha. R. P.; Siqueira, J. F. M. *Bioresour. Technol.* **1995**, *52*, 219-223.
212. Faveri, D. D.; Perego, P.; Converti, A.; Borghi, M. *Chem. Eng. J.* **2002**, *90*, 291-298.
213. Tran, A. V.; Chambers, R. P. *Biotechnol. Bioeng.* **1987**, *29*, 343-351.
214. Maddox , I. S. In: Roehr, M. Ed.; *Biotechnology, Vol 6. Products of Primary Metabolism*, VCH Publishers, Weinheim, Germany, **1996**, pp 269-291.
215. Magee, R. J.; Kosaric, N. *Adv. Appl. Microbiol.* **1987**, *32*, 89-161.
216. Gong, C. S.; Cao, N.; Tsao, G. T. In: Saha, B. C.; Woodward, J. Eds.; *Fuels and Chemicals from Biomass*, American Chemical Society, Washington, D. C., **1997**, pp 280-293.
217. Yu, E. K. C.; Saddler, J. N. *Trends Biotechnol.* **1985**, *3*, 100-104.
218. Kosaric, N.; Velikonja, J. *FEMS Microbiol. Rev.* **1995**, *16*, 111-142.
219. Saha, B. C.; Bothast, R. J. *Appl. Microbiol. Biotechnol.* **1999**, *52*, 321-326.
220. Nilegaonkar, S. S.; Bhosale, S. B.; Dandage, C. N.; Kapadi, A. H. *J. Ferment. Bioeng.* **1996**, *82*, 408-410.
221. Ui, S.; Okajima, Y.; Mimura, A.; Kanai, H.; Kudo, T. *J. Ferment. Bioeng.* **1997**, *84*, 185-189.
222. Kosaric, N.; Magee, R. J.; Blaszczyk, R. *Chem. Biochem. Eng.* **1992**, *6*, 145-152.
223. Jansen, N. B.; Flickinger, M. C.; Tsao, G. T. *Biotechnol. Bioeng.* **1984**, *26*, 362-368.
224. Syu, M. J. *Appl. Microbiol .Biotechnol.* **2001**, *55*, 10-18.
225. Garg, S. K.; Jain, A. *Bioresour. Technol.* **1995**, *51*, 103-109.
226. Faulds, C. B.; Bartolome, B.; Williamson, G. *Ind. Crops Prod.* **1997**, *6*, 367-374.
227. Lesage-Meessen, L.; Delattre, M.; Haon, M.; Thibault, J. F.; Colonna Ceccaldi, B.; Brunerie, P.; Asther. M. *J. Biotechnol.* **1996**, *50*, 107-113.

Pretreatment

Chapter 2

Pretreatment for Enzymatic Hydrolysis of Used Newspaper

Sung Bae Kim and Jin Won Chun

Division of Applied Chemical Engineering and ERI, Gyeongsang National University, Jinju 660–701, Korea

Wastepaper is one of the plentiful and low cost feedstockes for making bioethanol. A pretreatment method suitable for wastepaper was developed using ammonia with nonionic surfactants (NP series) and/or hydrogen peroxide. The enzymes used were commercially available cellulase and β-glucosidase, and a mixture of both enzymes was used to alleviate the end-product inhitition by cellobiose. In order to avoid that ink and additives contained in newspaper are solubilized or reacted with ammonia in high temperature, newspaper was pretreated with aqueous ammonia solution on a shaking bath at 40°C using 130 strokes/min for 3 h. The enzymatic digestibility of newspaper, office paper, and corrugated paper at enzyme loading 30 IFPU/g glucan was found to be 70%, 85%, and 88%, respectively. The lowest digestibility value of newspaper means that it is the hardest substrate for enzymatic hydrolysis among them. Effect of surfactant in ammonia solution on digestibility was found to be very significant. Digestibility achieved by the treatment with ammonia and NP-5 was the same as the simultaneous treatment with ammonia, hydrogen peroxide, and NP-5, which means that hydrogen peroxide does not need if surfactant is added in ammonia solution. In the ammonia and NP-5 treatment, the effect of input stage of NP-5 on digestibility was also investigated. It was found that surfactant addition in pretreatment stage was a little more effective on cellulose hydrolysis than surfactant addition in enzymatic hydrolysis.

Introduction

Since the first oil crisis in the 1970s, oil prices are steadily going up because of limitations in the discovery of new oil fields. Before its depletion, alternative sources of energy need to be developed in the near future. A significant part of this effort has been devoted to the utilization of renewable resources, particularly in biomass. Biomass has been shown to have a considerable promise as a raw material for liquid fuels such as ethanol that can replace gasoline. Lignocellulosic materials such as agricultural residues, wood, and municipal solid waste could be a raw material for bioethanol. Among those resources, wastepaper is one of the plentiful and low cost feedstocks for making bioethanol (1). Especially in Korea, any biomass except wastepaper is not enough to make it into energy. Moreover, the possibility for converting waste biomass to energy source may serve as a means of waste disposal.

Typically, wastepaper constitutes half of municipal solid waste, and newspaper alone 14% of the waste (2). Generally, over 50% of wastepaper is relatively well recycled. In the past, most of it was used only once and then landfilled or incinerated. Even recycled wastepaper can be used only two to three times before the fibers become unacceptably short. Currently only relatively long fibers obtained after deinking process are recycled for paper production, but the residues constituting mostly short fibers are discarded.

Cellulose, the major component of wastepaper, can be transformed into fermentable sugars by enzymatic hydrolysis. Newspaper is mostly derived from softwood and exhibits low enzymatic digestibility because of its high lignin content and dense structure. Additionally, chemicals such as fillers, ink, and other additives make it difficult to hydrolyze enzymatically. Thus, effective pretreatment is needed to increase enzymatic hydrolysis of newspaper.

Of numerous studies on the pretreatment of lignocellulosic biomass, there have been very limited studies on partially delignified biomass such as wastepaper (3-8). The methods used in most of these studies, however, were the same as those used on woody and herbaceous materials. Since the paper was made after treating wood physically and/or chemically in pulping process, wastepaper does not require the severe pretreatment developed for other lignocellulosic biomass. Our previous studies revealed that an ammonia-hydrogen peroxide mixture proved to be very effective in pretreating newspaper (9,10). Surfactant as well as ammonia and hydrogen peroxide was used in this study, because it can aid in reducing the adhesion of the ink to the fibers (11).

In this chapter we discussed our previous studies briefly and presented ongoing studies about the pretreatment and enzymatic digestibility of used newspaper.

Materials and Methods

Substrates and Surfactants

A mixture of three newspaper issued in Korea was used as substrate. Office paper and corrugated paper were arbitrarily selected from our lab. Paper sample was cut into approx. 0.5×0.5 cm pieces. The nonionic surfactants (TCI, Japan) used in this study were listed in Table 1.

Table I. Nonionic surfactants

Surfactant name	*Chemical composition*	*Appearance*	*EO* content*	*HLB***
NP-5	Polyethylene Glycol Mono-4-nonylphenyl Ether	Oily liquid	5 mol	10.0
NP-10		Oily liquid	10 mol	13.3
NP-20		White solid	20 mol	16.0

*EO = ethylene oxide, **HLB = hydrophile-lipophile balance

Pretreatment

Pretreatment was performed on a reciprocating shaking water bath. Five grams of substrate were added to a 500 mL autoclavable bottle with 100 g of ammonia-hydrogen peroxide solution. The concentration of each component was expressed as wt% based on the total amount of the solution. Then 0.5~1.5 wt% of a surfactant was added to this solution. The concentration of the surfactant was calculated as wt% based on the substrate (5 g). The bottle was placed for 3 h on a shaker operating at 40 °C and 130 strokes/min. After pretreatment, the wet solid was washed with deionized water until neutral and then it was separated into two portions. One was oven-dried overnight at 105°C to measure the moisture content and subsequently, the weight loss in pretreatment. It was further subjected to composition analysis. The other was stored in a refrigerator to carry out the enzymatic digestibility test.

Enzyme and Digestibility Test

Commercial cellulase and β-glucosidase (Novo Nordisk, Bagvard, Denmark) supplied from Novozymes Korea Ltd were used for digestibility test. A mixture of Celluclast (80 IU or international filter paper units [IFPU]/mL) and Novozym 188 (792 cellobiase units [CBU]/mL) was used with a ratio of 4 IU of Celluclast/CBU Novozym to alleviate the end-product inhibition by cellobiose.

Enzymatic digestibility of pretreated substrate was performed in duplicate according to National Renewable Energy Laboratory (NREL) standard procedure no. 009 (12). The amount of solid required to give 0.5 g glucan in 50 mL was added to 250-mL flask. The buffer solution was 0.05 M citrate, pH 4.8 and the cellulase enzyme loading was 30 IFPU/g glucan. The content of the flask was preheated to 50°C before the enzyme was added and the flask was placed on a shaking bath that operated at 50°C and 90 strokes/min. Samples were taken periodically and analyzed for glucose using HPLC. The glucose content after 72 h of hydrolysis was used to calculate the enzymatic digestibility.

Analytical Methods

The solid biomass sample was analyzed for moisture, sugars, klason lignin, and ash by NREL standard procedures no.001~005 (12). Sugars were measured by HPLC (Thermo Separation Products) using Bio-Rad HPX-87H (condition; 0.6 mL/min, 65°C, 0.005M H_2SO_4). Since this column does not resolve xylose, mannose, and galactose, the combined value of XMG is used in this article. All sugar contents of solid residue were expressed as wt% based on the original oven-dried untreated biomass and expressed as glucan, xylan, mannan, and galactan equivalents.

Results and Discussion

We investigated factors that affect enzymatic hydrolysis of newspaper, such as ash content, substrate size, and ink (10). Results showed that ink had a significant effect on enzymatic digestibility, whereas ash content and substrate size had a very small effect. Therefore, it was concluded that ink removal from newspaper is prerequisite for enzymatic hydrolysis.

Pretreatment by Percolation Process

This process has shown a proven performance when treating oak and corn stover in ammonia and/or hydrogen peroxide solution (13,14). The main advantage of this process is its ability to attain high sugar yield without causing significant sugar decomposition. Newspaper was pretreated at 170°C, 60 min, and 1 mL/min with 10 wt% and 20 wt% ammonia solution (10). The results found in this study were quite different from our previous results (15) obtained when oak was pretreated in the same condition. Percent solid remaining of oak was 70%, but that of newspaper was about 90%. When 10% of ammonia was used, the removal percent of cellulose, hemicellulose, and klason lignin in oak was 5%, 33%, and 45%, respectively, but that of each component in newspaper was 1%, 10%, and 23%, respectively. Such low removals in three major components can be explained by the fact that easily removable portion in each component was already removed during pulping process.

Digestibilities at both ammonia concentrations, however, were almost same or 15% less than that of untreated sample. Here, untreated sample means a substrate that is soaked in water for 3 hr at room temperature. This means that no pretreatment effect can be expected even though a significant amount of lignin was removed at each ammonia condition.

After the pretreatment, sample was recovered, washed, and then dried for composition analysis. The appearance of newspaper sample was found to be pretty hard. It was thought that some sticky material was dissolved in ammonia solution, bound to solid residues, and became hard film after drying. It was not observed when prereating hardwood and herbaceous materials (13,14). This might be due to some additives from paper or binders contained in ink. These materials could be solubilized in high temperature and high pH solution (16). Actually, pH of 10% ammonia solution is approx. 11.5. And then, they were dispersed and made a film on fiber surface, resulting in the interference of enzyme access to fibers. Therefore, it could be concluded that percolation process, which has shown a good performance in hardwood and herbaceous substrate treatment, was not suitable for newspaper pretreatment.

Pretreatment by Batch Process

Ammonia-hydrogen Peroxide Treatment

When it was considered that ink and additives added in paper production could be solubilized or reacted with ammonia in high temperature, newspaper should have been pretreated under 100 °C. Since ammonia alone was not effective in pretreating newspaper at low temperature, hydrogen peroxide was added to ammonia solution as an oxidant. In addition, hydrogen peroxide can help ink removal to break chemical cross-linkage between the binders of the ink (11).

With the concept above, the pretreatment process was devised that can remove ink and swell the substrate easily. In this study (9), we investigated parameters that affect enzymatic digestibility when treating newspaper in ammonia-hydrogen peroxide mixture on a shaking bath. Optimum condition was found to be 40°C, 3h, 130 strokes/min, and 4 wt% ammonia-2 wt% hydrogen peroxide. In this condition, the digestibility obtained was almost 90% of theoretical or 25% higher than digestibility of untreated or 4 wt% ammonia treated substrates when enzyme loading was 60 IFPU/g glucan. Such an increase in digestibility probably depends on ink removal and substrate swelling because the appearance of the ammonia-hydrogen peroxide treated sample was very different from that of the ammonia-treated sample. The volume of the former was about 1.5 times bigger than the latter's volume. Also a significant amount of dark-colored ink components was observed in the upper portion of the bottle, whereas a very small amount of ink was observed in ammonia-treated sample.

Enzymatic Digestibilities of Wastepapers

In order to compare the enzymatic digestibilities of different wastepapers, these papers were arbitrary selected from our lab: newspaper, office paper, and corrugated paper. Their typical initial compositions are listed in Table 2. Newspaper has the highest hemicellulose and klason lignin, and the lowest glucan and ash among them. It can be expected from the fact that newspaper is made from mainly groundwood pulp with a small amount of fillers. Office paper has the lowest hemicellulose and lignin and highest glucan and ash contents.

Table II. Initial composition of newspaper, office paper, and corrugated paper

Substrate	*Glucan(%)*	*XMG(%)*	*Klason lignin(%)*	*Ash(%)*
Newspaper	58.7	13.9	14.4	5.8
Office paper	66.1	8.5	2.3	16.5
Corrugated paper	62.1	11.8	11.0	10.0

Before measuring enzymatic digestibility, each substrate was treated with a 4 wt% ammonia and 2 wt% hydrogen peroxide solution and the results were listed in Table 3. Noticeable point in this table is the solid remaining after the treatment: newspaper, office paper, and corrugated paper lost 4%, 18%, and 15% of original weight, respectively. This means that lots of materials in office and corrugated papers were removed from this pretreatment.

Table III. The composition of solid residue of newspaper, office paper, and corrugated paper after pretreatment with 4 wt% ammonia and 2 wt% H_2O_2

Substrate	*Solids remaining(%)*	*Glucan(%)*	*XMG(%)*	*Klason lignin(%)*
Newspaper	96.1	57.8	12.8	14.2
Office paper	81.9	63.4	8.3	2.0
Corrugated paper	85.0	55.8	9.5	8.7

Enzymatic digestibility was measured at enzyme loading, 30 IFPU/ g glucan, and the results are shown in Fig. 1. Digestibility of newspaper, office paper, and corrugated paper was 70, 85, and 88%, respectively. These values were much higher than those of untreated samples, meaning that wastepaper should be pretreated before enzymatic hydrolysis. The digestibilities of office and corrugated papers at enzyme loading, 30 IFPU/g glucan, were almost equal to that of newspaper hydrolyzed at 60 IFPU/g glucan (9). Therefore, it is said that newspaper is the toughest substrate for enzymatic hydrolysis.

Most office paper is made from chemically pulped fibers and has much less lignin content. Lower lignin content in office paper may result in higher digestibility. The highest digestibility value was, however, achieved with corrugated paper even it has almost four times lignin content than office paper. In Korea, corrugated paper is generally made from recycled pulp obtained through deinking pulping process. Both mechanical and chemical actions during this process could change the structure of cellulose chain. Otherwise it is difficult to explain why the digestibility of corrugated paper is higher than that of office paper.

Surfactant Effect

Surfactants have been used in deinking process of recycled paper to promote ink detachment from fibers and help the dispersion of detached ink in process water (17). Their typical dosages are 0.25-1.5 wt% based on dry paper weight. Surfactant, also, improves cellulose hydrolysis when it is added with enzyme because it helps cellulase to desorb easily from the cellulose surface after

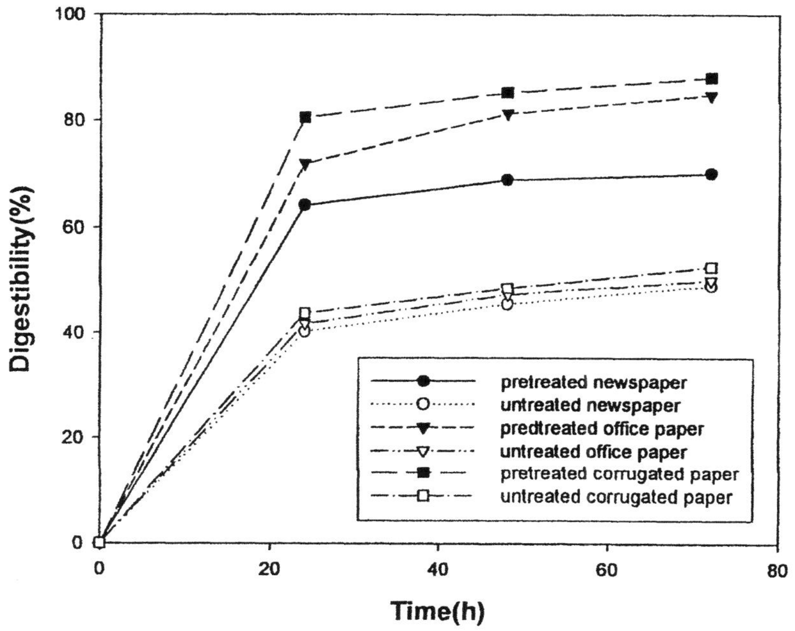

Figure 1. Enzymatic digestibility of newspaper, office paper and corrugated paper treated with 4 wt% ammonia and 2 wt% H_2O_2 solution.

hydrolysis reaction (18). Since nonionic surfactants are most common in deinking (11), NP series surfactants were selected as listed in Table 1.

Fig. 2 shows the effect of hydrogen peroxide and nonionic surfactant (NP-5) on enzymatic digestibility when newspaper was pretreated with 4 wt% ammonia solution. In this figure, control sample means a substrate pretreated with a 4 wt% ammonia and 2 wt% hydrogen peroxide solution. In the case of control run, the dark-colored band was observed in the upper portion of the bottle. This band was, however, disappeared when NP-5 was added. It was apparent that the dark-colored components were emulsified by surfactant, so these components were evenly distributed among the fibers. A most notable point is digestibility achieved by (ammonia+NP-5) run, which has the same as the (ammonia+H_2O_2 +NP-5) run. If the digestibility of (ammonia+H_2O_2+NP-5) run is compared with that of control, which uses the same concentrations of ammonia and hydrogen peroxide without NP-5, NP-5 effect on digestibility is very obvious. This means that hydrogen peroxide does not needed if NP-5 is added in ammonia solution. Hydrogen peroxide is almost ten times more expensive than ammonia. If hydrogen peroxide does not need in this pretreatment, a lot of money can be saved and consequently brings a substantial reduction of pretreatment cost.

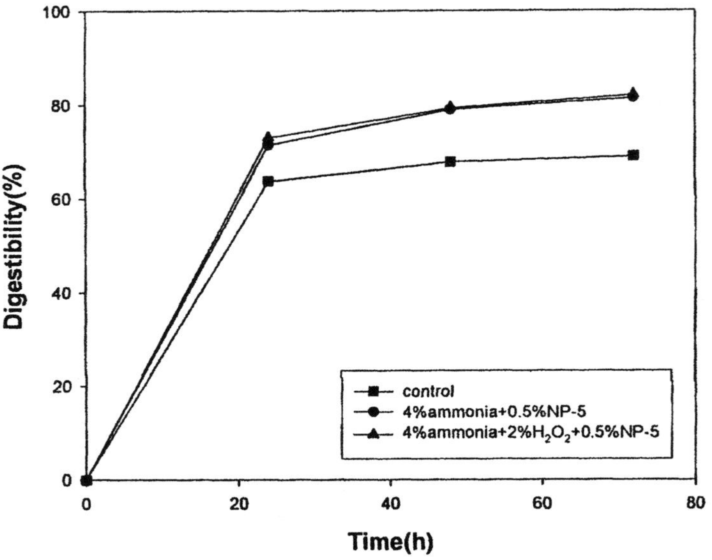

Figure 2. Effect of H_2O_2 and/or NP-5 on enzymatic digestibility of newspaper.

To find out NP-5 effect on digestibility, solid residues recovered after pretreatment were analyzed and listed in Table 4. From this table, it shows that hydrogen peroxide or NP-5 in ammonia solution did not affect that much on the composition of solid residue. Such a high digestibility caused by NP-5 must not be due to residual solid components because the compositions of both samples are almost same. Pretreated sample in our study is washed with water to remove chemicals and detached ink particles. In this stage detached ink particles are dispersed in the aqueous phase and some of them redeposit on the fiber surface. This redeposition may interfere enzyme access to substrate. After washing the pretreated sample with water, we found much less ink particles in the NP-5-contained sample than those in the control sample through microscope. But we cannot explain this digestibility enhancement by NP-5 clearly yet.

Fig. 3 shows the effect of NP-5 loading on digestibility when substrate was treated in 4 wt% ammonia solution. NP-5 loading was not affected on digestibility, so 0.5 wt% was selected for further experiments.

Fig. 4 shows the effect of HLB value on digestibility when substrate is treated in 4 wt% ammonia solution. The hydrophile-lipophile balance (HLB) represents the relative attraction of an emulsifier for water and oil. Generally, values below 7 refer to hydrophobic agents and values above 7 refer to hydrophilic ones. The optimum HLB value depends on ink composition (17). For washing deinking, the values are usually above 10. As shown in Fig. 4, digestibility seems to decrease a little if HLB value is increased. NP-5 (HLB 10.0) shows the highest digestibility among them.

Table IV. Effect of hydrogen peroxide or surfactant in 4 wt% ammonia solution on the composition of solid residue of newspaper

Pretreatment	*Solids remaining(%)*	*Glucan(%)*	*XMG(%)*	*Klason lignin(%)*
4 wt% ammonia	96.8	57.6	12.8	14.2
4 wt% ammonia + 2 wt% H_2O_2	97.0	57.5	12.3	14.2
4 wt% ammonia + 0.5 wt% NP-5	94.4	56.0	12.4	14.2

So far, surfactant was added in pretreatment stage with ammonia solution. Since surfactant can enhance digestibility if it is added in hydrolysis stage with enzyme, the effect of input stage of NP-5 on digestibility was investigated. As shown in Fig. 5, the digestibility obtained when surfactant was added in pretreatment stage was approximately 3% higher than digestibility obtained when surfactant in hydrolysis stage. This difference does not seem too much, but it is said that surfactant addition in pretreatment stage is a little more effective on cellulose hydrolysis than that in enzymatic hydrolysis.

Conclusions

Pretreatment condition for wastepaper requires much milder condition than those for woody and herbaceous materials. In order to increase enzymatic digestibility of newspaper, the method that adds nonionic surfactant into ammonia or ammonia-hydrogen peroxide solution was found to be very effective. We have believed from our previous studies that enzymatic digestibility can be increased only when ammonia is used with hydrogen peroxide together. But if ammonia was used with nonionic surfactant, the same digestibility was achieved without hydrogen peroxide. This means that ammonia-surfactant pretreatment does not need hydrogen peroxide and consequently brings a substantial reduction of pretreatment cost. It was also found that newspaper was the hardest substrate for enzymatic hydrolysis among newspaper, office paper, and corrugated paper.

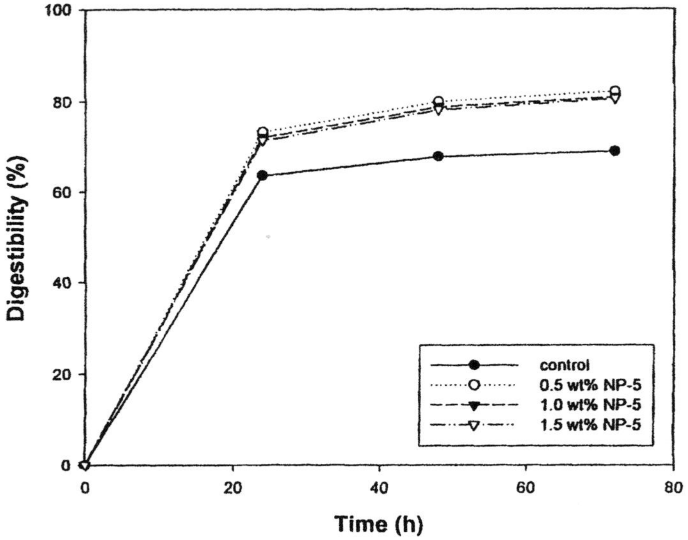

Figure 3. Effect of NP-5 loading in 4 wt% ammonia solution on enzymatic digestibility of newspaper.

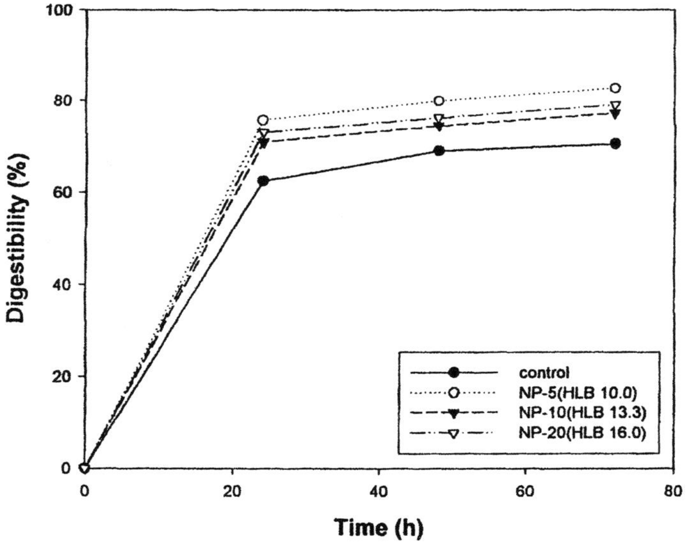

Figure 4. Effect of HLB value in 4 wt% ammonia solution on enzymatic digestibility of newspaper. (surfactant loading = 0.5 wt%)

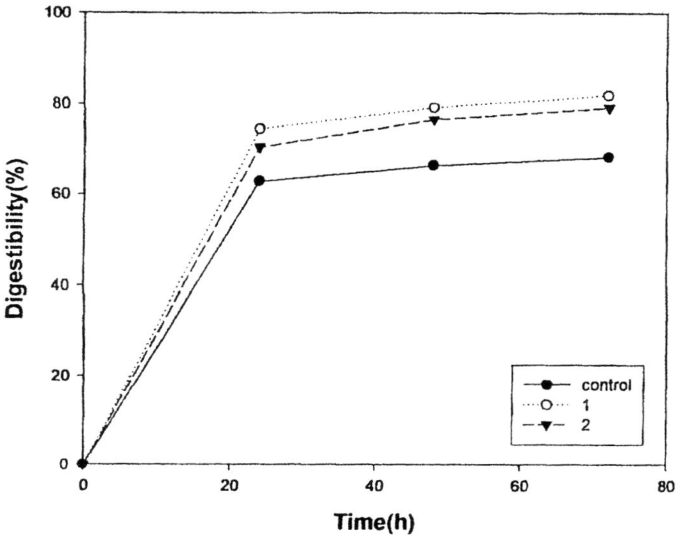

Figure 5. Effect of input stage of NP-5 on enzymatic digestibility of newspaper. (1: 4 wt% ammonia and NP-5 add together in pretreatment stage, 2: 4 wt% ammonia adds in pretreatment stage and then NP-5 in hydrolysis stage.)

References

1. Wyman, C. E. *Bioresour. Technol.* **1994,** 50, 3-16.
2. Scott, C. D.; Davison, B. H.; Scott, T. C.; Woodward, J.; Dees, C.; Rothrock, D. S. *Appl. Biochem. Biotechnol.* **1994,** 45/46, 641-653.
3. Khan, A. W.; Labrie, J.; McKeown, J. *Radiat. Phys. Chem.* **1987,** 29, 117-120.
4. Zheng, Y.; Lin, H. M.; Tsao, G. T. *Biotechnol. Prog.* **1998,** 14, 890-896.
5. Holtzapple, M. T.; Lundeen, J. E.; Sturgis, R.; Lewis, J. E.; Dale, B. E. *Appl. Biochem. Biotechnol.* **1992,** 34/35, 5-21.
6. Kim, J. S.; Lee, Y. Y.; Park, S. C. *Appl. Biochem. Biotechnol.* **2000, 84-86,** 129-139.
7. Capek-Menard, E.; Jollez, P.; Chornet, E.; Wayman, M.; Doan, K. *Biotechnol. Lett.* **1992,** 14, 985-988.
8. Nikolov, T.; Bakalova, N.; Petrova, S.; Benadova, R.; Spasov, S.; Kolev, D. *Bioresour. Technol.* **2000,** 71, 1-4.
9. Kim, S. B.; Moon, N. K. *Appl. Biochem. Biotechnol.* **2003,** 105-108, 365-373.

10. Moon, N. K.; Kim, S. B. *Korean J. Biotechnol. Bioeng.* **2001,** 16, 446-451.
11. Lassus, A. In *Recycled Fiber and Deinking;* Gottsching, L., ed.; Papermaking Science and Technology and Book 7; Gummerus Printing: Finland, 2000; pp 241-264.
12. National Renewable Laboratory, Golden, CO., Chemical Analysis and Testing Standard Procedures, 1996.
13. Kim, S. B.; Lee, Y. Y. *Appl. Biochem. Biotechnol.* **1996,** 57/58, 147-156.
14. Kim, S. B.; Um, B. H.; Park, S. C. *Appl. Biochem. Biotechnol.* **2001,** 91-93, 81-94.
15. Huh, S. J.; Kim, S. B.; Park, S. C. *Korean J. Biotechnol. Bioeng.* **1999,** 14(3), 1-7.
16. Allen, L. H.; Sithole, B. B.; Lapointe, C. L.; Truitt, R. *J. Pulp and Paper Sci.* **1997,** 23(4), 157-160.
17. Borchardt, J. K.; Recycling Paper; *Kirk-Othmer Encyclopedia of Chemical Technology;* John Wiley & Sons, Inc., USA, 2000.
18. Helle, S. S.; Duff, S. J. B.; David, G. C. *Biotechnol. Bioeng.* **1993,** 42, 611-617.

Chapter 3

Ethanol from Lignocellulosic Materials: Pretreatment, Acid and Enzymatic Hydrolyses, and Fermentation

Mohammad J. Taherzadeh[1,2] and Claes Niklasson[1]

[1]Department of Chemical Engineering and Environmental Science, Chalmers University of Technology, SE–412 96, Göteborg, Sweden
[2]Department of Chemical Engineering, Isfahan University of Technology, Isfahan, Iran

Ethanol production from lignocellulosic materials is a major global task in producing liquid fuel by sustainable processes. The structure of the lignocellulose is usually opened by dilute-acid hydrolysis or steam explosion in a pretreatment step, while the resulting cellulose and hemicelluloses can be cleaved to the monomers (sugars) by acid or enzymatic hydrolyses. The hydrolyzates are then fermented to ethanol by using baker's yeast or other microorganisms. The acid hydrolysis suffers from a number of inhibitory by-products including furans, phenolic compounds and carboxylic acids, whereas the enzymatic one is still expensive and slow. Very good progress has been made within the last two decades on development of pretreatments, hydrolyses, fermentation techniques and recombinant microorganisms. These advances are briefly reviewed here.

Introduction

Lignocellulosic materials represent an abundant, and largely unused, source of raw materials for the production of fuel ethanol. These materials can be obtained at a low cost from a variety of resources, *e.g.* forest residues, municipal solid waste, waste paper and crop residue resources *(4)*. The amounts of the three major groups of polymers, i.e. cellulose, hemicellulose and lignin, depend on the type of material. Garotte *et al.* *(5)* present a compilation of compositions of lignocelluloses of different hardwoods, softwoods and agricultural residues reported in publications. The hardwoods (white birch, aspen, red maple, *Eucalyptus*, *Populus* and oak) contain 39%-54% cellulose, 14%-37% hemicellulose and 17%-30% lignin. The corresponding values for softwoods (pines and firs) are 41%-50% cellulose, 11-27% hemicellulose and 20-30% lignin. The composition of different agricultural residues varies widely. For instance, wheat straw may consist of up to 50% cellulose, whereas the cellulose content reported for sunflower seed hulls is only 24%.

The basic steps necessary for obtaining fermentable sugars include a pretreatment, followed by one or several hydrolysis steps, in which the actual hydrolysis of the polymers into monomeric sugars takes place. The sugars are then fermented to ethanol by using several types of microorganisms. These three steps are discussed separately below.

Pretreatment

The purpose of the pretreatment is primarily to open up the structure of the material to facilitate access to the cellulose structure. However, a small amount of sugars from the hemicellulose may be formed already during the pretreatment process. The pretreatment process can be carried out with a variety of methods *(6, 7)*:

1. Physically by *e.g.* ball-milling, two-roll milling, colloid milling, hammer milling, vibro energy milling, high pressure steaming, extrusion, expansion, pyrolysis, and high energy radiation.
2. Chemically by *e.g.* alkalis (NaOH, NH_3, $(NH_3)_2SO_4$), acids (H_2SO_4, HCl, H_3PO_4), gases (ClO_2, NO_x, SO_2), oxidizing agents (H_2O_2, O_3), cellulose solvent (Cadoxen – ethylene diamine and water, or CMCS – sodium tartarate, ferric chloride, sodium sulfite and sodium hydroxide solution) and solvent extraction by ethanol-water, benzene-ethanol, ethylene glycol, or butanol-water.
3. Biologically by *e.g.* enzymes or fungi.

However, not all of these methods are desired for pretreatment of the lignocellulosic materials because of technical or economic feasibility. In some cases, a method is used to increase the efficiency of another method. For instance, milling could be applied to create a better steam explosion by reducing the chip size *(8)*. Furthermore, it should be noticed that the selection of pretreatment method should be compatible with the selection of hydrolysis. For example, if acid hydrolysis is to be applied, a pretreatment with alkali may not be beneficial. The most commonly applied pretreatment methods will be discussed here.

Steam Explosion

Steam explosion is among the most frequent methods for the pretreatment of lignocellulosic materials. The method involves treatment of the chipped biomass with high-pressure saturated steam followed by a rapid reduction of the steam pressure to obtain an explosive decompression. Either high temperature and low residence time (*e.g.* 270°C and 1 min) or lower temperature and longer residence time (*e.g.* 175°C and 30 min) could be applied for an optimal pretreatment *(9)*. Early processes were running at high pressure (220-270°C) and short residence time (40-90 s), whereas recent investigators use lower temperature (190-200°C) and a longer residence time of about 10 min *(10)*. However, the type of lignocellulose, residence time, temperature, chip size and humidity are important factors in optimizing the treatment conditions of a steam explosion process *(11)*. The variation in these factors results in various degrees of degradation of lignin, hemicellulose and cellulose *(12, 13)*. Gregg and Saddler *(14)* give examples of recommended time and temperature/pressure to be used for various feedstocks.

Dilute-acid Pretreatment

It is known that the addition of any kind of acid or gases that results in acidic conditions for steam explosion, such as H_2SO_4, SO_2 (or H_2SO_3), or CO_2 (or H_2CO_3), will successfully improve the enzymatic hydrolysis, decrease the formation of inhibitory compounds, and lead to a better removal of hemicellulose *(15)*. The effects are almost similar for hardwoods, softwoods and herbaceous materials. SO_2 presumably diffuses into the wood chips as sulfurous acid, H_2SO_3, which turns into sulfuric acid during steaming *(16)*. Sulfuric acid is the most extensively studied, apparently because of its low price and high efficiency *(17)*. In addition to the factors affecting steam explosion, the acid concentration plays an important role in dilute-acid pretreatment. There are a number of examples of the treatment conditions for various feedstocks in the literature. As examples we may mention 220°C, 30 s residence time, and 1% H_2SO_4 for the treatment of sugar cane bagasse *(18)* and 210°C, 2 min residence time and 0.175% H_2SO_4 for the treatment of *Eucalyptus grandis (19)*. More severe conditions (*e.g.* higher acid concentration, higher temperature and/or longer residence time) will result in degradation of cellulose *(2)*, which will be discussed later in more detail.

Ammonia Fiber (or Freeze) Explosion (AFEX)

The concept of AFEX is similar to steam explosion, except in the type of gas: lignocelluloses are exposed to ammonia instead of water vapor. In a typical treatment, one kg dry biomass is exposed to 1-2 kg liquid ammonia at high temperature (*e.g.* 50-90°C) and high pressure for a period of time (*e.g.* 30 min), and then the pressure is swiftly reduced *(7, 20)*. The efficiency of the AFEX method in cellulose decrystallization, with increased accessible surface area, dramatically increases lignocellulose susceptibility to enzymatic attack in the hydrolysis step *(21)*. In spite of acid pretreatment, AFEX does not dissolve hemicellulose, which is an advantage for the enzymatic hydrolysis *(22)*. Furthermore, no inhibitory compounds for the fermentation step are formed *(23)*, and the method is not affected by the chip size so that milling is unnecessary. There are many adjustable parameters in the AFEX process: ammonia loading, water loading, temperature, time, blowdown pressure, and number of treatments *(21)*. A lower lignin content of the lignocellulose results in a higher efficiency of the AFEX-pretreated enzymatic hydrolysis. While 90% hydrolysis of cellulose and hemicellulose of Bermuda grass (5% lignin) was obtained, the efficiencies for the hydrolysis of newspaper (18-30% lignin) and aspen chips (25% lignin) were less than 40% and 50%, respectively *(24)*.

Alkaline Pretreatment

Some bases such as NaOH, $Ca(OH)_2$, NH_4OH can be used for pretreatment of the lignocelluloses *(25)*. Dilute sodium hydroxide with concentration of 0.8 to 50 g/l *(26)* or 5-10 g NaOH/100 g feedstock *(6)* has primarily been used for this purpose. The mechanism behind this method is believed to be saponification of intermolecular ester bonds crosslinking xylan hemicelluloses and other components *(7)*. Consequently, the base causes swelling, leading to an increased internal surface area, decrease in the degree of polymerization, decrease in the crystallinity, separation of structural linkages between lignin and carbohydrates, and disruption of the lignin structure *(6)*. NaOH showed more effective results in increasing the susceptibility of the residual corn husk towards enzymes, compared to sulfuric and phosphoric acids, yielding 83-96% reducing sugars *(26)*.

Hydrolysis

Hydrolysis is the second step in production of ethanol from lignocellulosic materials. It involves cleaving the polymers of cellulose and hemicellulose. The cellulose usually contains only glucan, whereas hemicellulose contains polymers of several sugars such as mannan, xylan, glucan, galactan and arabinan. Consequently, the main hydrolysis product of cellulose is glucose, whereas the hemicellulose gives rise to several pentoses and hexoses. However, softwood hemicellulose is mainly composed of mannose, whereas the dominant sugar in hardwood hemicellulose is xylose *(2)*. The hydrolysis can be obtained chemically or enzymatically.

Chemical Hydrolysis

Chemical hydrolysis involves exposure of lignocellulosic materials to a chemical for a period of time at a specific temperature, and results in sugar monomers from cellulose and hemicellulose polymers. Acids are dominantly applied in chemical hydrolyses. Sulfuric acid is the most investigated acid *(27)*, although other acids such as HCl *(28)* have also been used. Acid hydrolyses can be divided into two groups: (a) concentrated-acid hydrolysis and (b) dilute-acid hydrolysis.

Concentrated-acid hydrolysis

Hydrolysis of lignocellulose by concentrated sulfuric or hydrochloric acids is a relatively old process. Braconnot in 1819 first discovered that cellulose could be converted to fermentable sugar by concentrated acid *(29)*. Concentrated-acid processes are generally reported to give higher sugar and ethanol yield, compared to dilute-acid processes. However, dilution and heating of the concentrated acid during the hydrolysis process make it extremely corrosive. Therefore, the process requires either expensive alloys or specialized non-metallic constructions, such as ceramic or carbon-brick lining. The high investment and maintenance costs have greatly reduced the commercial potential for this process. Furthermore, the environmental impact strongly limits the application of hydrochloric acid *(30)*.

Dilute-acid hydrolysis

Among the chemical hydrolysis methods, dilute-acid hydrolysis is probably the most commonly applied. It is a method that can be used either as a pretreatment preceding enzymatic hydrolysis, or as the actual method of hydrolyzing lignocellulose to the sugars *(31)*. The first established dilute-acid hydrolysis process was probably the Scholler process. This was a batch process, in which the wood material was kept in 0.5% sulfuric acid at 11-12 bar for approximately 45 minutes *(32)*. Nowadays, almost all dilute-acid hydrolysis processes are performed in a batch mode with a retention time of a few minutes. However, there have been some studies concerning continuous hydrolysis in plug flow reactors *(33)*. A recent study *(2)* presents data for one-stage dilute-acid hydrolysis, where 0.5% sulfuric acid was used at temperatures of 188-234°C and a retention time of 7 minutes. A major part of the hemicellulose (more than 80%) could be hydrolyzed by dilute-acid hydrolysis at temperatures less than 200°C, but the maximum overall glucose yield occurred at a hydrolysis temperature higher than 220°C. This is due to the larger recalcitrance of cellulose to hydrolysis. In no case was a better yield than 40% of the theoretical glucose yield from glucan obtained (Figure 1).

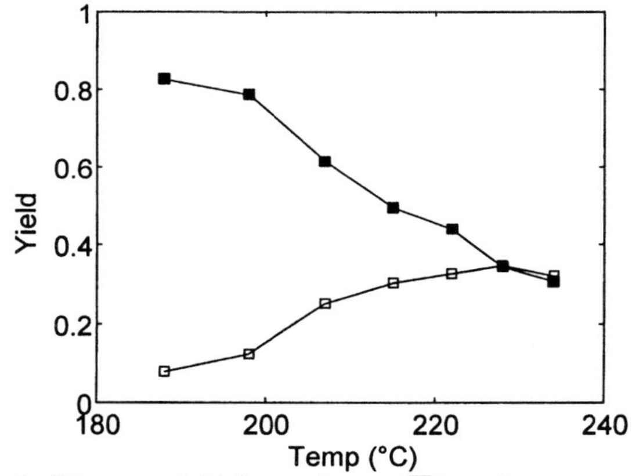

Figure 1. Glucose yield from glucan (□) and mannose yield from mannan (■) in a one-step dilute-acid hydrolysis of (25% dry weight) spruce as a function of hydrolysis temperature (1).

To avoid degradation of monosaccharides at high temperatures, the dilute-acid hydrolysis can be carried out in two (or more) stages. In the first stage, which should be carried out at relatively mild conditions, hemicellulose is converted to sugar monomers. It is considered as equivalent to a dilute-acid pretreatment step. In the second stage, the residual solid is hydrolyzed at more severe conditions allowing cellulose to be hydrolyzed *(34)*. When dilute-acid hydrolysis is used for pretreatment of lignocellulosic materials, considerably milder conditions than those mentioned above are applied. In a one-stage pretreatment, a temperature between 140 and 170 °C can be used, but two treatments at about 120°C for a longer time may also be used *(35)*.

The main drawback of the acid hydrolysis processes is the formation of undesirable by-products. This not only lowers the yield of sugars, but several of the by-products severely inhibit the formation of ethanol in the fermentation process. Potential inhibitors are furfural, 5-hydroxymethylfurfural (HMF), levulinic acid, acetic acid, formic acid, uronic acid, 4-hydroxybenzoic acid, vanillic acid, vanillin, phenol, cinnamaldehyde, formaldehyde, etc. *(1, 36)*. Some inhibitors, such as terpene compounds, are initially present in the wood, but apparently most of the inhibitors are formed in the hydrolysis process.

Enzymatic Hydrolysis

Enzymatic hydrolysis of cellulose and hemicellulose can be carried out by highly specific cellulase and hemicellulase enzymes (glycosylhydrolases). This group includes at least 15 protein families and some subfamilies *(37)*. A complete cellulase system consists of three classes of enzymes: 1,4-β-D-glucan cellobiohydrolases, endo-1,4-β-D-glucanases and 1,4-β-D-glucosidases. While the first and second enzymes cleave the cellulose to cellobiose, it is further split to glucose by the third enzyme. β-glucosidase is not a cellulase, but its action is very important to complete depolymerization of cellulose to glucose. Since hemicellulose contains different sugar units, the hemicellulytic system is more complex and involves at least endo-1,4-β-D-xylanases, exo-1,4-β-D-xylosidases, endo-1,4-β-D-mannanases, β-mannosidases, α-glucuronidases, acetyl xylan esterases, α-L-arabinofuranosidases, and α-galactosidases *(38)*. Several species of bacteria including *Clostridium, Cellumonas, Bacillus, Thermomonospora, Ruminococcus, Bacteriodes, Erwinia, Acetovibrio, Microbispora*, and *Streptomyces* and fungi including *Tricoderma, Penicillium, Fusarium, Humicola, Phanerochaete* and *Schizophillum* spp. are able to produce cellulases and hemicellulases *(7, 37)*. Among these microorganisms, *Trichoderma* was mentioned as the most efficient cellulose-hydrolysing organism *(17)*. The maximum cellulase and β-glucosidase activities occur at 40-60°C and a pH of 4.0 to 5.0. However, the optimal condition may change with the hydrolysis residence time *(39, 40)*. The order of magnitude for the incubation time is 1-3 day.

There are several advantages and disadvantages of dilute-acid and enzymatic hydrolyses compared to each other (Table I). Enzymatic hydrolysis is carried out under mild conditions, whereas high temperature and low pH result in corrosive conditions for the acid hydrolysis. While it is possible to obtain a cellulose hydrolysis of close to 100% by enzymatic hydrolysis after a pretreatment *(41)*, it is difficult to achieve such high yield with the acid hydrolysis. Furthermore, the previously mentioned inhibitory compounds are formed during acid hydrolysis, whereas this problem is not so severe for enzymatic hydrolysis.

Enzymatic hydrolysis has its own problems in comparison to the dilute-acid hydrolysis. A hydrolysis of several days is necessary for enzymatic hydrolysis *(39)*, whereas a few minutes are enough for the acid hydrolysis *(2)*. The prices of the enzymes are still very high and could be decreased dramatically by increasing the enzyme-specific activity *(42)*. Another problem of the enzymatic hydrolysis is that the sugars released inhibit the enzymes during hydrolysis. In order to overcome this latter problem, simultaneous saccharification and fermentation (SSF) was invented, as distinct from a separate hydrolysis and fermentation (SHF). In SSF, the glucose produced by the hydrolysis is consumed immediately by the fermenting microorganism, which avoids end-product inhibition of β-glucosidase. However, since the optimal temperatures of the hydrolysis and fermentation are 45-50°C and 30°C, it is difficult to obtain the entire process at optimum conditions *(17)*.

Table I. A comparison between dilute-acid and enzymatic hydrolyses

Comparing variable	*Dilute-acid hydrolysis*	*Enzymatic hydrolysis*
Mild hydrolysis conditions	No	Yes
High yield of hydrolysis	No	Yes
Avoiding formation of inhibitory by-products	No	Yes
Avoiding product inhibition during hydrolysis	Yes	No
Low cost of hydrolysis	Yes	No
Short time of hydrolysis	Yes	No

Fermentation of the hydrolyzates

Fermentation of the lignocellulosic hydrolyzates is more difficult than the well-established processes of ethanol production from *e.g.* molasses and starch. Hydrolyzates contain a broader range of inhibitory compounds, where the composition and the concentration of these compounds depend on the type of lignocellulosic materials and the chemistry and nature of the pretreatment and hydrolysis processes. Secondly, the hydrolyzates of hemicelluloses contain not only hexoses but also pentoses, where xylose is the dominant sugar in the hydrolyzates from hardwood hemicelluloses. Therefore, the fermenting microorganism should be able to produce ethanol from the hydrolyzates with a high yield and productivity, withstand potential inhibitors, and produce ethanol from pentoses, as well as being safe for humans. Baker's yeast (*Saccharomyces cerevisiae*) is the most commercially used microorganism for ethanol production, but it cannot take up xylose. This section is devoted to the effect of the inhibitory compounds, different fermentation techniques, and the xylose-fermenting microorganisms.

The effects of inhibitory compounds on the fermentation

The by-products mentioned earlier inhibit the fermentation by different mechanisms. As a function of conditions and method of hydrolysis, different by-products will dominate. Most likely, a combination of the action of several substances is the reason for observed inhibition *(43)*. Furthermore, it is not only the quantitatively dominant inhibitors which determine the fermentability of a hydrolyzate. The toxicity of a hydrolyzate is found to differ from that of a synthetic medium with the same amount of the major hydrolyzate inhibitors added, indicating the importance of other inhibitors present in small amounts *(44)*.

Carboxylic acids

Acetic acid, formic acid and levulinic acid are the most common carboxylic acids found in the hydrolyzates. Acetic acid is not only a by-product of hydrolysis *(45)* but is also a well-known by-product in fermentation *(46)*. Acetic acid is mainly formed from acetylated sugars in the hemicellulose, which are cleaved off already at mild hydrolysis conditions. Therefore, the acetic acid yield in the hydrolysis does not significantly depend on the severity of the hydrolysis process *(2)*. Hydrolysis of hardwoods (alder, aspen and birch) at 198-234°C, 0.5 g/l H_2SO_4, and 33% wood dry materials for 7 min resulted in approximately 10 g/l acetic acid, whereas the hydrolysis of softwoods (pine and spruce) produced 3 g/l acetic acid at similar conditions *(2)*.

Acetic acid is sometimes mentioned as an important inhibitor *(47, 48)*. Since acetic acid has a dissociation constant of 4.75 in water, it will be partly dissociated at a pH of 5-5.5 (typical values for the fermentation). It is generally accepted that the effect of the undissociated part of the acid is larger than the effect of the dissociated part *(49)*. The undissociated carboxylic acids can diffuse through the cell membrane *(50, 51)*. Since the intracellular pH is higher than that of the extracellular medium *(52)*, the undissociated acid which has diffused into the cell is partly dissociated into acetate and hydrogen ions, thereby potentially lowering the intracellular pH *(53, 54)*. The optimum extracellular pH for growth of *S. cerevisiae* is about 5 *(55)*, although growth is possible down to a pH as low as 2.5. Without added acetic acid, there is no clear effect on the specific growth rate of the yeast cells down to a pH of 3.5 *(3)*. However, a clear effect on both $Y_{ATP/X}$ and the minimum pH of growth can be observed in the presence of acetic acid in the medium (Figure 2).

Formic and levulinic acids are other weak carboxylic acids (dissociation constants 3.75 and 4.4 respectively), which are found in hydrolyzates. Formic acid is most likely formed from degradation of HMF *(56, 57)*, although other parallel formation routes are possible. Formic acid is a stronger inhibitor than acetic acid *(57)*, and acts inhibitory to the fermentation process above a concentration of about 1 g/l *(58)*. Levulinic acid is a degradation product of HMF *(56)*, and was shown to have a negative effect on fermentability of the hydrolyzates *(57)*. However, due to the low concentration of formic and levulinic acid normally found in hydrolyzates, they are probably of secondary importance with respect to inhibitory effects.

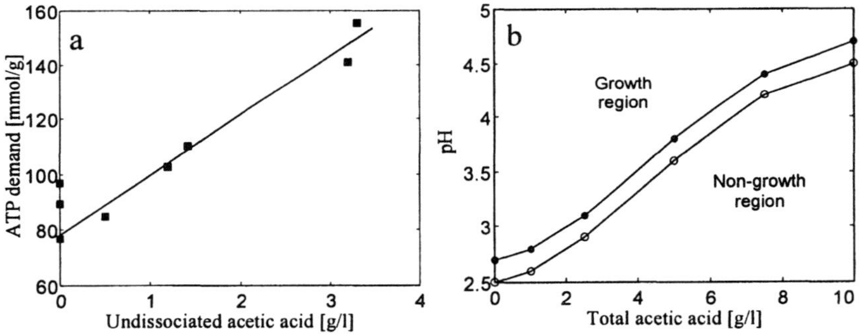

Figure 2. (a) The ATP demand for the growth of S. cerevisiae in media containing different concentration of undissociated acetic acid, and (b) region of anaerobic growth of S. cerevisiae as a function of medium pH and total acetic acid concentration. Under the lower line no growth occurs and above the upper line $\mu \geq 0.2\ h^{-1}$ *(Reproduced from reference (3). Copyright 1997 Elsevier)*

Furans

Furfural and HMF are the only furans usually found in hydrolyzates in significant amounts *(2, 59)*. The inhibitory effects of these furans have previously not been quite clear, and partly contradictory results have been reported in the literature (see *e.g.* *(60-68)*). However, studies in the past few years have provided a better understanding of the physiological effects of these inhibitors *(69-74)*.

Furfural has been found to inhibit the *in vitro* activity of several important enzymes in the primary carbon catabolism such as hexokinase, aldolase, phosphofructokinase, triosephosphate dehydrogenase and alcohol dehydrogenase. Of these, triosephosphate dehydrogenase and alcohol dehydrogenase appeared to be the most sensitive enzymes *(64)*. However, the inhibition of certain non-glycolytic enzymes, such as pyruvate dehydrogenase and aldehyde dehydrogenase, is even more severe *(72)*. Consequently, cell growth is more sensitive to the presence of furfural than the ethanol production from glucose. Interestingly, furfural can be converted by the yeast, forming mainly furfuryl alcohol and furoic acid. This conversion has been reported not only for *Saccharomyces cerevisiae*, but also for a number of other yeasts of the genera *Torulopsis, Pichia* and *Rhodotorula (70, 75-77)*. The yeast *S. cerevisiae* has a capacity to convert furfural at a maximum specific conversion rate of 0.6 g/g·h. However, at that conversion rate, cell growth stops. The maximum conversion rate with a maintained cell growth is approximately 0.15 g/g·h. The clearly dominating product is furfuryl alcohol, and less than 1% of the furfural is converted to furoic acid *(70)*. In addition to furfuryl alcohol and furoic acid, an acyloin product of furfural has been reported *(1)*. This metabolite, 3-(2-furfuryl)-2-hydroxy-2-methyl-3-oxo-propanoic acid, is produced when the cells are growing on glucose and a high concentration of furfural (above 2 g/l) is present in the medium.

HMF is not as severely toxic to *S. cerevisiae* as furfural *(69)*. This is in line with the observation that the *in vitro* inhibition of the enzymes pyruvate dehydrogenase and aldehyde dehydrogenase is smaller by HMF than by furfural. On the other hand, the conversion rate of furfural is about 4 times faster than the conversion rate of HMF. This means that HMF remains much longer than furfural in the medium, and consequently, the effects of HMF last longer than those of furfural. Furthermore, when furfural and HMF are present simultaneously in the medium, the conversion rates of both compounds are lowered. This results in the presence of inhibitors for a longer time in the medium, and in a higher toxicity for the cells *(69)*. The main products of HMF conversion by *S. cerevisiae* were shown to be hydroxymethyl-furfuryl alcohol, 5-hydroxymethyl-furan-carboxylic acid and a condensation product of HMF and pyruvic acid. Each of these compounds could be the dominant product of HMF depending on the cultivation condition.

Phenolic compounds

A large number of phenolic compounds have been found in hydrolyzates *(78, 79)*. However, reported concentrations are normally a few milligrams per liter *(57)*. This could be due to the low water solubility of many of the phenolic compounds, or a limited degradation of lignin during the hydrolysis process. Among the phenolic compounds, less heavily substituted phenolics are probably the most inhibitory materials in the hydrolyzates *(43, 80)*. The inhibition effects of some of the phenolic compounds are summarized in Table II. Like furans, many of the phenolic substances can be converted by microorganisms. For instance, vanillin, hydroxybenzaldehyde, and syringaldehyde are assimilated by *S. cerevisiae* in the fermentation process *(81)*, and conversion of vanillin to vanillyl alcohol by *Klebsiella pneumoniae* has also been reported *(80)*.

Table II. The inhibition effects of some phenolic compounds

Name	*Concentration (g/l)*	*Inhibition effect*	*Reference*
Phenol	1	Strong	*(82).*
Vanillin	5	Strong	*(81),*
Vanillin	1	Weak	*(81, 82)*
Vanillic acid	<1	No effect	*(83)*
Hydroxybenzaldehyde	0.5	Some inhibition	*(81)*
Syringaldehyde	<0.5	No effect	*(81)*
Syringaldehyde	1.5	Almost strong	*(81)*
4-hydroxybenzoic acid	2	No effect	*(84)*

Fermentation Techniques

Several of the inhibitory compounds found in hydrolyzates can be biotransformed, or, in a few cases, even be fully metabolized by yeast. Conversion occurs for several of the carboxylic acids, furans and phenolic compounds. This suggests that continuous *in situ* detoxification of the hydrolyzate during the fermentation may be possible. However, this requires a suitable mode of operation, and the bioconversion of inhibitors must be taken into account in the design of the process.

Batch cultivation

Batch cultivation is the least demanding method of fermentation with respect to requirements of the control system. After initial pH adjustment the hydrolyzate is used as it is, with the exception of possible additions of minerals or other supplementary nutrients needed, and only temperature and pH need to be controlled during the process. In a completely non-inhibiting hydrolyzate, exponential growth will be obtained after inoculation of the bioreactor. If the hydrolyzate is slightly inhibiting, there will be a relatively long lag phase, during which part of the inhibitors are converted (*e.g.* *(2, 62)*). However, if the hydrolyzate is severely inhibiting, no conversion of the inhibitors will occur, and neither cell growth nor fermentation will occur (see Figure 3). A slightly inhibiting hydrolyzate can thus be detoxified during batch fermentation. However, too high concentration of the inhibitors will cause a complete inactivation of the metabolism. *In-situ* detoxification in batch fermentation may be improved by increasing the initial cell density *(67)*, by adaptation of the cells to the medium before inoculation *(65, 85)*, by choosing optimal reactor conditions to minimize the effects of inhibitors *(3)*, or by genetic modification of the microorganism used to increase its tolerance against the inhibitors.

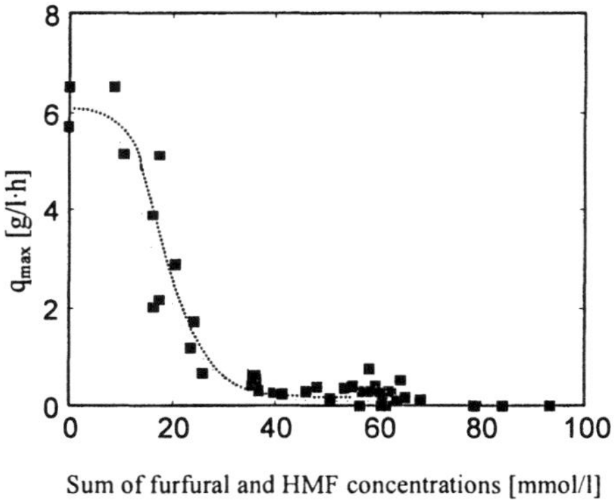

Figure 3. The maximum ethanol production rate (q_{max}) in batch cultivations of S. cerevisiae (10 g/l) on a large number of different hydrolyzates containing various amounts of inhibitors (Reproduced from reference (2). Copyright 1997 American Chemical Society)

Fed-batch cultivation

Fed-batch technique is a promising method for the fermentation of dilute-acid hydrolyzates, and its application for this purpose has recently been studied *(44, 86-88)*. The basic concept behind the success of this technique is the capability of *in situ* detoxification by the cells. Since the yeast has a limited capacity for the conversion of the inhibitors, the achievement of a successful fermentation strongly depends on the feed rate of the hydrolyzate. At a too high feed rate using an inhibiting hydrolyzate, both ethanol production and cell growth can be expected to stop, whereas at a very low feed rate, the hydrolyzate may still be converted but at a very low productivity, which was experimentally confirmed *(44)*. Consequently, there should exist an optimum feed rate (or dilution rate) specific to a particular hydrolyzate. If the hydrolyzate is only slightly inhibitory, a high feed rate can be applied. On the other hand, for a severely inhibiting hydrolyzate, a low feed rate is necessary to prevent build-up of the concentration of inhibitors in the bioreactor to levels which completely stop the cellular metabolism. Since the optimum feed rate is difficult to determine beforehand, and, furthermore, may vary during the cultivation, it would be advantageous to control the feed rate on-line (Figure 4). The variable upon which to base the control should preferably be easy to measure and be measurable on-line, and should be directly related to the fermentation. Carbon dioxide evolution rate (CER) is a good candidate to meet these criteria. A feed rate control based on measured changes of carbon dioxide evolution rate (CER) and different algorithms has recently been developed, which shows good performance in fermentation of fermentable or inhibiting dilute-acid hydrolyzates *(44, 86-88)*.

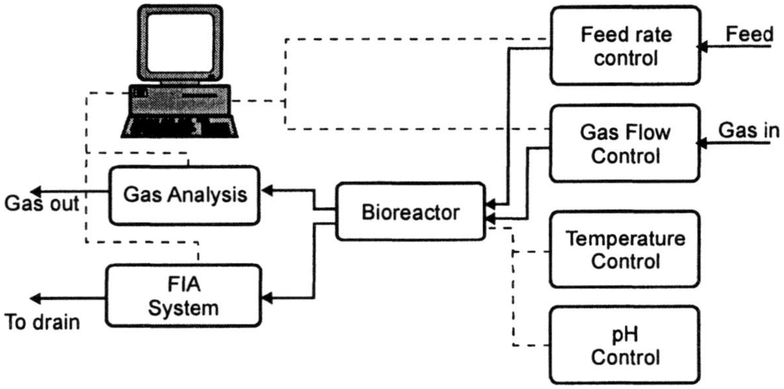

Figure 4. A schematic picture of on-line control of fed-batch cultivation (1).

Continuous cultivation

Continuous cultivation is the third mode of operation for fermentation of hydrolyzates. In spite of a number of potential advantages in terms of productivity, this method has not been much developed in the fermentation of the acid hydrolyzates. The major drawback of the continuous fermentation is that inhibitors present in the medium will limit the specific growth rate of the cells. This will result in the wash-out of the bioreactor, unless a very low dilution rate is applied, giving a very low productivity. Furthermore, at a very low dilution rate the conversion rate of the inhibitors can be expected to decrease due to the decreased specific growth rate of the biomass. Thus, wash-out may occur even at very low dilution rate. Chung and Lee *(67)* faced this problem when they tried to continuously ferment acid hydrolyzates. They reported 90% decrease in the viable cell number after only three residence times. A straightforward continuous cultivation is therefore not an option, but some kind of cell retention is necessary to avoid wash-out. Options include recirculation of the biomass or cell retention by filtration or immobilization of the biomass. Lee *et al.* *(89)* suggested a continuous cultivation with total cell retention (using an internal filter system with pore size of 2 µm). These authors obtained higher ethanol productivity of *S. cerevisiae* fermenting enzymatically hydrolyzed oak wood, but the cell concentration did not increase due to effects of toxic materials.

In a recent work, immobilized cells of *S. cerevisiae* entrapped in Ca-alginate were used to continuously ferment hydrolyzate *(90)* (see Table III). It was concluded that the method could be successfully applied to dilute-acid hydrolyzates, giving a several-fold increase in the fermentation rate compared to a suspended cell process. However, the method still needs further improvement in order to ferment more inhibiting hydrolyzate. Probably, a partial detoxification is still needed in addition to cell immobilization for highly inhibiting hydrolyzates.

Table III. Results of continuous cultivation of a spruce hydrolyzate at steady state conditions by free and immobilized S. cerevisiae at different dilution rates (90).

Parameter	*Immobilized cells*	*Immobilized cells*	*Immobilized cells**	*Immobilized cells*	*Free cells*	*Free cells*
Dilution rate (h^{-1})	0.3	0.5	0.5	0.6	0.1	0.2
Glucose consumption (%)	86	83	83	79	92	Wash out
Mannose consumption (%)	72	67	67	55	63	
Furfural conversion (%)	98	97	97	97	100	
HMF conversion (%)	60	66	68	60	66	
Residual glucose (g/l)	1.4	1.7	1.8	2.2	0.60	
Ethanol yield§ (g/g)	0.45	0.47	0.48	0.47	0.44	
Glycerol yield§ (g/g)	0.04	0.05	0.03	0.04	0.035	

*Furfural concentration in the hydrolyzate was manually increased to 2.5 g/l
§All the yields are reported in grams per gram of consumed glucose and mannose

Pentose Fermentation

Saccharomyces cerevisiae does not have genes encoded for xylose reductase and xylitol dehydrogenase *(91)*. Therefore, it cannot produce ethanol from xylose. There have been several attempts to introduce these genes into *S. cerevisiae* and to overexpress some other genes such as xylulokinase (*e.g.* *(92)*). However, the yeast still suffers from low ethanol yield and productivity *(93)*. There are several other naturally occurring and recombinant microorganisms reported to produce ethanol from xylose. Olsson and Hahn-Hägerdahl *(94)* presented a list of ethanol-producing bacteria, yeast and fungi from xylose. Among the naturally occurring organisms, we can name yeasts *Candida, Pichia Schizosaccharomyces, Kluveromyces* and *Pachysolen* species, fungi *Fusarium, Mucor, Monilia* and *Paecilomyces* species, and bacteria *Clostridium, Bacillus, Bacteroides, Thermoanaerobacter* and *Ervinia* species. Among these microorganisms, *Candida shehatae, Pichia stipitis* and *Fusarium oxysporum* resulted in a high yield (>0.45 g ethanol/g xylose) and a reasonable productivity (>0.17 g/l.h).

Several attempts have been made to genetically modify other microorganisms than *S. cerevisiae* in order to produce ethanol from both hexoses and pentoses. The method principally involves addition and expression of all the genes which are not present in the genome of the microorganisms and are necessary to run the pathway from the sugars to ethanol. For example, the genes for xylose isomerase, xylulokinase, transketolase and transladolase were inserted into bacteria *Zymomonas mobilis* which resulted in an ethanol-producing organism from xylose *(95)*. So far, the most successful recombination has probably been made by Ingram and his group, where they added the genes for pyruvate decarboxylase and alcohol dehydrogenase to *Escherichia coli (96)*. The resulting recombinant was able to produce 41 g/l ethanol from 80 g/l xylose with a productivity of 0.87 g/l.h. Furthermore, they made a similar recombination to *Klebsiella oxytoca,* which is natively able to ferment cellobiose and cellotriose, in order to eliminate the need for β-glucosidase. The strain was further recombinated to contain endoglucanase *(97)*.

Conclusion

The process of producing ethanol from lignocellulose is not a simple task. It has made good progress in the last two decades, but still needs work to become commercialized. Pretreatment can be combined with acid hydrolysis, while enzymatic hydrolysis may be combined with fermentation if a suitable organism (recombinant or natural) can be applied. Hydrolyzates contain quite a number of hydrolysis by-products which are inhibitory to fermentation by *S. cerevisiae* and the other organisms. However, the yeast and some of the organisms are able to degrade or convert many of these inhibitors to less inhibitory compounds, provided that a suitable mode of fermentation is applied. A necessary requirement is to avoid high concentrations of the inhibitors in the culture, which can be achieved by fed-batch or continuous cultivations with a high cell content in the bioreactor.

REFERENCES

1. Taherzadeh M. J. PhD Thesis.Chalmers University of Technology. Göteborg, Sweden, 1999.
2. Taherzadeh M. J.; Eklund R.; Gustafsson L.; Niklasson C.; Lidén G. *Ind. Eng. Chem. Res.* **1997**, *36*, 4659-4665.

3. Taherzadeh M. J.; Niklasson C.; Lidén G. *Chem. Eng. Sci.* **1997**, *52*, 2653-2659.
4. Wiselogel A.; Tyson S.; Johnson D. In: *Handbook on bioethanol: production and utilization.* Wyman C. E., editor. Washington DC: Taylor & Francis; 1996. p. 105-118.
5. Garrote G.; Dominguez H.; Parajo J. C. *Holz Als Roh-und Werkst.* **1999**, *57*, 191-202.
6. Fan L.; Lee Y.; Gharpuray M. *Adv. Biochem. Eng. Biotechnol.* **1982**, *23*, 158-183.
7. Sun Y.; Cheng J. *Bioresource Technol.* **2002**, *83*, 1-11.
8. Ballesteros I.; Oliva J. M.; Navarro A. A.; Gonzalez A.; Carrasco J.; Ballesteros M. *Appl. Biochem. Biotechnol.* **2000**, *84-6*, 97-110.
9. Nesse N.; Wallick J.; Harper J. M. *Biotechnol. Bioeng.* **1977**, *19*, 323-336.
10. Wright J. D. *Chem. Eng. Prog.* **1988**, *84*, 62-74.
11. Duff S.; Murray W. *Bioresource Technol.* **1996**, *55*, 1-33.
12. Josefsson T.; Lennholm H.; Gellerstedt G. *Holzforschung.* **2002**, *56*, 289-297.
13. Karina M.; Tanahashi M.; Higuchi T. *Mokuzai Gakkaishi.* **1992**, *38*, 166-172.
14. Gregg D.; Saddler J. N. *Appl. Biochem. Biotechnol.* **1996**, *57/58*, 711-727.
15. Carrasco J. E.; Saiz M. C.; Navarro A.; Soriano P.; Saez F.; Martinez J. M. *Appl. Biochem. Biotechnol.* **1994**, *45-6*, 23-34.
16. Weil J.; Westgate P.; Kohlmann K.; Ladisch M. *Enzyme Microb. Technol.* **1994**, *16*, 1002-1004.
17. Galbe M.; Zacchi G. *Appl. Microbiol. Biotechnol.* **2002**, *59*, 618-628.
18. Morjanoff P. J.; Gray P. P. *Biotechnol. Bioeng.* **1987**, *29*, 733-741.
19. Emmel A.; Mathias A. L.; Wypych F.; Ramos L. P. *Bioresource Technol.* **2003**, *86*, 105-115.
20. Dale B. E.; Leong C. K.; Pham T. K.; Esquivel V. M.; Rios I.; Latimer V. M. *Bioresource Technol.* **1996**, *56*, 111-116.
21. Holtzapple M. T.; Jun J. H.; Ashok G.; Patibandla S. L.; Dale B. E. *Appl. Biochem. Biotechnol.* **1991**, *28-9*, 59-74.
22. Vlasenko E. Y.; Ding H.; Labavitch J. M.; Shoemaker S. P. *Bioresource Technol.* **1997**, *59*, 109-119.
23. Dale B. E.; Henk L. L.; Shiang M. *Dev. Ind. Microbiol.* **1985**, *26*, 223-33.
24. McMillan J. D. In: *ACS Symposium Series (Enzymatic Conversion of Biomass for Fuels Production).* Himmel M. E.; Baker J. O.; Overend R. P., editors.; 1994. p. 292-324.
25. Rodriguezvazquez R.; Diazcervantes D. *Bioresource Technol.* **1994**, *47*, 159-164.
26. Barl B.; Biliaderis C. G.; Murray E. D.; Macgregor A. W. *J. Sci. Food Agric.* **1991**, *56*, 195-214.

27. Harris E.; Beglinger E.; Hajny G.; Sherrard E. *Ind. Eng. Chem.* **1945**, *37*, 12-23.
28. Hashem A. M.; Rashad M. M. *Egypt J. Food Sci.* **1993**, *21*, 171-180.
29. Sherrard E.; Kressman F. *Ind. Eng. Chem.* **1945**, *37*, 5-8.
30. Katzen R.; Madson P. W.; Monceaux D. A. In: *The alcohols textbook.* Lyons T. P.; Murtagh J. E.; Kelsall, editors.: Nothingham University Press; 1995. p. 37-46.
31. Qureshi N.; Manderson G. *Energy Sources, vol: 17: 241-265.* **1995**, *17*, 241-265.
32. Faith W. *Ind. Eng. Chem.* **1945**, *37*, 9-11.
33. Church J.; Wooldridge D. *Ind. Eng. Chem. Prod. Res. Dev.* **1981**, *20*, 371-378.
34. Harris J.; Baker A.; Zerbe J. *Energy biomass wastes , vol. 8: 1151-1170.* **1984**, *8*, 1151-1170.
35. Torget R.; Hsu T. A. *Appl. Biochem. Biotechnol.* **1994**, *45-6*, 5-22.
36. Larsson S. PhD thesis.Lund University of Technology. Lund, 2000.
37. Rabinovich M. L.; Melnik M. S.; Boloboba A. V. *Appl. Biochem. Microbiol.* **2002**, *38*, 305-321.
38. Jorgensen H.; Kutter J. P.; Olsson L. *Anal. Biochem.* **2003**, *317*, 85-93.
39. Tengborg C.; Galbe M.; Zacchi G. *Biotechnol. Prog.* **2001**, *17*, 110-117.
40. Ortega N.; Busto M. D.; Perez-Mateos M. *Int. Biodeterior. Biodegrad.* **2001**, *47*, 7-14.
41. Ogier J. C.; Ballerini D.; Leygue J. P.; Rigal L.; Pourquie J. *Oil & Gas Science and Technology-Revue De L Institut Francais Du Petrole.* **1999**, *54*, 67-94.
42. Sheehan J. S.; Himmel M. E. *Agro Food Ind. Hi-Tech.* **2001**, *12*, 54-57.
43. Clark T.; Mackie K. *J. Chem. Tech. Biotechnol.* **1984**, *34B*, 101-110.
44. Taherzadeh M. J.; Niklasson C.; Lidén G. *Bioresource Technol.* **1999**, *69*, 59-66.
45. Schneider H. *Enzyme Microb. Technol.* **1996**, *19*, 94-98.
46. Oura E. *Process Biochem.* **1977**, *12*, 19-21,35.
47. Ferrari M. D.; Neirotti E.; Albornoz C.; E. S. *Biotechnol. Bioeng.* **1992**, *40*.
48. Kusumegi K.; Yoshida H.; Tomiyama S. *J. Ferm. Bioeng.* **1998**, *85*, 213-217.
49. Thomas K. C.; Hynes S. H.; Ingledew W. M. *Appl. Environ. Microbiol.* **2002**, *68*, 1616-1623.
50. Verduyn C.; Postma E.; Scheffers W. A.; van Dijken J. P. *Yeast.* **1992**, *8*, 501-517.
51. Gottschalk G. *Dechema-Monographs, vol.105: 43-53.* **1987**, *105*, 43-53.
52. Dombek K. M.; Ingram L. O. *Appl. Environ. Microbiol.* **1987**, *53*, 1286-1291.
53. Pampulha M. E.; Loureiro V. *Biotechnol. Lett.* **1989**, *11*, 269-274.

54. Pampulha M.; Loureiro-Dias M. *FEMS Microbiol. Lett.* **2000**, *184*, 69-72.
55. Verduyn C.; Postma E.; Scheffers W. A.; van Dijken J. P. *J. Gen. Microbiol.* **1990**, *136*, 405-412.
56. Sjöström E. *Wood chemistry: fundamentals and applications*. 2 ed, Academic Press; 1993.
57. Larsson S.; Palmqvist E.; Hahn-Hägerdal B.; Tengborg C.; Stenberg K.; Zacchi G., et al. *Enzyme Microb. Technol.* **1999**, *24*, 151-159.
58. Ingram M.; Mossel D. A. A.; de Lange P. *Chemistry and Industry, 63-4.* **1955**, 63-64.
59. Tran H. C. PhD Thesis.University of California. Berkeley, 1986.
60. Fireoved R. L.; Mutharasan R. *Ann. N. Y. Acad. Sci.* **1986**, *469(Biochem. Eng. 4)*, 433-446.
61. Azhar A. F.; Bery M. K.; Colcord A. R.; Roberts R. S. *Dev. Ind. Microbiol.* **1982**, *23*, 351-360.
62. Sanchez B.; Bautista J. *Enzyme Microb. Technol.* **1988**, *10*, 315-318.
63. Zauner E.; Bronn W. K.; Dellweg H.; Tressl R. *Branntweinwirtschaft.* **1979**, *119*, 158-160,163.
64. Banerjee N.; Bhatnagar R.; Viswanathan L. *Eur. J. Appl. Microbiol. Biotechnol.* **1981**, *11*, 224-228.
65. Banerjee N.; Bhatnagar R.; Viswanathan L. *Enzyme Microb. Technol.* **1981**, *3*, 24-28.
66. Azhar A. F.; Bery M. K.; Colcord A. R.; Roberts R. S.; Corbitt G. V. *Biotechnol. Bioeng. Symp.* **1981**, *11*, 293-300.
67. Chung I. S.; Lee Y. Y. *Biotechnol. Bioeng.* **1985**, *27*, 308-315.
68. Boyer L. J.; Vega J. L.; Klasson K. T.; Clausen E. C.; Gaddy J. L. *Biomass Bioenergy.* **1992**, *3*, 41-48.
69. Taherzadeh M. J.; Gustafsson L.; Niklasson C.; Lidén G. *Appl. Microbiol. Biotechnol.* **2000**, *53*, 701-708.
70. Taherzadeh M. J.; Gustafsson L.; Niklasson C.; Lidén G. *J. Biosci. Bioeng.* **1999**, *87*, 169-174.
71. Taherzadeh M. J.; Gustafsson L.; Niklasson C.; Lidén G. *J. Biosci. Bioeng.* **2000**, *90*, 374-380.
72. Modig T.; Lidén G.; Taherzadeh M. J. *Biochem. J.* **2002**, *363*, 769-776.
73. Horváth I. S.; Taherzadeh M. J.; Niklasson C.; Lidén G. *Biotechnol. Bioeng.* **2001**, *75*, 540-549.
74. Sarvari Horvath I.; Franzen C. J.; Taherzadeh M. J.; Niklasson C.; Liden G. *Appl. Environ. Microbiol.* **2003**, *69*, 4076-86.
75. Morimoto S.; Hirashima T.; Matutani N. *J. Ferment. Technol.* **1969**, *47*, 486-490.
76. Villa G. P.; Bartroli R.; Lopez R.; Guerra M.; Enrique M.; Penas M., et al. *Acta Biotechnol.* **1992**, *12*, 509-512.
77. Diaz de Villegas M. E.; Villa P.; Guerra M.; Rodriguez E.; Redondo D.; Martinez A. *Acta Biotechnol.* **1992**, *12*, 351-354.

78. Larsson S.; Quintana-Sainz A.; Reimann A.; Nilvebrant N. O.; Jonsson L. J. *Appl. Biochem. Biotechnol.* **2000,** *84-6,* 617-632.
79. Palmqvist E.; Hahn-Hägerdal B. *Bioresource Technol.* **2000,** *74,* 25-33.
80. Nishikawa N.; Sutcliffe R.; Saddler J. *Appl. Microbiol. Biotechnol.* **1988,** *27,* 549-552.
81. Delgenes J.; Moletta R.; Navarro J. *Enzyme Microb. Technol.* **1996,** *19,* 220-225.
82. Pfeifer P. A.; Bonn G.; Bobleter O. *Biotechnol. Lett.* **1984,** *6,* 541-546.
83. Ando S.; Arai I.; Kiyoto K.; Hanai S. *J. Ferment. Technol.* **1986,** *64,* 567-70.
84. Palmqvist E.; Grage H.; Meinander N. Q.; Hahn-Hagerdal B. *Biotechnol. Bioeng.* **1999,** *63,* 46-55.
85. Qureshi N.; Manderson G. J. *J. Ind. Mocrobiol., vol.8: 45-52.* **1991,** *7,* 117-122.
86. Taherzadeh M. J.; Niklasson C.; Lidén G. *Biotech. Bioeng.* **2000,** *69,* 330-338.
87. Nilsson A.; Taherzadeh M. J.; Lidén G. *J. Biotechnol.* **2001,** *89,* 41-53.
88. Nilsson A.; Taherzadeh M. J.; Lidén G. *Bioprocess Biosys. Eng.* **2002,** *25,* 183-191.
89. Lee W.; Lee J.; Park B.; Kim M.; Park S.; Chang H. *Korean J. Chem. Eng.* **1996,** *13,* 453-456.
90. Taherzadeh M. J.; Millati R.; Niklasson C. *Appl. Biochem. Biotechnol.* **2001,** *95,* 45-57.
91. Kötter P.; Ciriacy M. *Appl. Microbiol. Biotechnol.* **1993,** *38,* 776-783.
92. Gardonyi M.; Jeppsson M.; Liden G.; Gorwa-Grausland M. F.; Hahn-Hagerdal B. *Biotechnol. Bioeng.* **2003,** *82,* 818-824.
93. Johansson B.; Hahn-Hagerdal B. *FEMS Yeast Res.* **2002,** *2,* 277-282.
94. Olsson L.; HahnHagerdal B. *Enzyme Microb. Technol.* **1996,** *18,* 312-331.
95. Zhang M.; Eddy C.; Deanda K.; Finkestein M.; Picataggio S. *Science.* **1995,** *267,* 240-243.
96. Ohta K.; Beall D. S.; Mejia J. P.; Shanmugam K. T.; Ingram L. O. *Appl. Environ. Microbiol.* **1991,** *57,* 893-900.
97. Ingram L. O.; Aldrich H. C.; Borges A. C. C.; Causey T. B.; Martinez A.; Morales F., et al. *Biotechnol. Prog.* **1999,** *15,* 855-866.

Chapter 4

Recent Progress in Transformation of Lignocellulosics to Fuels and Chemicals by Supercritical Water Technology

Katsunobu Ehara[1,2] and Shiro Saka[1]

[1]Graduate School of Energy Science, Kyoto University, Yoshida-Honmachi, Sakyo-ku, Kyoto 606–8501, Japan
[2]Current address: Kyoto University as New Energy and Industrial Technology Development Organization (NEDO) Fellow, Kyoto University, Kyoto, Japan

Lignocellulosics were treated in supercritical water (>374 °C, >22.1 MPa), and fractionated into the water-soluble portion, precipitates, methanol-soluble portion and supercritical water-insoluble residue. The water-soluble portion contained carbohydrate-derived hydrolyzed products, dehydrated products, fragmented products and organic acids. The precipitates were found to be glucan which is insoluble in ordinary water but soluble in supercritical water. On the other hand, methanol-soluble portion was found to be derived from lignin, and mainly consisted of condensed-type linkages of lignin due to the preferential degradation of the ether linkages of lignin. From these findings, the supercritical water treatment was concluded to be appropriate to use as a pretreatment for ethanol producing system that is followed by enzymatic saccharification or dilute acid hydrolysis and fermentation.

Due to the global warming by the excessive use of fossil resources, biomass resources will become more important in the future as alternatives to fossil resources. The features of biomass are renewable, carbon-neutral and abundant. However, these resources have not been utilized efficiently and their non-used portions are often wasted (*1*). Therefore, it is essential to develop a technology for efficient utilization of biomass resources.

For the conversion of biomass resources into useful chemicals and bio-energy, saccharification of cellulosic resources followed by fermentation is one of the directions. For this purpose, two major methods exist for hydrolysis of carbohydrates, by acid catalyst (*2*) and enzymatic saccharification (*3*). Although these processes have been refined and considerably developed, they have at least the following drawbacks; the former has a corrosion problem of the reactor by acid, while the latter involves the use of enzyme expensive for its preparation (*4*). Besides these approaches, supercritical water (>374 °C, >22.1 MPa) technology has recently received increasing attentions.

Supercritical water behaves very differently from water under normal condition as follows (*5*): The ionic products of water (Kw) is about three to four orders of magnitude larger in its supercritical state than that in an ordinary condition. Therefore, supercritical water can act as an acid catalyst. Furthermore, the dielectric constant of supercritical water is in a range between 10 and 20 since water in ordinary condition is about 80 in its dielectric constant. Thus, the hydrophobic substances can be solvated with the supercritical water.

This chapter, therefore, reviewed recent progress in transformation of lignocellulosics to fuels and useful chemicals by the supercritical water technology. Based on the obtained results, the appropriate process for ethanol production using supercritical water technology was discussed.

Treatment Systems and Separation

Supercritical Fluid Biomass Conversion Systems

Several types of systems such as batch, semi-flow and flow are established to treat various biomass resources. The batch-type system is the simplest, and can endure a relatively higher pressure compared with the flow-type system. However, this system requires some time to heat up ordinary water to supercritical condition and cool down supercritical water to ordinary condition. Accordingly, excessive decomposition of target products is often found inevitably (*6*). For the semi-flow system, the sample is fixed in a reaction tube and water is continuously delivered into the reaction tube. This system permits elution of only water-soluble products from reaction tube, and is often used to extract hemicelluloses from lignocellulosics (*7*). Thus, compared with the batch-type system, the decomposition of target products is restrained (*8*).

Figure 1 shows our flow-type supercritical water biomass conversion system. This system can cover a range up to 45 MPa in pressure and 450 °C in temperature, and can be divided into slurry pumping section, solvent-pumping section, supercritical treatment tube and cooling section.

For the slurry pumping section, the slurried sample with water (2.0 - 8.0wt %) is routinely stirred by the circulating pump. This can keep its concentration constant, which prevents the sample from precipitating. The sample is injected into the reaction tube of 1.9 cm, and pressurized by the slurry injector up to an appropriate pressure. The sample is then mixed with the supercritical water prepared by the solvent pumping section. After the reactant had passed through the reaction tube, an appropriate amount of the water is injected into the reactant to quench the reaction and its mixture is further cooled by passing through the cooler.

The immediate heating and cooling system enable us to regulate the residence time within 0.1 s, and prevent target products from excessive decomposition. Detailed information of our batch-type and flow-type systems is described elsewhere (*9, 10*).

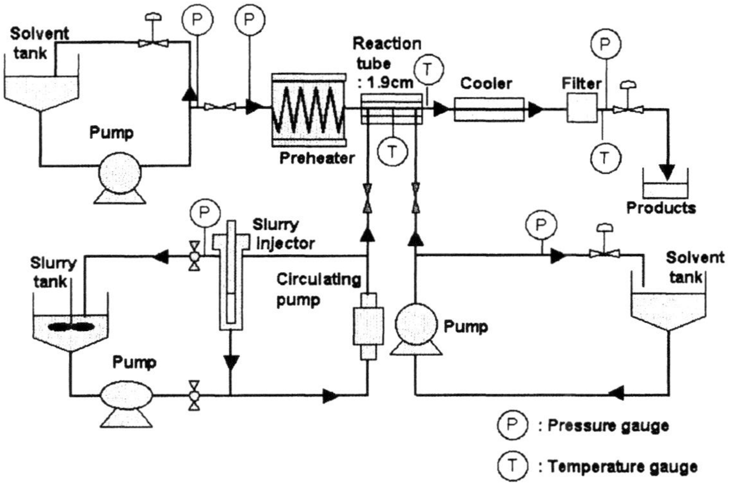

Figure 1. Schematic diagram of flow-type supercritical water biomass conversion system.

Separation of Treated Lignocellulosics

Figure 2 shows the separation scheme of the lignocellulosics treated in supercritical water. The supercritical water-soluble portion is retrieved by filtration from the supercritical water-insoluble portion, immediately after the treatment. During standing for 12 hours, the former generates the precipitates and oily substances due to the change of dielectric constant of water from the supercritical water to the ordinary one. They are filtrated to separate from the water-soluble portion. The precipitates and oily substances are, then, extracted with methanol, and also separated to precipitates and methanol-soluble portion by filtration.

Consequently, the water-soluble portion, water-insoluble portion (precipitates), methanol-soluble portion (oily substances) and methanol-insoluble portion are obtained from the treated sample. The latter two portions are not obtained when cellulose is treated appropriately in supercritical water. On the other hand, lignin sample such as milled wood lignin (MWL) is only separated to the latter two portions. Therefore, the former and latter two portions mainly consist of carbohydrate-derived and lignin-derived products, respectively (*11*).

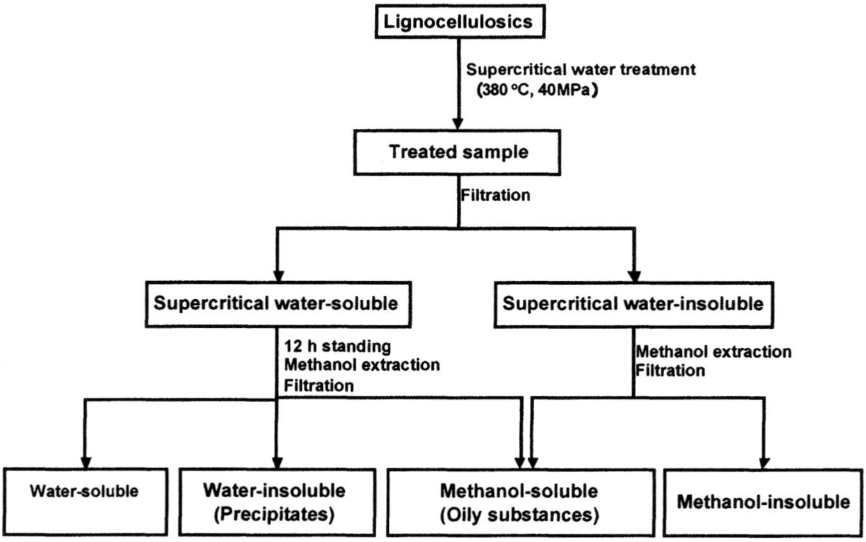

Figure 2. Separation of lignocellulosics treated in supercritical water.

Decomposition of Carbohydrates

Characterization of Water-Soluble Portion

Cellulose was decomposed by supercritical water and separated to water-soluble portion. This portion contained cello-oligosaccharides (oligomers) and glucose as a hydrolyzed product, while levoglucosan, 5-hydroxymethyl furfural (5-HMF) and furfural were found as dehydrated products. In addition, fragmented products such as methylglyoxal, glycolaldehyde, dihydroxyacetone and erythrose were also found (*10*).

In addition to these compounds, furfural, which was removed during the evaporation process to condense the water-soluble portion, was apparently present. Furthermore, low-molecular weight organic acids, such as pyruvic acid, lactic acid, glycolic acid, levulinic acid and acetic acid, were identified by HPLC system equipped with electric conductivity detector.

Hemicelluloses were also decomposed by supercritical water and separated to water-soluble portion. There were expected hydrolyzed products such as glucose and mannose from glucomannan, and xylose from xylan. On the other hand, lignin was surely decomposed by supercritical water but not separated to this portion (*11*).

To characterize the oligosaccharides in detail, matrix assisted laser desorption ionization-time of flight mass spectrometric (MALDI-TOFMS) analysis of this portion was carried out. As a result, it was clarified that this portion contains cello-oligosaccharides up to cello-dodecaose (Degree of polymerization; DP = 12), and the reducing ends of some oligosaccharides were found to be decomposed to levoglucosan, erythrose and glycolaldehyde (*10*).

Characterization of Precipitates

The formation of the precipitates after supercritical water treatment of cellulose is reported by some research groups (*12, 13*). However, their distribution of the DP had not been known yet. Thus, GPC chromatographic study was performed after derivatization of the precipitates by phenylisocyanate. From the chromatogram, the DP of the precipitates was estimated to be in a range between 13 and 100.

On the other hand, once the precipitates were dried, X-ray diffractometry showed that they were clearly crystallized to be cellulose II which is usually prepared as regenerated cellulose after its dissolving (*10*).

From these results, we concluded that the precipitates are glucan which is insoluble in ordinary water but soluble in supercritical water (*10*).

Decomposition Pathway of Cellulose

Table I shows chemical composition of cellulose treated in supercritical water. The total yield of hydrolyzed products, which are polysaccharides (precipitates), oligosaccharides, glucose and fructose, is 75.6 % for 0.12 s treatment time, and decreased when the treatment time is prolonged. On the other hand, the yields of dehydrated products, fragmented products and organic

Table I. Chemical composition of cellulose treated in supercritical water at 380 °C and 40 MPa using flow-type system

		Yield (%)		
	Product name	*0.12 s*	*0.24 s*	*0.48 s*
Hydrolyzed				
	Polysaccharides	31.2	12.1	n.d.**
	Oligosaccharides	41.1	37.6	7.5
	Glucose	2.8	8.4	8.9
	Fructose	0.5	3.4	10.0
(Subtotal)		(75.6)	(61.5)	(26.4)
Dehydrated				
	Levoglucosan	0.2	2.4	3.7
	5-HMF	0.1	1.9	7.3
	Furfural	n.d.**	1.0	3.4
(Subtotal)		(0.3)	(5.3)	(14.4)
Fragmented				
	Erythrose	0.7	0.9	1.9
	Methylglyoxal	0.6	3.0	6.7
	Glycolaldehyde	2.5	6.7	14.6
	Dihydroxyacetone	0.2	1.6	2.4
(Subtotal)		(4.0)	(12.2)	(25.6)
Organic acid				
	Pyruvic acid	n.d.**	0.5	0.7
	Glycolic acid	1.3	2.9	3.9
	Lactic acid	0.7	3.2	4.0
	Formic acid	0.3	1.6	2.5
	Acetic acid	0.1	0.7	1.0
(Subtotal)		(2.4)	(8.9)	(12.1)
Others*		17.7	12.1	21.5
Total		100.0	100.0	100.0

Note: * Others consist of unidentified products, gasified products and water derived from dehydration of saccharides. ** Not detected.

acids are increased when the treatment time is prolonged. Thus, shorter treatment time is favorable to obtain the hydrolyzed products which can be used as a feedstock of ethanol production.

The supercritical water, which is low in viscosity and has a high diffusibility, can be liable to diffuse into cellulose macromolecules. In addition, free water around cellulose microfibril would behave as supercritical water. Therefore, cellulose microfibril would be released into the supercritical water. Furthermore, the crystalline structure of cellulose would be possibly broken down because hydrogen bonds of cellulose would be cleaved under supercritical condition, as observed in water (*14*). That is to say, cellulose macromolecules would be degraded both internally and externally by supercritical water (*15*).

From these lines of evidence, we propose a pathway of cellulose decomposition by hydrolysis, dehydration and fragmentation in Fig. 3. Cellulose is hydrolyzed to polysaccharides whose DP is in a range between 13 and 100. Subsequently, polysaccharides are further hydrolyzed to oligosaccharides with DP between 2 and 12, whose reducing end of glucose is dehydrated or fragmented to levoglucosan, erythrose and glycolaldehyde. These dehydrated and fragmented oligosaccharides are, further, hydrolyzed to glucose, which is isomerized to fructose. The resultant hexoses are then further decomposed, if the treatment is prolonged, to levoglucosan, 5-HMF, erythrose, glycolaldehyde, methylglyoxal, and dihydroxyacetone by way of dehydration or fragmentation, (*10, 16, 17*). The dehydrated and fragmented products are further oxidized to low-molecular-weight organic acids such as pyruvic, glycolic, lactic, formic and acetic acid.

It is very important to control this reaction pathway to obtain the targeted products in high yield. Therefore, basic and applied researches have been conducted by some research groups. We reported that the crystal structure of cellulose II is easier to be hydrolyzed to glucose than that of cellulose I (*18*). The flow-type system can hydrolyze cellulose with minimizing pyrolyzed products compared with batch-type system (*10*). Sasaki et al. reported kinetic study of cellulose decomposition using some model compounds. They suggested that cellulose macromolecules would possibly be diffused into the supercritical water (*12*), and clarified that fragmentation of glucopyranoside increased with decreasing treatment pressure (*19*). The extraction of hemicelluloses from lignocellulosics has also been carried out by subcritical water using a semi-flow system (*7, 20*). Furthermore, addition of catalyst such as H_2O_2, NaOH in the supercritical water is also conducted to obtain the low-molecular-weight organic acids (*21, 22*).

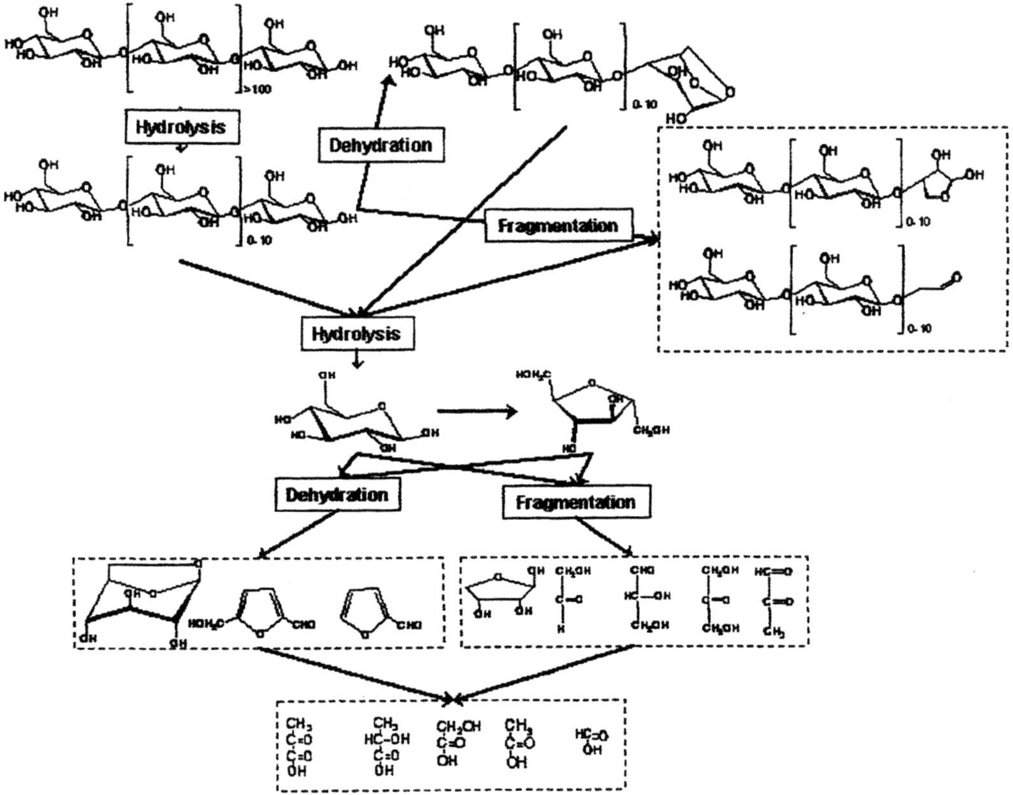

Figure 3. The proposed pathway of cellulose decomposition as treated in supercritical water.

Decomposition of Lignin

Decomposition of Lignin Model Compounds

To simulate the reaction of lignin, dimeric ß-O-4 (non-condensed) and biphenyl (condensed) types lignin model compounds were treated with supercritical water. As a result, ß-O-4 type lignin model compound was readily cleaved in their ether linkages, whereas biphenyl type compounds were stable during supercritical water treatment. The same result was also obtained when the non-phenolic model compounds were treated in supercritical water. This result clearly indicated that the ß-O-4 linkages of lignin are easily cleaved under supercritical water treatment, whereas the 5-5 linkages of lignin are rather stable (*23*).

Characterization of Methanol-Soluble Portion

As above-mentioned, the methanol-soluble portion (oily substances) was only obtained from the sample containing lignin, and the yield of that was close to the lignin content of the sample when the appropriate condition was selected. Therefore, methanol-soluble portion mainly consists of lignin-derived products (*11*).

Table II shows characteristic features of the methanol-soluble fraction from Japanese cedar (*Cryptomeria japonica* D. Don). The number of the phenolic hydroxyl groups of untreated lignin in Japanese cedar was determined to be 16.7 per 100 units of phenylpropane (C_6-C_3) residues. However, the methanol-soluble portion was found to have more phenolic hydroxyl groups than untreated lignin. On the other hand, the total yield of nitrobenzene oxidation products from methanol-soluble portion was lower than that of untreated lignin.

The phenolic hydroxyl groups are formed by splitting of ether linkages of lignin. This result is consistent with the lignin model compound study described above. Because alkaline nitrobenzene oxidation products would mainly be derived from ether linkages of lignin, it is a measure of the non-condensed types of the lignin linkages present. Therefore, this result suggests that the methanol-soluble portion consists of the condensed-type linkages of lignin. This is due to the preferential degradation of ether linkages in supercritical water (*23*).

Monomeric and Dimeric Products in Methanol-Soluble Portion

The molecular weight of the methanol-soluble portion was found to be less than 2000. Thus, it was expected that this portion contains lignin-derived monomeric, dimeric, trimeric and some oligomeric products (*24*).

Table II. Comparative analysis of untreated lignin and methanol-soluble portion from Japanese cedar after the supercritical water treatment (400 °C, 115 MPa, 8 s, Batch-type system)

	Phenolic hydroxyl groups (Number/100 of C_6-C_3)	*Nitrobenzene oxidation products (mmol/g of lignin)*
Untreated lignin	16.7	31.3
Methanol-soluble	1.9	0.5

Figure 4 shows a total ion chromatogram of the methanol-soluble portion from Japanese cedar as treated in supercritical water. All identified products in Fig. 4 had guaiacyl nuclei (2-methoxyphenol), except for 5-HMF. 5-HMF was a contaminant derived from cellulose and separated predominantly into the water-soluble portion (*10, 11, 18*). Eugenol, propylguaiacol, *cis*-iso-eugenol, *trans*-iso-eugenol, propioguaiacone, guaiacylacetone, 2-methoxy-4-(1-hydroxypropyl)-phenol, 2-methoxy-4-(prop-1-en-3-one)phenol, trans-coniferylaldehyde and feruric acid were composed of phenylpropane units (C_6-C_3). These products would be derived through the cleavage of ether linkages of lignin as above-mentioned.

In addition to C_6-C_3 units, C_6-C_2 and C_6-C_1 units of products were also detected. The C_6-C_2 units were ethylguaiacol, vinylguaiacol, homovanillin, acetoguaiacone and homovanillic acid, while the C_6-C_1 units were methylguaiacol and vanillin. In our previous study of lignin model compound in the supercritical water, the cleavage between $C_ß$ and $C_γ$ ($C_ß$/ $C_γ$) bond in the C_6-C_3 unit was observed (*23*). Therefore, the existence of these products suggests that the cleavage between $C_ß$/ $C_γ$ and $C_α$/ $C_ß$ linkages of lignin take place in supercritical water. Sato also reported that the dealkylation of propyl chain of alkyl phenols took place in supercritical water (*25*).

Furthermore, four dimeric lignin-derived products were estimated to be biphenyl-type (5-5), diphenylethane-type (ß-1), stilbene-type (ß-1) and phenylcoumaran-type (ß-5) compounds. These products were considered to be more stable than those that have ether linkages such as ß-O-4 and α-O-4 because preferential degradation of ether linkages of lignin occurs during supercritical water treatment (*24*).

Supercritical Water Process for Efficient Use of Biomass

Lignocellulosics consist of chemical components such as cellulose, hemicelluloses and lignin. Therefore, it is essential to depolymerize these polymers at once in supercritical treatment of water. The obtained results clearly demonstrate that cellulose is hydrolyzed to be polysaccharides (DP 13-100),

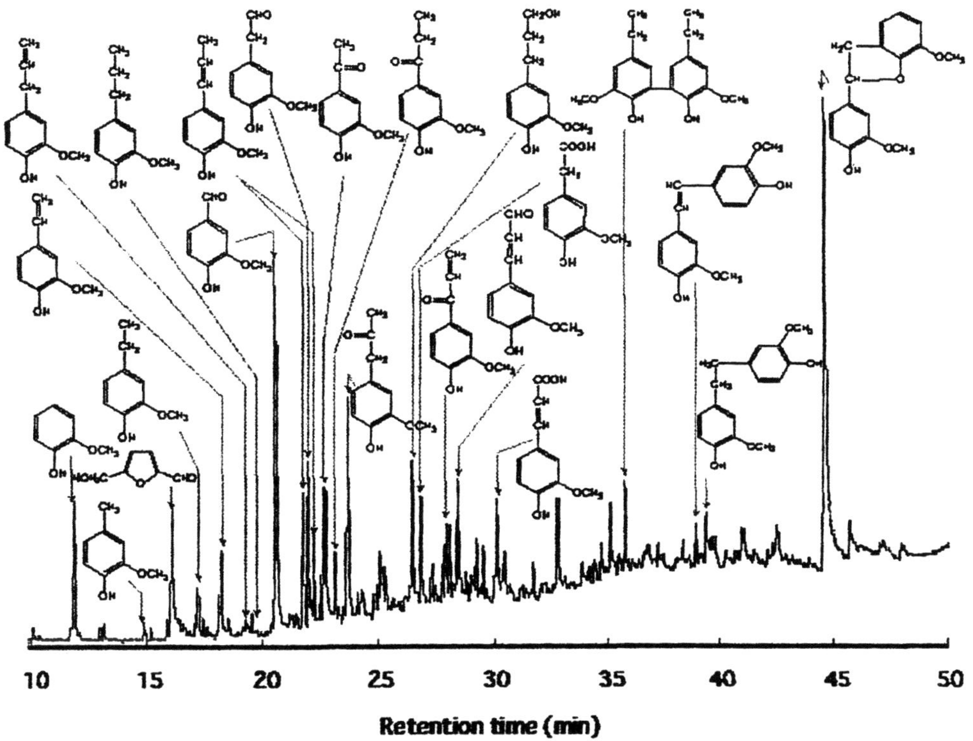

Figure 4. Various aromatic compounds from Japanese cedar identified in methanol-soluble portion as treated in supercritical water.
(380°C, 100 MPa, 8 s, Batch-type system)

oligosaccharides (DP 2-12) and glucose with a trace of the fragmented and pyrolyzed products such as 5-HMF and furfural under an appropriate condition. At the same time, ether linkages of lignin are readily cleaved to produce various aromatic products. Therefore, the supercritical water treatment can be appropriate for the pretreatment of ethanol production if it is combined with the subsequent enzymatic saccharification or diluted acid hydrolysis and fermentation. We and Sasaki et al. reported that polysaccharides and oligosaccharides prepared from cellulose by supercritical water treatment was effectively hydrolyzed to glucose by dilute sulfuric acid (*26*) or cellulase (*27*).

Figure 5 shows our proposed system which is the ethanol producing process using supercritical water technology. This process consists of supercritical water treatment followed by the separation of the products of the water-soluble and methanol-soluble portions. For water-soluble portion derived from cellulose and hemicelluloses, enzymatic saccharification or diluted acid hydrolysis is carried out to produce monosaccharides of glucose, mannose and xylose. The obtained monosaccharides are, then, fermented to ethanol by genetically engineered yeast. For methanol-soluble portion, various aromatic compounds derived from lignin would be converted into the value-added products by hydrogenolysis, etc. For example, they are fuels as octane number enhancer, and chemicals such as benzene, phenol and pyrocatechol and their related products.

This process can provide some advantages for ethanol production, compared with acid hydrolysis process; because of the extremely rapid reaction, rather small and simple plant may be designed without use of acid catalyst. In addition, lignin-derived products are not contaminated as in sulfuric acid lignin in the acid hydrolysis process, so that they can be appropriate to be converted into the value-added aromatic products. Thus, the overall process would lead to the efficient utilization of the whole lignocellulosics.

To realize this process, not only basic but also applied studies are necessary concerned with chemical, physical, biological and industrial approaches. As a result, biomass resources will be effectively used as alternatives of fossil resources.

Acknowledgements

This research has been carried out under the program of the Research for the Future (RFTF) of The Japan Society of the Promotion of Science (JSPS-RFTF97P01002), the COE program in the 21st Century of "Establishment of COE on Sustainable-Energy System" and by a Grant-in-Aid for Scientific Research (B)(2) (No.12460144, 2001.4-2003.3) from the Ministry of Education, Culture, Sports, Science and Technology, Japan.

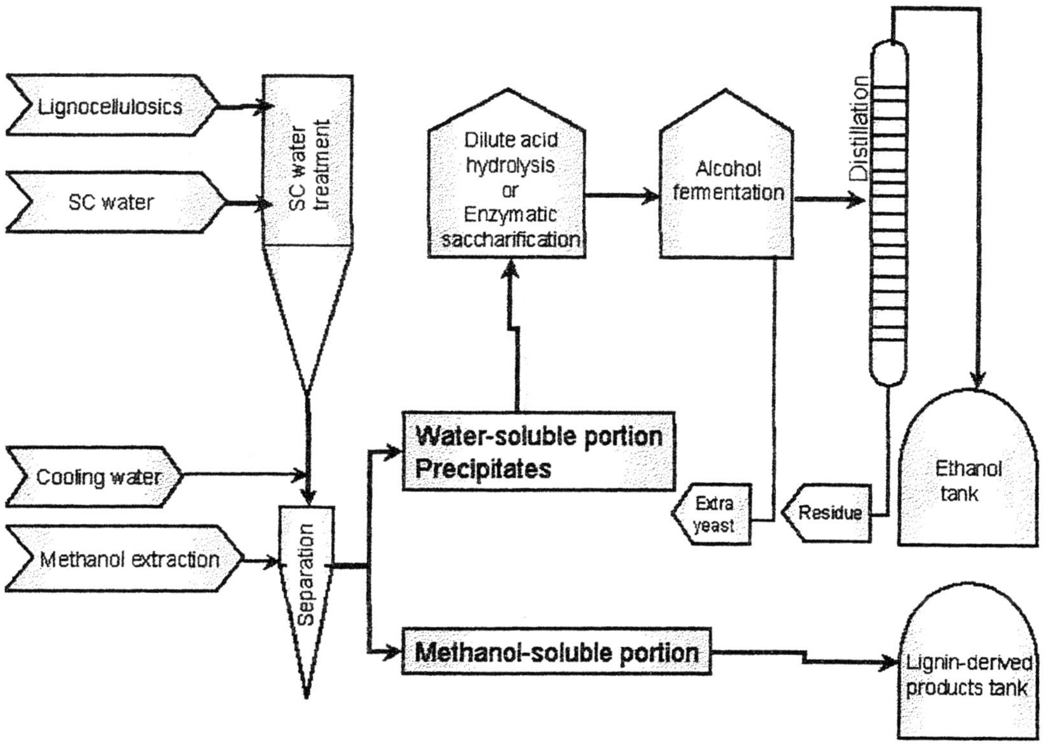

Figure 5. Schematic process for ethanol production from lignocellulosics using supercritical water treatment.

References

1. Minami, E.; Saka, S. Quantity of Biomass in Japan. *Biomass, Energy, Environment (in Japanese)*; Saka, S. Ed.; IPC: Tokyo, 2001; pp 61-103.
2. Ehara, K.; Saka, S. Acid Hydrolysis. *Biomass, Energy, Environment (in Japanese)*; Saka, S. Ed.; IPC: Tokyo, 2001; pp 251-260.
3. Chang, V. S.; Holtzapple, M. T. *Appl. Biochem. Biotechnol.* **2000**, *84-86*, 5-37.
4. Goldstein, I. S. *Tappi* **1980**, *63*, 141-143.
5. Holzapfel, W. B. *J. Chem. phys.* **1969**, *50*, 4424-4428.
6. Saka, S. Supercritical Fluid. *Biomass, Energy, Environment (in Japanese)*; Saka, S. Ed.; IPC: Tokyo, 2001; pp 291-313.
7. Ando, H.; Sakaki, T.; Kokusho, T.; Shibata, M.; Uemura, Y.; Hatate, Y. *Ind. Eng. Chem. Res.* **2000**, *39*, 3688-3693.
8. Sakaki, T.; Shibata, M.; Sumi, T.; Yasuda, S. *Ind. Eng. Chem. Res.* **2002**, *41*, 661-665.
9. Kusdiana, D.; Minami, E.; Ehara, K.; Saka, S. Development of the Batch-Type and Flow-Type Supercritical Fluid Biomass Conversion Systems. *12th European Conference and Technology Exhibition on Biomass for Energy, Industry and Climate Protection.*: Amsterdam, 2002; pp789-792.
10. Ehara, K.; Saka, S. *Cellulose* **2002**, *9*, 301-311.
11. Saka, S.; Konishi, R. Chemical Conversion of Biomass Resources to Useful Chemicals and Fuels by Supercritical Water Treatment. *Progress in Thermochemical Biomass Conversion*; Bridgwater, A. V. Ed.; Blackwell Sci.: Oxford, 2001; pp 1338-1348.
12. Sasaki, M.; Fang, Z.; Fukushima, Y.; Adschiri, T.; Arai, K. *Ind. Eng. Chem. Res.* **2000**, *30*, 2883-2890.
13. Sumi, T.; Sakaki, T.; Ohba, H.; Shibata, M. *Rapid Commun. Mass Spectrom.* **2000**, *14*, 1823-1827.
14. Lindsay, W. T. Hydrogen Bonding and Water Structure. *The ASME Handbook on Water Technol. for Thermal Power Systems*; Cohen, P. Ed.; ASEME: N.Y., 1981; pp 371.
15. Saka, S. *Mokuzai Kougyou* (in Japanese). **2001**, *56*, 105-110.
16. Kabyemela, B. M.; Adschiri, T.; Malaluan, R.; Arai, K. *Ind. Eng. Chem. Res.* **1997**, *36*, 1552-1558.
17. Kabyemela, B. M.; Adschiri, T.; Malaluan, R.; Arai, K. *Ind. Eng. Chem. Res.* **1999**, *38*, 2888-2895.
18. Saka, S.; Ueno, T. *Cellulose* **1999**, *6*, 177-191.
19. Sasaki, M.; Frukawa, M.; Minami, K.; Adschiri, T.; Arai, K. *Ind. Eng. Chem. Res.* **2002**, *41*, 6642-6649.

20. Garrote, G.; Dominquez, H.; Parajo', J. C. *Hollz als Roh- Werkstoff* **1999**, *57*, 191-202.
21. Quitain, A. T.; Faisal, M.; Kang, K.; Daimon, H.; Fujie, K. *J Hazardous Materials* **2002**, *B93*, 209-220.
22. Calvo, L.; Vallejo, D. *Ind. Eng. Chem. Res.* **2002**, *41*, 6503-6509.
23. Ehara, K.; Saka, S.; Kawamoto, H. *J. Wood Sci.* **2002**, *48*, 320-325.
24. Takada, D.; Ehara, K.; Saka, S. *J. Wood Sci.* **2003**, in press.
25. Sato, T.; Sekiguchi, G.; Saisu, M.; Watanabe, M.; Adschiri, T.; Arai, K. *Ind. Eng. Chem. Res.* **2002**, *41*, 3124-3130.
26. Nakata, T.; Miyafuji, H.; Ehara, K.; Saka, S. Bio-Ethanol from Cellulose with Supercritical Water Treatment Followed by Sulfuric Acid Hydrolysis. *Proceedings of Kyoto University 21st Century COE Program, The 1st International Symposium on Sustainable Energy System*: Kyoto, 2003; pp 174.
27. Sasaki, M.; Iwasaki, K.; Hamatani, T.; Adschiri, T.; Arai, K. *Kobunshi Ronbunshu* (in Japanese). **2001**, *58*, 527-532.

Chapter 5

Thermochemical Treatment, Separation, and Conversion of Corn Fiber to Ethanol

Charles Abbas, Kyle Beery*, Eric Dennison, and Pam Corrington

James Randall Research Center, Archer Daniels Midland Company, Decatur, IL 62521

Corn fiber is composed of approximately 20% starch, 35% hemicellulose, 18% cellulose, 11% protein, 3% oil, and 6% ash. The hemicellulose is primarily composed of xylose (~55%) and arabinose (~36%). Corn fiber can be fractionated using combinations of enzymatic and thermochemical (heat plus acid or base) methods. Many different hydrolysis schemes have been carried out on the corn fiber to obtain various industrially useful components. The hydrolysis step creates mainly oligosaccharides, with lower concentrations of monosaccharides and degradation products. These oligosaccharides are further hydrolyzed to monosaccharides by a secondary hydrolysis. In this experiment, the corn fiber was hydrolyzed at 30% solids at 140°C for 30 minutes followed, optionally, by an acid hydrolysis or enzyme hydrolysis. The most optimal hydrolysis method was the initial thermochemical hydrolysis method without the further acid or enzyme hydrolysis. The glucose and xylose monosaccharides present in the hydrolysates were fermented to ethanol by a recombinant strain of *Saccharomyces cerevisiae 424A* to produce up to 58 g/L of ethanol in the fermentation broth.

Introduction

A corn wet-milling biorefinery processes corn to such products as corn syrup, high fructose corn sweetener, corn oil, Vitamin E, food and industrial starches, beverage and industrial alcohol, carbon dioxide, corn gluten animal feed, corn gluten meal, and glucose. The glucose can be sold as is or it can be converted to other products by fermentation or catalytic conversion. Ethanol is the major fermentation product. The corn is subjected to steeping with sulfur dioxide to soften the corn and break down the starch-protein matrix. After steeping, the corn is milled to separate the germ from the starch and fiber. The fiber and starch are then milled and the fiber is separated and washed to remove residual starch. The fiber is dried and mixed with the corn gluten, and then corn steep liquor and stillage are sprayed on the corn fiber. The mixture is then dried and pelletized. The pelletized fiber is sold as corn gluten feed (CGF). As the demand for corn gluten feed is not anticipated to significantly increase over the next decade, an increase in corn wet-milling would cause an oversupply of CGF. Therefore, an additional use for the corn fiber stream is needed.

Corn fiber contains several components that, if separated, can be readily upgraded to value-added products using current commercial technology. Over 13,000 tons of corn fiber are produced per day in the U.S. and over 80% of that is exported (*1*). Corn fiber is an ideal feedstock, because it is abundant and requires no special harvesting or additional transportation costs as it is already in the plant. If the corn fiber volume can be reduced by a conversion process, the biorefinery would be able to process additional corn and also potentially increase the ethanol yield from each bushel of corn by up to 0.3 gallons/bushel (*2*), while exporting less corn gluten feed.

Several approaches to converting corn fiber to value-added products have been attempted. The main areas of research have consisted of treating a corn fiber slurry with acid, base, or enzymes. Acid hydrolysis of the corn fiber leads to degradation of the starch and hemicellulose fractions to oligosaccharides (*3-6*). The cellulose portion will also be hydrolyzed if the conditions are adequately severe. Generally, the research on acid hydrolysis follows a mild acid treatment with enzyme hydrolysis of any remaining carbohydrate polysaccharides or oligosaccharides. Base treatment is generally concerned with solubilization of the hemicellulose fraction for purification into corn fiber gum (*7-12*). However, ammonia fiber explosion facilitates hydrolysis of the carbohydrate fractions by reducing enzyme requirements by creating more surface area (*13-15*). Enzyme hydrolysis is almost always paired with an acid or base pretreatment. In this research, we examined dilute-acid hydrolysis of the corn fiber with a secondary acid hydrolysis step to hydrolyze the oligosaccharides to monosaccharides.

In the process described in the following, the corn fiber is subjected to an initial thermochemical hydrolysis step, in which the SO_2 contained in the corn fiber is utilized as an acid catalyst to hydrolyze the starch and hemicellulose polymers. The residual SO_2 contained in the corn fiber originates with the corn kernel steeping step. Corn fiber contains 35% hemicellulose, 18% cellulose, 17% starch, 11% protein, 6% ash, 5% galactan, 3% oil, 1% mannan, and 4% other materials. The starch and hemicellulose are hydrolyzed to oligosaccharides, and are further hydrolyzed to monosaccharides. Cellulose hydrolysis is not desirable as it is needed as a carrier for the corn steep liquor and stillage. The corn steep and stillage streams are valuable sources of protein for animal feed, but they need a carrier in order to be dried and utilized. The hydrolysis of the starch yields glucose, and the hydrolysis of the hemicellulose yields xylose, arabinose, galactose, and mannose. Other components of the hemicellulose are ferulic acid, coumaric acid, acetic acid, and glucuronic acid. The ferulic acid has been shown to be linked to the hemicellulose chain and forms diferulic linkages between the hemicellulose strands thereby facilitating the formation of the hemicellulose matrix (*16-18*). The glucose, xylose, and arabinose can be converted to ethanol or other fermentation products or catalytically converted to ethylene glycol, propylene glycol, and glycerol. The acid hydrolysis of the corn releases acetic acid from the hemicellulose polymer and also causes a fraction of the monosaccharides to degrade to hydroxymethyl furfural and furfural. These components have been reported to cause inhibition of the yeast during fermentation of corn fiber hydrolysates to ethanol (*19-29*). The residual fiber contains primarily cellulose and protein. Since the protein passes through the process unutilized, the volume reduction of the hydrolyzed fiber leads to a higher protein content on the corn gluten feed. This presumably leads to a higher value, since the corn gluten feed is sold on protein content.

Materials and Methods

Corn Fiber Hydrolysis Experiments

Initial Thermochemical Hydrolysis Method

Corn fiber (Archer Daniels Midland, Co., Decatur, IL) was treated by three methods in a custom-made, jacketed, 50 gallon rotary reactor to hydrolyze the starch and hemicellulose fractions of the fiber. The reactor was equipped with a 0.55 MPa pop-off valve. All of the methods included initially hydrolyzing a 30%

solids corn fiber slurry, and the remaining two methods added a 1% sulfuric acid hydrolysis or a glycohydrolyase enzyme hydrolysis step.

A 13.6 kg sample of corn fiber from the plant at 35% solids was mixed with 2.26 kg of deionized water to lower the solids content to 30%. The slurry was loaded into the 50 gallon reactor and hydrolyzed for 30 minutes at 140°C and 1 RPM. The reactor was vented and the hydrolyzed corn fiber was dewatered by processing with vacuum filtration through cheesecloth (Fisher Scientific, Pittsburgh, PA). The residual hydrolyzed corn fiber was washed twice with a total of 8 gallons of 95°C water and dewatered between washings.

The filtrates were pooled and concentrated using a vacuum evaporator to 21.3% dry solids. This concentrate was then subjected to a secondary acid hydrolysis step as follows: 1% sulfuric acid (Aldrich Chemicals, Milwaukee, WI) treatment at 121°C for 30 minutes to degrade the oligosaccharides to monosaccharides. A sample of the corn fiber filter cake was dried and weighed to determine the extent of corn fiber solubilization.

Acid Hydrolysis

A 14.0 kg sample of corn fiber from the plant at 36.0% solids was mixed with 2.80 kg of deionized water to lower the solids content to 30%. The slurry was loaded into the 50 gallon reactor and hydrolyzed for 30 minutes at 140°C and 1 RPM. The reactor was vented and 168 g of sulfuric acid (Aldrich Chemicals, Milwaukee, WI) were added to the slurry. The reactor was pressurized and run at 130°C for 45 minutes at 1 RPM. The reactor was vented and the hydrolyzed corn fiber slurry was dewatered by processing with vacuum filtration through cheesecloth. The residual hydrolyzed corn fiber was washed twice with 4 gallons of 95°C water each time and dewatered between washings.

The filtrates were pooled and concentrated using a vacuum evaporator to 33.9% dry solids. This concentrate was then subjected to the secondary acid hydrolysis step.

Enzyme Hydrolysis

A 12.7 kg sample of corn fiber at 36% solids was mixed with 2.6 kg of deionized water to lower the solids content to 30%. The slurry was loaded into the 50 gallon reactor and hydrolyzed for 30 minutes at 140°C and 1 RPM. The reactor was vented and 50 mL each of the following enzymes were added to the reactor: Dyadic International (Jupiter, Florida) hemicellulases: FoodCel Plus,

ViscoStar 150 L, ViscoStar CL CONC, Neutral Fungal Cellulase; and Genencor (Rochester, NY) Spezyme FRED α-amylase and Novozymes (Bagsvaerd, Denmark) Optidex L-400 amyloglucosidase. The temperature was kept at 60°C for 48 hours with the reactor rotating at 1 RPM. After the enzyme hydrolysis, the corn fiber slurry was dewatered by processing with vacuum filtration through cheesecloth. The residual hydrolyzed corn fiber was washed twice with 4 gallons of 95°C water each time and dewatered between washings.

The filtrates were pooled and concentrated using a vacuum evaporator to 26.6% dry solids. The concentrate was then subjected to the secondary acid hydrolysis step.

Secondary Acid Hydrolysis

In order to hydrolyze the oligosaccharides in the various corn fiber hydrolysates to monosaccharides, the hydrolysates were subjected to a secondary acid hydrolysis step. This step consisted of taking the concentrated hydrolysate and adding 1% H_2SO_4, followed by heating of the mixture to 121°C for 30 minutes in a stirred reactor

Hydrolysate Fermentation

The hydrolysate from the initial corn fiber hydrolysis was used as the feed for continuous fermentations using *Saccharomyces cerevisiae* r424A (*30-31*) in a set of three New Brunswick BioFlo III fermentors (Edison, New Jersey). The recombinant yeast was provided by Nancy Ho of the Laboratory of Renewable Resources Engineering at Purdue University, West Lafayette, IN. One mL of amyloglucosidase (EC 3.2.1.3, ADM, Decatur, IL) per liter of fermentation media was added to the fermentors at the start of the fermentation to hydrolyze any maltooligosaccharides remaining in the corn fiber hydrolysate. The initial fermentation media compositions in the fermentation vessels are shown in Table I, the remainder of the initial fermentation media was water. The initial fermentation volume was 1500 mL and the yeast inoculum was 10%. The fermentations were run at 31°C, pH 4.5, agitation on a single impeller at 150 rpm, and a working volume of 2100mL. The fermentation was run without air addition. The pH was controlled by ammonium hydroxide addition. The only feeds to the fermentors were corn fiber hydrolysate and ammonium hydroxide. Samples were taken periodically and the concentrations of sugars, organic acids, glycerol, and degradation products were determined by HPLC.

Table I: Initial Fermentation Vessel Compositions

	Dried Yeast Extract	*Hydrolysate*	*Corn Steep Liquor*
Fermentation Vessel 1	1.6%	33%	
Fermentation Vessel 2		33%	20%
Fermentation Vessel 3		33%	20%

Analytical Methods

Carbohydrates via Gas Chromatography

The carbohydrates in the hydrolysates and fermentation media: xylose, arabinose, fructose, glucose, mannose, galactose, maltose, maltotriose, and sucrose are analyzed by gas chromatography. The liquid samples are derivatized with Trimethylchlorosilane solution (Pierce Biotechnology, Inc., Rockford, IL) for detection. The GC was a Shimadzu GC-17A (Kyoto, Japan) with an injection volume of 1 μL and a temperature of 340°C. The column is a Phenomenex ZB-5 (Torrance, CA) and the detector is a flame ionization detector (FID) set at 340°C. The column carrier gas is helium with a flowrate of 1.1 mL/minute.

High Performance Liquid Chromatography Organic Acid Column

Organic acids, glucose, ethanol, hydroxymethyl furfural (HMF), and furfural concentrations were determined by high performance liquid chromatography (HPLC) using an organic acid column. The system was composed of all Shimadzu (Kyoto, Japan) equipment except for the column, which was a Bio-Rad HPX-87H column (Hercules, CA) kept heated to 65°C, and the solvent degasser, which was a Phenomenex Degassex DG-4400 (Torrance, CA). The injection volume is 10 μL and the mobile phase flowrate is 0.5 mL/min.

Results and Discussion

Corn Fiber Hydrolysis

Initial Thermochemical Hydrolysis Method

The effect of three high-solids hydrolysis methods on corn fiber were determined by the monomer sugar concentrations and the solubilization percentage. Generally, the hydrolysis of corn fiber proceeds, as severity of treatment increases, in the order of starch, hemicellulose branches, hemicellulose xylose backbone, and then cellulose (*4*).

The high-solids treatment in the jacketed reactor is unique, as most corn fiber hydrolyses in the literature are performed at solids concentrations below 15%, or at high solids in steam or ammonia explosion equipment. The high-solids hydrolysis is useful for keeping the concentrations of monosaccharides in the hydrolysate high.

Each of the three treatments were exposed to the same initial thermochemical hydrolysis at 140°C. In this case, the residual sulfur dioxide will act as an acid catalyst to hydrolyze the polysaccharides in the corn fiber. The results are shown in Table II. As can be seen, the larger part of the starch was hydrolyzed: 17.3% of the corn fiber was recovered as starch, where the starch content in corn fiber ranges from 15 to 20%. Approximately 72% of the theoretical 35 to 40% hemicellulose was hydrolyzed. This would be expected in a mild thermochemical hydrolysis. The main hemicellulose sugar present in the hydrolysate is arabinose, as that is the main component of the hemicellulose branch chains and is also hydrolyzed before the xylose backbone (*4, 16*). 42% of the total protein was solubilized by the treatment and small amount of organic acids were released or generated. The degradation products concentration is low at 0.1%. The low concentrations of organic acids and degradation products will keep the inhibition of the yeast fermentation to a minimum (*26, 27*).

Acid Hydrolysis

In this step, the corn fiber was subjected to the initial hydrolysis method followed by adding 1% sulfuric acid and hydrolyzing again for an additional 30 minutes at 140°C. The additional hydrolysis leads to a more severe treatment as can be seen from the degradation products concentrations in Table III. The overall solubilization liberates 17.7g/100g corn fiber more than the initial thermochemical hydrolysis method. The degradation product concentration is

Table II: Corn Fiber Hydrolysate Components - Initial Hydrolysis Method

	Per 100g Corn Fiber (%)
Starch – Hexoses	17.3
Hemicellulose - Pentoses and Hexoses	25.2
Degradation Products	0.1
Protein	4.6
Organic Acids	1.1
Total Mass	48.3

9.7g/100g corn fiber higher than the initial thermochemical hydrolysis, due to the more severe treatment. The sulfuric acid will degrade the oligosaccharides to monosaccharides and then further to hydroxymethyl furfural (HMF) and furfural. The organic acid concentration liberates 1.7g/100g corn fiber higher, because more acetic acid is released from the hemicellulose polymer by the sulfuric acid hydrolysis. Acetic acid, HMF and furfural in any concentrations will cause a decrease in the yeast specific growth rate and the specific ethanol production rate of *Saccharomyces cerevisiae* (*24, 26, 27*). At 2 g/L of both HMF and furfural or 10 g/L of acetic acid, the specific growth rate of *S. cerevisiae* drops to approximately zero (*24, 27*). Also, 5.2 g/100g corn fiber of additional protein is solubilized by the more severe treatment.

Table III: Corn Fiber Hydrolysate Components – Initial Hydrolysis Method Followed by Acid Hydrolysis of the Solid Fiber

	Per 100g Corn Fiber (%)
Starch – Hexoses	15.4
Hemicellulose – Pentoses and Hexoses	25.6
Degradation Products	9.5
Protein	9.8
Organic Acids	2.8
Other	2.8
Total Mass	66.0

Enzyme Hydrolysis

The corn fiber was subjected to the initial thermochemical hydrolysis at 140°C for 30 min, and then glycohydrolases were added at an approximately 1% level. The fiber was held at 60°C for 48 hours while rotating. Table IV shows the results of the enzyme hydrolysis. The results show only slightly higher levels of starch, organic acids and degradation products than the results from the initial hydrolysis method.

The pH of the corn fiber was not adjusted before addition of the enzymes. This was done so that the process would be similar to the manufacturing-scale process. The pH of the slurry is approximately 3.5 at this point in the process. The glycohydrolases utilized included α-amylase, amyloglucosidase, several hemicellulases, and a cellulase. Almost all of the starch was hydrolyzed to glucose by the initial thermochemical hydrolysis step, so the starch-degrading enzymes would not be able to generate a large concentration of glucose. The hemicellulases are general (mainly food-grade enzymes for the brewing industry) and not specific for corn fiber. The overall solubilization of the corn fiber was increased 8g total/100g corn fiber by the hemicellulase treatment over that of the initial thermochemical hydrolysis. The increase in solubilization by the starch enzymes and hemicellulases is accounted for partially by the additional 2 g organic acids/100g corn fiber, acetic acid being the main organic acid in the mixture. Acetic acid is formed when the acetyl groups on the side chains on the arabino-xylan backbone are released. The degradation products concentration is increased an additional 2.5g/100g corn fiber, and an additional 1.1g starch/100g corn fiber is liberated as glucose. Cellulase enzymes were not used, because the cellulose is needed for use as a carrier for the corn steep liquor and stillage from the corn wet-milling process. It has been noted previously that corn fiber arabinoxylan is not easily hydrolyzed by hemicellulase enzymes due to diferulic acid cross-linkages between ferulic acid molecules on separate hemicellulose chains (*5, 16*).

Table IV: Corn Fiber Hydrolysate Components – Initial Hydrolysis Method Followed by Enzyme Hydrolysis of the Solid Fiber

	Per 100g Corn Fiber (%)
Starch – Hexoses	18.4
Hemicellulose – Pentoses and Hexoses	24.6
Degradation Products	2.6
Protein	4.9
Organic Acids	3.1
Other	3.0
Total Mass	56.5

Secondary Acid Hydrolysis

The secondary acid hydrolysis step is important in preparing the hydrolysate as a fermentation media. In the initial hydrolysis step, the polysaccharides are hydrolyzed to oligosaccharides without generating a large concentration of monosaccharides. Table V shows the concentration of oligosaccharides in the extract and secondary acid hydrolyzed extract. The glucose concentration derived from the starch increases by over an order of magnitude during the secondary hydrolysis step. The monosaccharides derived from the hemicellulose increase less severely due to the liberation of monosaccharides from the side chains during the primary hydrolysis step, mainly arabinose. The organic acid concentration also increases due to the acetyl group liberation from hemicellulose oligomers.

The severity of the overall treatment increases when the secondary acid hydrolysis step is added. This can be seen by the increase in the degradation product concentration between the primary and secondary hydrolysis steps. As the monosaccharides are created, they are exposed to conditions that can cause degradation to HMF and furfural.

Fermentation

Due to the high percentage of glucose and xylose in corn fiber, it is necessary to have an organism that ferments both sugars with high performance simultaneously. The organism also needs to have a high resistance to inhibitors in the fermentation media as well as to the product ethanol. *Saccharomyces cerevisiae* has a high ethanol tolerance and a resistance to other common inhibitors in lignocellulosic hydrolysates (*24, 26, 27*). The *S. cerevisiae* r424A organism is genetically engineered by cloning the xylose reductase and xylitol dehydrogenase genes from *Pichia stipitis* and the xylulokinase gene from *S. cerevisiae* into *S. cerevisiae* (*30*). This allows the organism to convert xylose to ethanol.

Table VI shows the results of the three continuous fermentations completed with the r424A organism. The carbon source in the fermentation media was corn fiber hydrolysate produced by the initial hydrolysis method followed by secondary acid hydrolysis. The yeast metabolized over 60% of the total carbohydrates with a g ethanol/g carbohydrate yield between 0.50 and 0.55. The lower percentage of total carbohydrates metabolized is likely due to the continuous fermentation method, as shake flask fermentations, albeit at lower dry solids, metabolized up to 91% of the carbohydrates (*unpublished data*). The carbohydrate utilization will be optimized in future fermentations. The g/g yield is near theoretical, therefore the metabolized carbohydrate is converted solely to ethanol. The organic acids, ethanol and dextrose concentrations for a sample fermentation are shown in Figure 1. The ethanol concentration in the fermentor at the end of the run was 55g/L. The concentrations of the citric acid, lactic acid and glycerol increase substantially during the fermentation, which are normal byproducts of the ethanol fermentation by *Saccharomyces cerevisiae*. The volume of the fermentation was doubled with hydrolysate fed into the fermentor. These fermentations show that the organism can ferment the glucose and xylose from the corn fiber hydrolysate to ethanol without detoxification of the hydrolysate. The fermentation conditions were not optimized, so additional improvement in the fermentation is expected.

Table V: Effect of Secondary Acid Hydrolysis on Corn Fiber Hydrolysis

	Starch – Hexoses (g/L)	*Hemicellulose - Pentoses and Hexoses (g/L)*	*Degradation Products (g/L)*	*Protein (g/L)*	*Organic Acids (g/L)*	*Other (g/L)*
Extract (before secondary acid hydrolysis)	2.39	17.31	1.05	20.00	0.92	0.06
Secondary Acid Hydrolyzed Extract	35.80	43.10	2.77	24.00	2.52	0.04

Table VI: Ethanol Yields from Continuous Fermentation

Fermentor Number	*Fermentation Method*	*Ethanol Titer g/L*	*Percent Total Carbohydrate Consumed*	*Gram of Ethanol /gram of consumed carbohydrate*
1	Continuous	55 g/L	63%	0.51
2	Continuous	54 g/L	63%	0.50
3	Continuous	58 g/L	61%	0.55

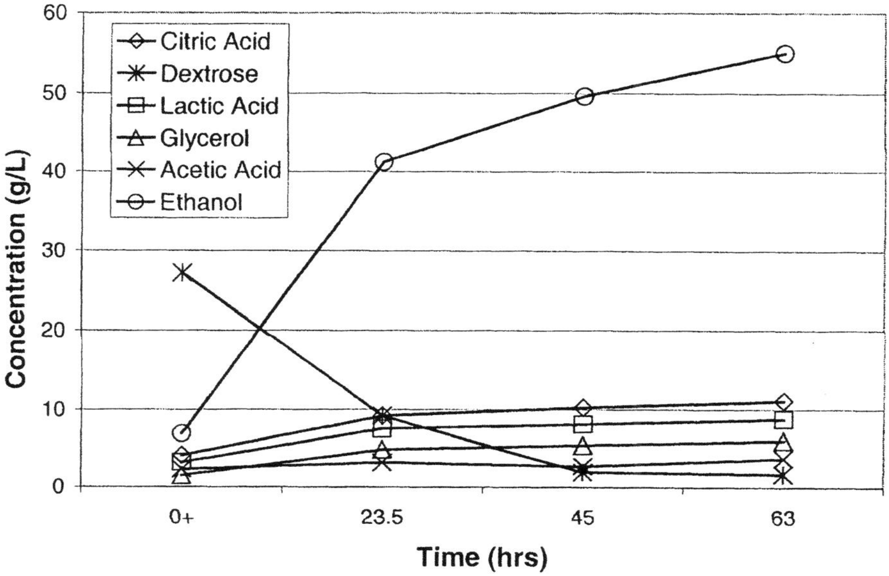

Figure 1. Corn Fiber Hydrolysate Fermentation with S. cerevisiae *r424A (fermentor 1).*

Conclusions

High-solids corn fiber hydrolysis is more likely to be commercially feasible than dilute corn fiber hydrolysis. This method keeps the concentration of sugars at high concentrations, which will increase the ethanol concentration, which is necessary for an economical ethanol distillation.

The most commercially feasible hydrolysis method is the initial hydrolysis at 140°C for 30 minutes. This method hydrolyzes most of the starch and 72% of the hemicellulose. This leaves the cellulose and a portion of the hemicellulose to act as a carrier for the corn steep liquor and stillage to make corn gluten feed. The additional acid hydrolysis step creates a large amount of degradation products, which would be inhibitory to the ethanol fermentation, without giving a greater monosaccharide concentration. The enzyme hydrolysis step is not feasible without enzymes that contain activities specifically for the corn fiber hemicellulose matrix. These types of enzymes are not commercially available, or would be prohibitively expensive, therefore, enzyme hydrolysis of corn fiber is not currently commercially feasible.

The fermentation of the corn fiber hydrolysate generated by the initial hydrolysis method proved to be successful in producing a high concentration of ethanol from glucose and xylose. The total carbohydrate conversion is lower than the best possible, but the fermentation method has not been optimized.

Acknowledgements

The authors wish to thank U.S. Department of Energy Jim Spaeth and Gerson Santos-Leon and DOE-OIT Mark Paster. The funding for this project is DOE/EE/OTT grant #DE-FC36-00G10596, A000 "Corn Fiber Separation and Subsequent Conversion to Fuels and Chemicals".

References

1. Corn Refiners Association website, *http://www.corn.org/web/shipprod.htm*
2. Gulati, M.; Kohlmann, K.; Ladisch, M.R.; Hespell, R.; Bothast, R.J. *Bioresour. Technol.* **1996**, *58*, 253-264.
3. Osborn, D.; Chen, L.F. *Starch*, **1984**, *36*, 393-395.
4. Grohmann, K.; Bothast, R.J. *Process Biochemistry,* **1997**, *32*, 405-415.
5. Saha, B.C.; Bothast, R.J. *Appl. Biochem. Biotechnol.* **1999**, *76*, 65-77.
6. Bura, R.; Mansfield, S.D.; Saddler, J.N.; Bothast, R.J. *Appl. Biochem. Biotechnol.*, **2002**, *98-100*, 59-72.
7. Antrim, R.L.;Harris, D.W. U.S. Patent 4,038,481, 1977.
8. Antrim, R.L.; Harris, D.W. U.S. Patent 4,104,463, 1978.
9. Doner, L.W.; Hicks, K.B. *Cereal Chem.* **1997**, *74*, 176-181.
10. Hespell, R.B. *J. Agric. Food Chem.* **1988**, *46*, 2615-2619.
11. Doner, L.W.; Sweeney, G.A.; Hicks, K.B. U.S. Patent 6,147,206, 2000.
12. Buchanan, C.M.; Buchanan, N.L.; Debenham, J.S.; Shelton, M.C.; Wood, M.D. U.S. Patent 6,388,069, 2002.
13. Dale, B.E.; Leong, C.K.; Pham, T.K.; Esquivel, V.M.; Rios, I.; Latimer, V.M. *Bioresour. Technol.* **1996**, *56*, 111-116.
14. Moniruzzaman, M.; Dale, B.E.; Hespell, R.B. Bothast, R.J. *Appl. Biochem. Biotechnol.* **1997**, *67*, 113-126.
15. Hespell, R.B.; O'Bryan, P.J.; Moniruzzaman, M.; Bothast, R.J. *Appl. Biochem. Biotechnol.***1997**, *62*, 87-97.
16. Saulnier, L.; Thibault, J-F. *.J Sci. Food. Agric.* **1999**, *79*, 396-402.
17. Lapierre, C.; Pollet, B.; Ralet, M-C.; Saulnier, L. *Phytochemistry* **2001**, *57*, 765-772.
18. Hosny, M.; Rosazza, J.P.N. *J. Nat. Prod.* **1997**, *60*, 219-222.
19. Lohmeier-Vogel, E.M.; Sopher, C.R.; Lee, H. *J. Ind. Microbiol. Biotechnol.* **1998**, *20*, 75-81.
20. Pampulha, M.E.; Louriero-Dias, M.C. *Appl. Microbiol. Biotechnol.* **1990**, *34*, 375-380.
21. Barber, A.R.; Hansson, H.; Pamment, N.B. J. *Ind. Microbiol. Biotechnol.* **2000**, *25*, 104-108.

22. Banerjee, N.; Bhatnagar, R.; Viswanathan, L. *European J Appl. Microbiol. Biotechnol.* **1981**, *11*, 226-228.
23. Pampulha, M.E.; Loureiro-Dias, M.C. *Appl. Microbiol. Biotechnol.* **1989**, *31*, 547-550.
24. Phowchinda, O.; Délia-Dupuy, M.L.; Strehaiano, P. *Biotechnol. Lett.* **1995**, *17*, 237-242.
25. Pampulha, M.E.; Loureiro, V. *Biotechnol. Lett.* **1989**, *11*, 269-274.
26. Palmqvist, E.; Grage, H.; Meinander, N.Q.; Hahn-Hägerdal, B. *Biotechnol. Bioeng.* **1999**, *63*, 46-55.
27. Taherzadeh, M.J.; Gustafsson, L.; Niklasson, C.; Lidén, G. *Appl. Microbiol. Biotechnol.* **2000**, *53*, 701-708.
28. Antoce, O.A.; Antoce, V.; Takahashi, K.; Pomohaci, N.; Namolosanu, I. *Am. J. Enol. Vitic.* **1997**, *48*, 413-422.
29. Persson, P.; Larsson, S.; Jönsson, L.; Nilvebrant, N-O.; Sivik, B.; Munteanu, F.; Thörneby, L.; Gorton, L. *Biotechnol. Bioeng.* **2002**, *79*, 695-700.
30. Ho, N. W-Y.; Chen, Z.; Brainard, A.P.; Sedlak, M. in *Green Chemical Syntheses and Processes;* Anastas, P.T.; Heine, L.G.; Williamson, T.C., Ed., ACS Symposium Series 767; American Chemical Society: Washington, D.C., 2000, pp 143-159.
31. Krishnan, M.S.; Xia, Y.; Ho, N. W-Y.; Tsao, G.T. in *Fuels and Chemicals from Biomass;* Saha, B.C., Woodward J., Ed., ACS Symposium Series 666; American Chemical Society Washington, D.C., 1997, pp 74-92.

Biodegradation

Chapter 6

Unconventional Relationships for Hemicellulose Hydrolysis and Subsequent Cellulose Digestion

Bin Yang, Matthew C. Gray, Chaogang Liu, Todd A. Lloyd, Suzanne L. Stuhler, Alvin O. Converse, and Charles E. Wyman

Thayer School of Engineering, Dartmouth College, Hanover, NH 03755

Cellulosic biomass can be pretreated with dilute sulfuric acid to recover high yields of sugars directly from hemicellulose and subsequently by enzymatic hydrolysis of the residual cellulose, and these sugars can be used to produce fuels and chemicals with unique and powerful economic, environmental, and strategic benefits. Pretreatment is costly for such biological routes, and we seek to better understand hydrolysis kinetics to support emerging applications and needed technology advances. Hemicellulose removal is affected by solids concentration and flow through the solids, contrary to customary first-order reaction models. We also measure a wide range of oligomer species, particularly at lower acid levels, that most kinetic models ignore or oversimplify, and a model is proposed to more accurately predict their behavior. The solubility of oligomers is also being measured by a new technique to better understand its effect on kinetics. New data suggests that lignin alterations during hemicellulose hydrolysis are important in improving enzymatic hydrolysis of cellulose.

Introduction

Biological production of fuels and commodity chemicals from inexpensive and abundant sources of cellulosic biomass can provide unparalleled environmental, economic, and strategic benefits *(1,2)*. For example, enzymatic hydrolysis of the cellulose in these materials to glucose followed by fermentation to ethanol is an economically attractive route to production of alternative liquid transportation fuels *(3,4,5,6,7)*. However, biomass must be pretreated to open up its structure prior to such biological operations so that high yields vital to economic success can be realized. Pretreatment is currently one of the most expensive steps in bioconversion routes, and advanced pretreatment technologies are needed to significantly reduce costs, improve cellulose digestibility, simplify upstream and downstream operations, and provide the potential for additional revenues from co-products *(8,9,10,11)*. Better knowledge of pretreatment systems would facilitate such advances and accelerate commercial applications by giving practitioners and financial organizations greater confidence in scale-up *(9)*.

Dilute Acid Hydrolysis

A number of organizations favor hemicellulose hydrolysis by dilute sulfuric acid for pretreatment because high sugar yields can be realized from hemicellulose during pretreatment and from cellulose in subsequent enzymatic hydrolysis of the solid residue *(11,12,13,14,15,16,17)*. Compared to water-only processes, sulfuric acid pretreatments increase the ratio of monomeric to oligomeric sugars and can produce monomeric sugar yields of up to 90% of the theoretical maximum *(19,20,21,22,23)*. Novel technologies based on flowing liquid through solid biomass have been shown to produce higher hemicellulose sugar recoveries, greater lignin removal, less inhibitors in the hydrolyzate liquid, and highly digestible cellulose when compared to conventional systems *(25,26,27,28,29)*. However, the high water consumption required for such flow systems results in excessive energy consumption for pretreatment and product recovery, and development of satisfactory equipment would be challenging. Nonetheless, understanding the cause of the enhanced performance of flow systems could lead to novel advanced pretreatments that reduce costs.

Data and kinetic models of dilute acid pretreatment are vital to provide a foundation for understanding hemicellulose hydrolysis and the cause of enhanced performance by flow system systems. Initial hemicellulose hydrolysis models were adapted from Saeman's first-order homogeneous kinetic model of cellulose hydrolysis in a dilute acid batch system *(30)* and later modified to include two different fractions of hemicellulose, one of which is more easily hydrolyzed than the other *(31)*. Oligomers were eventually included as reaction intermediates in a few studies *(32,33)*, but these models either ignored oligomers or treated them as only one or two discrete compounds. Although other models have been devised over the years *(34,35,36)*, they all evolve from the same first-order kinetic representation, and none reported in the literature is robust enough to adequately describe the changes in observed performance for different reactor configurations.

Jacobsen and Wyman studied the effect of solids loading on the total yield of xylose as both monomers and oligomers for hydrolysis of sugarcane bagasse in water at 200°C and found a statistically significant increase in yield by reducing solids loading from 1% to 0.5% *(37)*. Others have shown that sugar recovery increases with water content in batch systems *(38,39)*. Such results suggest that flowthrough operations benefit from the large amounts of water applied, but a complete explanation has not yet been established. In addition, the effect of solids concentration is not consistent with the Saeman model or any of its derivatives *(30,32)*.

To reconcile these differences, we have proposed that pretreatment models should combine reaction with mass transfer and solubility limitations so that the concentration gradient between the solid and the liquid phases is taken into account. Furthermore, we have applied depolymerization kinetics to account for the production of sugar monomers and oligomers observed during the hydrolysis of cellulosic biomass, particularly at low acid levels. Some aspects of these studies will be summarized in this chapter.

Enzymatic Hydrolysis of Pretreated Cellulose

Several structural and compositional attributes of biomass are thought to influence the enzymatic hydrolysis of cellulose to glucose including cellulose crystallinity, lignin levels, hemicellulose removal, accessible surface area of cellulose, and the presence of acetyl groups *(40,41)*. However, the complex structure of biomass makes it difficult to discern the relative importance of these features and their roles, and reducing one barrier to digestion can alter the importance of others. For example, removing hemicellulose also removes acetyl groups and usually changes the lignin left in the material making it difficult to ascertain which factor was most influential in improving performance. Various studies have reported that cellulose hydrolysis improves with increased lignin removal although differences were reported in the degree of lignin removal needed *(42,43,44)*. The ratio of syringyl to guaiacyl lignin groups was also shown to have important effects on enzymatic digestibility *(45)*. It is probable that one of the more significant factors is the impact of lignin on fiber swelling which in turn influences cellulose accessibility *(46,47)*. Lignin has also been claimed to depolymerize and then repolymerize in a different morphology during hemicellulose hydrolysis *(48,49)*.

The removal of lignin not only increases cellulose accessibility but also enhances cellulase availability. Lignin and its complexes physically and chemically resist enzymatic attack, and condensed lignin also has the ability to adsorb protein from aqueous solutions *(50)*. Thus, it appears that lignin removal not only opens up more space for enzymes but also reduces non-specific enzyme adsorption, improving the efficacy of hydrolysis *(41, 51,52,53,54)*.

On the other hand, several studies showed a direct relationship between cellulose digestion and hemicellulose removal. Knappert et al showed that removal of hemicellulose in poplar by dilute acid hydrolysis increased the susceptibility of cellulose to enzymes *(12)*. Grohmann et al also showed a direct relationship between hemicellulose removal and cellulose digestion and concluded that although lignin removal could enhance digestibility, it was not necessary to achieve good cellulose conversion *(55,56)*. However, some substrates required higher temperatures for effective cellulose hydrolysis at the same degree of hemicellulose removal, suggesting that hemicellulose is not the only factor impacting digestibility, while other studies do not support a role for hemicellulose in changing cellulose digestibility *(16,57,58)*. It is important to note that lignin could be altered at the high temperatures typically employed for hemicellulose hydrolysis, thereby impacting cellulose digestion rates *(16)*.

As part of our pretreatment research, we have evaluated the digestibility of corn stover cellulose following pretreatment and found that it changes significantly with pretreatment reactor configuration. Furthermore, we have observed that these differences in performance can be related to modification or removal of lignin as well as removal of hemicellulose, and this chapter will include a summary of some of these findings along with our results for hemicellulose hydrolysis.

Materials and Methods

Sample Preparation

The National Renewable Energy Laboratory (NREL) in Golden, Colorado graciously provided corn stover from a large lot they obtained from Harlan, Iowa and maintained at controlled conditions. This material was milled to pass through a 2 mm opening and then screened to obtain a –420 +250 μm fraction which was stored in plastic Ziploc bags and kept in a freezer (-20°C) for all tests.

The composition of the corn stover was determined through application of NREL LAP procedures 001, 002, and 012 *(59,60,61)*, and this substrate was found to contain 37.8% glucan, 21.3% xylan, 1.6% arabinan, 3.8% mannan, 1.4% galactan, 17.8% lignin, 7.8% ash, and 5.8% moisture by weight.

Reactors

For water-only tests, batch reactors were constructed by cutting stainless steel or Hastelloy C276 tubing (0.5" OD x 0.035" wall thickness) into 4" sections that were fitted with Swagelok couplings and removable threaded end caps, giving a total reactor volume of 9.5 mL. For dilute acid tests, Hastelloy C276 tubing was used, and Teflon plugs were inserted in the ends to protect the stainless steel caps as utilized by researchers at Auburn University *(35)*.

Work was also performed with a mixed batch reactor to gain insight into the effects of agitation on batch systems and the possible influence of fluid velocity on hemicellulose removal. For these tests, a 1-L Parr bomb constructed of Carpenter-20 (Parr Instruments, Moline, IL) was fitted with a flat-blade impeller on a one-piece shaft and operated at varying speeds using a Parr DC motor drive (A1750HC, Parr Instruments, Moline, IL) *(62)*.

Two flowthrough systems were employed in this work: a smaller reactor (1/2-in ID × 1.84-in length with an internal volume of 3.6 mL) and a larger reactor (1/2-in ID × 6-in length with an internal volume of 14.3 mL). All reactor parts were obtained from Maine Valve and Fitting Co., Bangor, ME. A 1/8-in stainless steel thermocouple (Omega Engineering Co., Stamford, CT) was installed at the outlet of the reactor to monitor temperature, and 316 stainless steel tubing was used as a preheating coil (1/4-in OD × 0.35 inch wall), to connect the reactor with other components of the system, and as a cooling coil (1/8-in OD × 0.028 inch wall). The preheating coil was long enough to allow the incoming water to reach the desired temperature before it entered the reactor. To operate the flowthrough unit, about 2 grams of corn stover was loaded into the reactor, and the reactor was then connected to the system. Distilled water at room temperature was pumped through the reactor to purge air, completely wet the biomass in the reactor, and pressurize the system to the set pressure. Then, the reactor and preheating coil were submerged for 2 minutes in a 4-kW fluidized sand bath (model SBL-2D, Techne Co., Princeton, NJ) set at a temperature of 100°C and then moved to a second sand bath set at the target reaction temperature. Flow was maintained until the desired reaction time, at which point the reactor and preheating coil were transferred to an ice water bath to stop the reaction *(63,64)*.

Solubility Measurements

The solubility of monomeric and oligomeric sugars was measured by an inline refractive index detector (AFAB Enterprises, Eustis, FL) attached to a 1" stainless steel tee with a plug at the bottom and a rubber stopper to cover the top. During operation, this device was loaded with a known mass of water and sugar, placed on a programmable hotplate/stirrer, and covered with styrofoam insulation to minimize heat losses. The RI signal rapidly increased with temperature as the sugars dissolved and more slowly once they were all in solution. Thus, the solubility could be determined from the total amount of sugar and water added and the temperature at which the slope of the RI signal versus temperature changed. Solubilities were also measured by a standard method in which a vial containing a sugar/water mixture was submerged in a constant temperature water bath and samples taken at equilibrium were analyzed by HPLC and gravimetric methods *(65)*.

Enzymes

Cellulase (Spezyme CP, Lot No. 301-00348-257, Genencor, Palo Alto, CA) at 28 FPU/mL supplemented with β-glucosidase (Novozyme 188 at 250 β-glucosidase IU/mL, Sigma, St. Louis, MO) at a ratio of 1:1.75FPUase:CBUase was used for all hydrolysis experiments. Enzymatic treatments were performed at 60 FPU/g cellulose with the activity as calculated by adding the activities of both the Spezyme CP and Novozyme 188 *(66)*.

Enzymatic Hydrolysis

Enzymatic hydrolysis of pretreated cellulose from both the batch and flowthrough systems was conducted at a 2% solids concentration (g dry weight/100 mL) in 50 mM acetate buffer (pH 4.8) containing 40 μg/mL tetracycline and 30 μg/mL cycloheximide. Flasks were pre-incubated at 50°C in water using an orbital shaker bath (3540, Barnstead International, Dubuque, IA) at 150 rpm for 10 minutes, and the enzymes were added to start the hydrolysis after acclimation. Aliquots of 0.5 mL were taken at different times (0, 4, 24, 48, 72 h), immediately chilled on ice, and centrifuged at 5000 G for 10 min. Total sugar analyses were carried out on the resultant supernatants.

Analytical Procedures

Sugar and acid insoluble lignin content of solids were determined using the Klason lignin procedure published as NREL LAP 003 and 014 *(67,68)*. The sugar concentrations were measured with a high performance liquid chromatography system (Waters 2695, Milford, MA) equipped with a pulsed refractive index detector (Waters 2410, Milford, MA). The column was equilibrated with deionized water at a flow rate of 0.6 mL/min. An Aminex HPX-87P (Bio-Rad, Sunnyvale, CA) column was used for sugar separations.

Results and Discussion

Effects of Flow Rate on Corn Stover

Total solids removal

As presented in Table 1 and Figure 1, our results show that the rate of total solids removal increased with both temperature and flow rate for both water-only and dilute acid pretreatment of corn stover. However, for water-only pretreatment, about 94-97% of the total material was accounted for in all runs at 180°C after 16 minutes, suggesting little of this loss in mass was due to decomposition at this temperature. However, the overall mass recovery decreased with increasing temperature, especially for batch and low flow rate runs, indicating that more mass was lost at these operating conditions. These results show that increasing flow rate increases removal of solids, supporting the idea that at longer residence times and higher temperatures, dissolved sugars decomposed to furfural and other volatile degradation products. It is expected that most of the overall loss in mass was due to dissolved xylan, as much less glucan was dissolved over this range of temperatures and times (Table 1).

Xylan removal

Xylan removal increased with flow rate and temperature for both water-only and very dilute acid pretreatment, as illustrated by Figure 1. As can be seen from this figure, flowthrough pretreatment of corn stover with hot water at a flow rate of 10 mL/min removes more xylan at the same temperature and time than

Table 1. Total mass dissolution (TMD), total mass remaining (TMR), xylan remaining, and glucan remaining in the solid residue for water-only pretreatment of corn stover.

Temp. (°C)	*Flow rate (mL/min)*	*Time (min)*	*TMD (%)*	*TMR (%)*	*OMB[1] (%)*	*Xylan remaining (%)*	*Glucan remaining (%)*
180	0	12	13.3	82.7	96.0	91.3	102.3
		16	18.1	78.9	97.0	80.0	101.5
180	10	12	27.0	66.5	93.5	60.8	
		16	32.4	62.3	94.7	42.5	99.4
200	0	12	22.5	63.5	86.0	40.1	101.0
		16	23.4	62.2	85.6	29.9	99.9
200	1	12	28.4	56.6	85.0	36.4	-
		16	32.7	53.0	85.7	29.4	99.5
200	10	12	51.2	48.5	99.7	14.1	99.3
		16	52.4	43.3	95.7	9.5	98.5
220	0	12	28.7	55.2	83.9	13.5	-
		16	30.2	53.1	83.3	8.7	98.8

OMB[1]: overall mass balance = TMD + TMR

batch pretreatment of stover with 0.05 wt% sulfuric acid. Although it has been postulated that autohydrolysis of hemicellulose is catalyzed by acetic and other organic acids released during the breakdown of hemicellulose *(69)*, this mechanism cannot fully explain flowthrough reactor behavior in which hemicellulose solubilization is enhanced by flow without acid addition, especially at high flow rates. At such high flow rates, the acids will have little time to act before they are swept from the reactor, and the large volumes of water used also decrease the concentration of organic acids. In addition, other studies demonstrated that adding supplemental acetic acid did not accelerate the solubilization of hemicellulose during pretreatment of biomass with hot water, suggesting that the organic acids released are not the primary hydrolytic agent *(70,71)*.

Almost all of the dissolved xylan was in oligomeric form for water-only pretreatment of corn stover over a temperature range of 160-220°C. Addition of very small amounts of acid accelerated hydrolysis of xylose oligomers to monomers and increased the monomeric fraction, but the largest portion of the overall xylan dissolved remained in oligomeric form. These results demonstrate that oligomers are important intermediates during hemicellulose hydrolysis, especially at no acid or very low acid conditions. The yield of oligomers also increased tremendously with flow rate, suggesting that flow rate accelerates solubilization of hemicellulose. Together, these studies demonstrate that factors other than acid concentration and temperature impact hemicellulose hydrolysis, and although a detailed mechanism responsible for this behavior is still under investigation, we postulate that mass transfer plays an important role in hemicellulose hydrolysis, with this effect being especially important at no acid or very low acid conditions *(37,63)*.

Lignin removal

Our results show the amount of lignin removed increased with flow rate and a positive correlation was observed between xylan and lignin removal for flowthrough pretreatment, as shown in Figure 2. Consistent with observations by others that lignin and hemicellulose are covalently linked in native materials, we expect that hemicellulose oligomer-lignin compounds are released initially. In addition, these hemicellulose oligomer-lignin materials are expected to be soluble at high temperatures and can therefore be swept from a flowthrough system, especially with the large volume of water at high flow rates. However, when held at reaction conditions, the hemicellulose oligomer-lignin compounds are expected to break down to separate sugar oligomer and lignin fragments that can in turn form sugar monomers and monomeric lignin, respectively, depending on temperature, acid concentration, and residence time. The lignin species formed have a limited solubility and can form other low solubility products through condensation reactions. Thus, for batch runs with longer residence times, the overall result is dramatically lower lignin removal. For intermediate flow rates, some of the hemicellulose oligomer-lignin compounds are removed while the remaining portions have enough time to react to insoluble lignin-based products. Such a mechanism would explain the relationship between xylan and lignin removal shown in Figure 2 and why lignin removal is impacted by flow rate and temperature *(63)*.

Effect of Solids Concentration

The effect of solids loading on hemicellulose hydrolysis was studied by Jacobsen and Wyman *(37)* and Stuhler *(71)*. Water-only batch tube experiments with corn stover were performed at 200°C for 15 minutes at solids loadings of 5 and 21%, and a number of performance criteria including total solids remaining and total soluble xylan oligomer yields were evaluated *(71)*. Statistically significant differences in several results were observed at a 95% confidence level, as summarized in Figure 3.

It is interesting to note that xylan remaining in the solids decreases with increasing solids loading, which is opposite of what would be expected if solubility were limiting. A more likely reason for this is that the increased hydronium ion concentration at higher solids loadings accelerates sugar release. On the other hand, solubilized oligomers and total mass removal decrease while monomers and furfural increase with increased solids loading, consistent with this hypothesis.

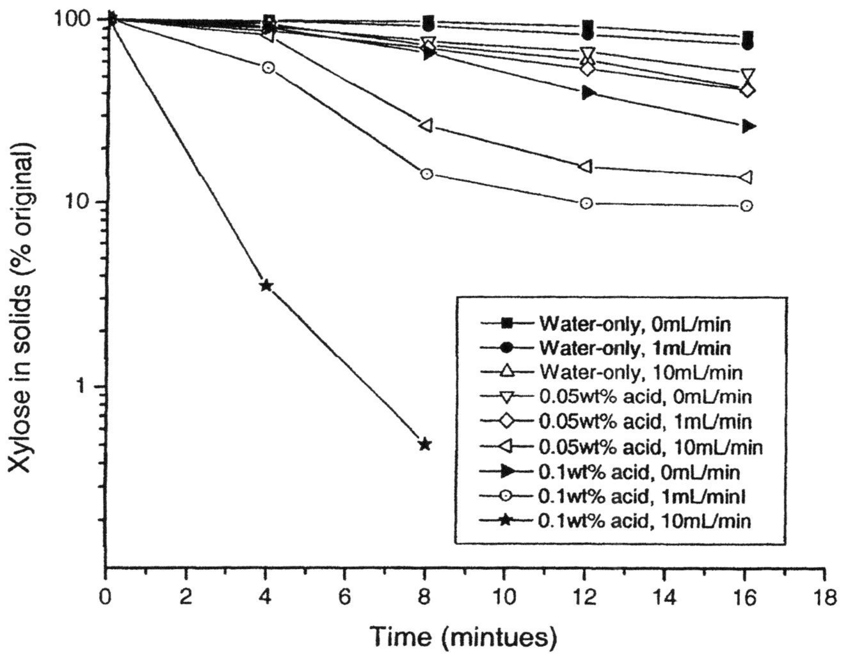

Figure 1. Effect of flow rate and acid concentration on xylose remaining in the solids at 180°C

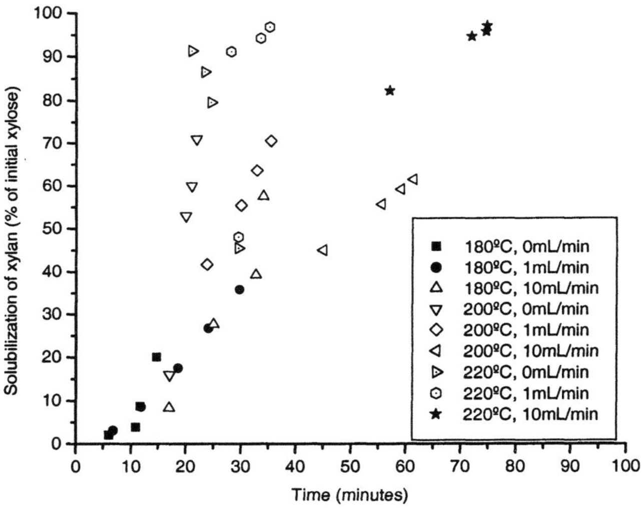

Figure 2. The effect of flow rate and temperature on the relationship between solubilization of xylan and lignin for water-only hydrolysis.

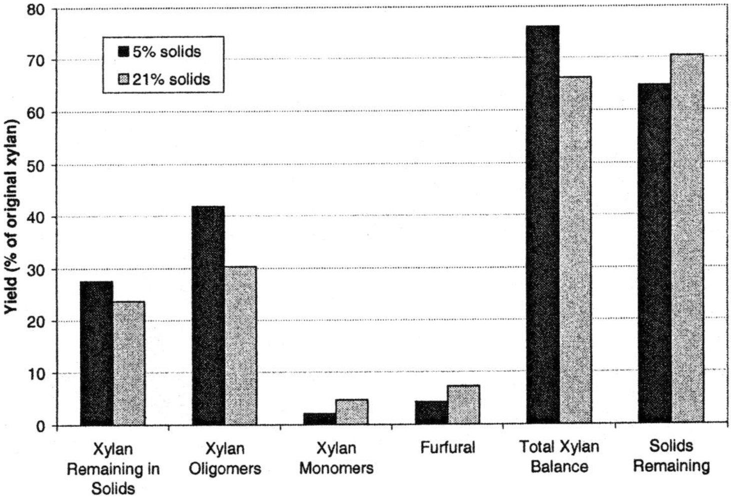

Figure 3. Effect of solids concentration on xylan hydrolysis yields after 15 minutes at 200°C.

The significant decrease in the xylan balance with increasing solids has two possible explanations: the degradation reactions are more complete (as evidenced by the higher furfural content) or higher concentrations of other solubilized components such as lignin remove more sugars from solution through condensation reactions. It is most likely that the observed effect results from a combination of these two mechanisms.

Traditional hemicellulose hydrolysis kinetic models cannot account for a change in hemicellulose sugar yields with solids concentration and suffer from inconsistencies that bring into question their mechanistic accuracy. Thus, although current models can be useful for a given flow regime, their ability to describe different systems such as flowthrough reactors on a consistent basis is unproven *(29)*.

Modeling Mass Transfer Effects

A model which includes the effects of mass transfer on hemicellulose hydrolysis has been developed by Brennan and Wyman with the following equations applied to describe a batch system *(62)*.

$$\frac{dH}{dt} = -k_1 H \qquad (1)$$

$$\frac{dX_n}{dt} = k_1 H - k_d A\left(\frac{X_n}{V_p} - \frac{X_s}{V_t}\right) \qquad (2)$$

$$\frac{dX_s}{dt} = k_d A\left(\frac{X_n}{V_p} - \frac{X_s}{V_t}\right) - k_4 X_s \qquad (3)$$

$$\frac{dD}{dt} = k_4 X_s \qquad (4)$$

Equation [1] expresses the reaction of hemicellulose, *H*, in the solid phase over time *t* to form oligomers X_n, where k_1 is a rate constant with units of time^{-1}.

Equation [2] is the differential equation describing the amount of soluble but still undissolved oligomers inside the particle, X_n, by accounting for their accumulation by chemical reaction and removal by mass transfer across a diffusive boundary layer. k_d is a mass transfer coefficient (length^2/time), X_s is the amount of xylan oligomers dissolved in the bulk solution (moles or mass), A is the surface area of a solid particle (length^2), V_p is the volume of the solid phase (length^3), and V_t is the total volume of the bulk solution (length^3). Equation [3] is the differential balance describing mass transfer of solubilized xylan into the bulk solution and its degradation to furfural, tars, etc. with k_4 being the degradation rate constant. Equation [4] is the differential balance for the chemical degradation of solubilized xylan to degradation products, D. Together, Equations 1-4 account for the effects of both mass transfer and chemical reaction during hemicellulose hydrolysis, a process that has been perceived to be completely reaction-controlled in the past.

Table 2 summarizes the mass transfer coefficients determined for this model for batch tubes, stirred batch, and flowthrough reactor configurations and shows that the diffusive mass transfer coefficient k_d increases in this order of reactor type. However, one would expect the mass transfer coefficient to follow such a pattern as flow is increased. On the other hand, although the rate constants for conventional models based on only chemical reaction can also be fit to data from these three reactor types, rate constants for these models should only depend on temperature, and these variations would not be expected. Thus, coupling mass transfer to reaction appears to provide a more meaningful explanation for the effects of flow on performance, but further work is needed to fully develop and evaluate this approach.

Oligomer Solubility

One of the factors that we need to integrate into our mass transfer model is the solubility limits of the sugars and oligomers involved and how they are influenced by the presence of other species. However, there is surprisingly little information available in the literature on the solubility of pure sugars in a two component (i.e., single sugar/water) mixture and almost none for sugar oligomers. For this reason, the in-situ solubility device described earlier was developed for measuring oligomer solubilities and was tested using β-cyclodextrin, an inexpensive starch oligomer of limited solubility *(72,73)*. Data was collected at various solids levels using both in-situ and water bath methods, with the results shown in Figure 4. The 95% confidence intervals on the data from this device intersect those of the literature data at every temperature except for at 55°C, where it is slightly higher. The in-situ method had a tighter confidence range than the water bath method, possibly because the latter method requires filtering a solution and removing it from a hot environment before analysis while the former method measures solubility instantly. We are now investigating using a much smaller apparatus to collect data with significantly less of the expensive oligomers needed for this research.

Table 2. Mass transfer coefficient k_d (cm^2/sec) determined for various reactor configurations for corn stover treated at 180 °C.

Reactor Type	*Water-Only*	*0.05wt% H_2SO_4*	*0.1wt% H_2SO_4*
Batch Tube	0.0025	-	0.0036
Stirred Batch	0.0058	-	-
Flowthrough – 1 ml/minute	0.004	0.0075	0.02
Flowthrough – 10 ml/minute	0.011	0.025	0.15

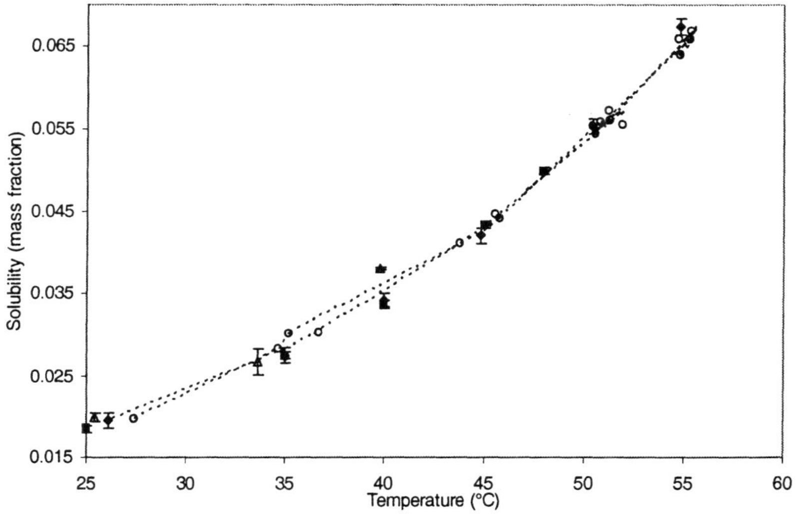

Figure 4. Solubilities of β-cyclodextrin obtained with the in-situ device (□), compared to isothermal water bath data (□)and literature values (◆ (72) and ■(73)). The dotted lines are 95% confidence intervals for the in-situ data.

An important goal of this element of our research is to develop tools to predict the solubility of oligomers at pretreatment temperatures at which it is difficult to collect solubility data. In this regard, we were able to estimate the experimental solubility data of low molecular weight xylo-oligomers using the ideal solubility law. However, many of the physical parameters needed in this model are unrecorded and had to be estimated. Subject to this limitation, the solubilities for xylobiose through xylohexaose at room temperature were predicted to be over 1% by mass and increased to greater than 55% at 120°C. This projection implies that solubility is not expected to be a limiting factor in water-only hydrolysis for oligomers with a degree of polymerization less than 6 *(65)*. However, additional data and modeling are needed to substantiate this finding and extend the method to higher molecular weight oligomers.

Oligomer Production, and Hydrolysis

We also studied the autohydrolysis of xylan at 200°C in hot water to help clarify the behavior of hemicellulose hydrolysis. After 5 minutes, around 37% of the potential xylose was present as soluble oligomers, as determined by post-hydrolysis of the resulting liquid fraction. These oligomers appeared to be high molecular weight compounds with a degree of polymerization (DP) greater than 10 because they could not be analyzed directly on our ion-moderated partition (IMP) chromatography column. However, after reaction for 10 minutes, most of the oligomers had a DP of less than ten and could be measured by the IMP column *(74)*.

To understand how the soluble oligomers react in solution, the kinetics of water-only hydrolysis of pure oligomers of DP 1 to 5 was followed at 200°C. Disappearance of each of these oligomer species could be described well by first-order homogeneous kinetics. It was further found that the decomposition rate constants of xylopentaose and xylotetraose were found to be similar. The rate constants of xylotriose and xylobiose were about two-thirds of the rate of xylotetraose and xylopentaose. Xylose decomposed at about half the rate of xylobiose. The formation of shorter oligomers from each of these pure oligomers was also tracked, and it was found that the release of lower DP species from the higher DP oligomers could not be accurately described unless degradation (products other than xylose or its DP 2 to 5 oligomers) of a significant fraction of the oligomers was integrated into the kinetic pathway. This result could help account for why yields are low for hemicellulose hydrolysis without acid addition.

Depolymerization Model

A primary limitation of traditional kinetic models is their inability to describe the time course distribution of oligomeric species of varying chain lengths that were observed in the experiments described in the previous sections. Because of the polymeric nature of hemicellulose, we would expect a distribution of hemicellulose fragments with varying chain lengths, and by adapting an approach developed by Simha *(75),* the hydrolysis of corn stover hemicellulose was modeled as a depolymerization reaction *(35).* Assuming that all bonds in the xylan chain are cleaved at the same rate, the instantaneous distribution of depolymerization fragments can be expressed by the following differential equation:

$$\frac{dN_j}{dt} = 2k_h \sum_{i=j+1}^{n} N_i - k_h(j-1)N_j \qquad (5)$$

or

$$N_j = N_n^0(1-\alpha)^{(j-1)}\alpha[2+(n-j-1)\alpha] \quad (6)$$

in which k_h is the hydrolysis rate constant, N_n^0 is the initial concentration of polymer of initial length n, N_j = molar concentration of a fragment of length j, and $\alpha = 1 - e^{-k_h t}$.

Figure 5a reveals that this depolymerization model shows trends consistent with data but is not particularly accurate. However, a modified model was developed in which the right-hand side of Equation [5] was multiplied by a first-order reactivity term, $a = e^{-kt}$, with a being the reactivity, k_a a proportionality constant, and t the time, and application of this model did a much better job of describing the time course of xylan hydrolysis, as shown in Figure 5b. The fact that declining bond reactivity describes the data well suggests that the original assumption that all xylan bonds are broken equally at random is not valid. An obvious next step is the determination of individual bond energies within the hemicellulose molecule and incorporation of this information into a modified model.

(a)

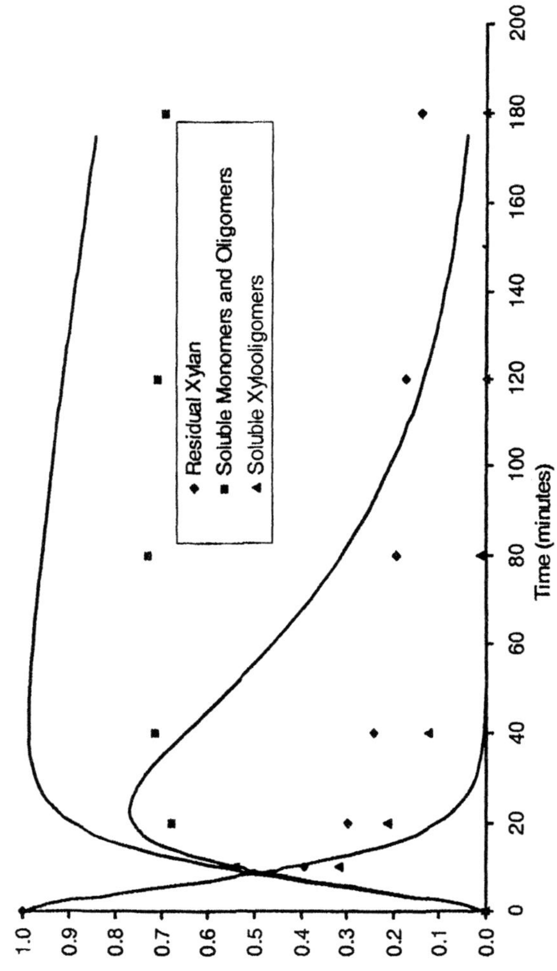

Residual Xylan
Soluble Monomers and Oligomers
Soluble Xylooligomers
Fraction of potential xylose (X/X0)
Time (minutes)
1.0
0.9
0.8
0.7
0.6
0.5
0.4
0.3
0.2
0.1
0.0
0
20
40
60
80
100
120
140
160
180
200

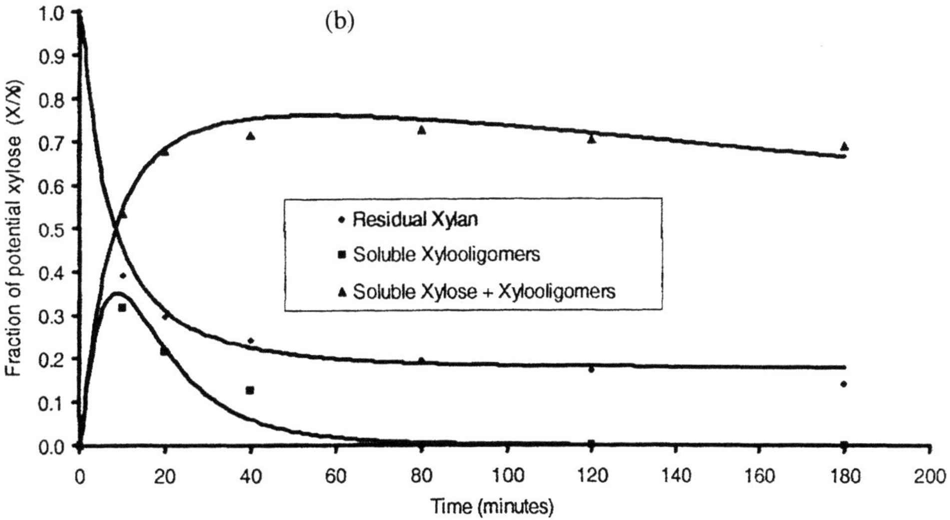

Figure 5. Comparison of an unmodified and modified depolymerization model applied to batch hydrolysis of corn stover at 140 °C with 0.5% sulfuric acid added.

The Relationship between Digestibility and Hemicellulose and Lignin Removal

Figure 6a shows the relationship of enzymatic digestibility of cellulose to xylan removal for both water-only and very dilute acid hydrolysis pretreatment in flowthrough and batch reactors. When no acid was used, less xylan removal was required to achieve the same enzymatic digestibility of cellulose in the solid residue from a flowthrough reactor as from a batch system. For example, removal of only about 20% of the xylan in a flowthrough reactor without addition of acid achieved a cellulose digestibility of approximately 55% while 78% of the xylan had to be removed in a batch system to achieve about the same yield. Alternatively, the digestibility of cellulose for corn stover pretreated in a flowthrough system was about 20% higher than that from the batch system at the same level of xylan removal. Because digestibilities peak at high xylan removal, this difference decreased when xylan removal was greater than about 85%.

Figure 6a also indicates that less xylan removal is needed for the flowthrough reactor to achieve the same level of enzymatic digestibility as a batch system when very dilute sulfuric acid is added. However, the relative advantage of flowthrough operation decreased with addition of sulfuric acid, and flowthrough reaction gave about a 10% greater enzymatic digestibility compared to the batch approach, with the differences again declining at high xylan removal levels. Nonetheless, acid addition improved the extent of enzymatic digestion of cellulose for both flowthrough and batch operations.

While the enzymatic digestibility of cellulose could be directly related to xylan removal for both batch and flowthrough operations, its relationship to lignin removal was not as consistent, as shown in Figure 6b. In particular, lignin removal tended to peak at about 20% in batch reactors with acid added and at about 30% when corn stover was treated without acid. Nonetheless, cellulose digestibility was as high as 90% for batch operations. On the other hand, cellulose digestibility was found to increase almost linearly with lignin removal for corn stover pretreated by the flowthrough reactor with or without added sulfuric acid, and the digestibility reached nearly 100% at the maximum of about 75% lignin removal observed for flowthrough pretreatment. In addition, while adding acid to the batch system decreased lignin removal but increased digestibility, the relationships between lignin removal and cellulose digestibility were quite similar for flowthrough operation whether acid was added or not.

Coupling the relationship between removal of xylan and lignin in Figure 2 with the digestibility results in Figure 6 leads us to conclude that adding dilute acid to biomass may enhance lignin solubilization, but for batch operations that lignin may precipitate as different species that interfere less with enzyme action than native lignin. The result would be that acid addition improves cellulose digestion relative to an uncatalyzed batch reactor while not being as effective as flowthrough systems that remove much more solubilized lignin before it can condense. Furthermore, it could be that lignin modification is more important than xylan dissolution and that the latter provides a convenient marker of lignin alterations that improve cellulose digestibility. This leads us to believe that lignin modification is important to enhance the digestibility of cellulose and that lignin removal provides even greater benefits. Of course, removing lignin also reduces non-productive binding of cellulase, increasing its efficiency during hydrolysis, and we believe it likely that enhancing accessibility of cellulose and reduction of nonproductive binding of enzyme both improve performance. Unfortunately, it is difficult to prove this mechanism, but further research is planned to clarify these possibilities *(64)*.

Conclusions

Our results show that increasing flow rate in a flowthrough reactor significantly enhanced xylan removal for pretreatment of corn stover with just hot water or when acid was added. In addition, yields of xylose monomer and oligomers increased with decreasing solids concentration in batch systems. This variation in xylose removal with flow rate and solids concentration is not consistent with predictions from first-order homogeneous kinetic models that incorporate only the effects of temperature, acid concentration, and time. Furthermore, these variations cannot be attributed to acetic acid released from biomass during hydrolysis as this effect should decrease, not increase, with flow rate. Our preliminary models show that coupling chemical reaction of xylan to soluble oligomers in the solids with mass transfer of these oligomers into solution followed by further reaction once in solution could more appropriately account for the change in reaction rate with flow rate and solids concentration. A novel apparatus is being applied to gather new data on oligomer solubility to better understand whether solubility limitations could also play an important role in this sequence. We have measured a range of oligomer chain lengths during xylan hydrolysis, and soluble oligomers were projected to directly degrade to nonsugars such as furfural as well as form lower DP species. Although existing models do not consider oligomer reactions in any detail, we have found that a depolymerization model can describe their basic features provided consideration is given to a change in reactivity as the reaction progresses. Further research targets enhancing our data and bringing these observations about oligomer behavior into a single analytical framework that can explain the effects of solids concentration and flow rate on hemicellulose hydrolysis on a consistent basis.

Lignin removal was limited to about 30% for a batch reactor without acid addition and dropped to about 20% when dilute sulfuric acid was used, and the enzymatic digestibility of cellulose in the residual solids increased with xylan removal for both, consistent with observations by others. On the other hand, introducing flow of liquid through the solids enhanced removal of lignin

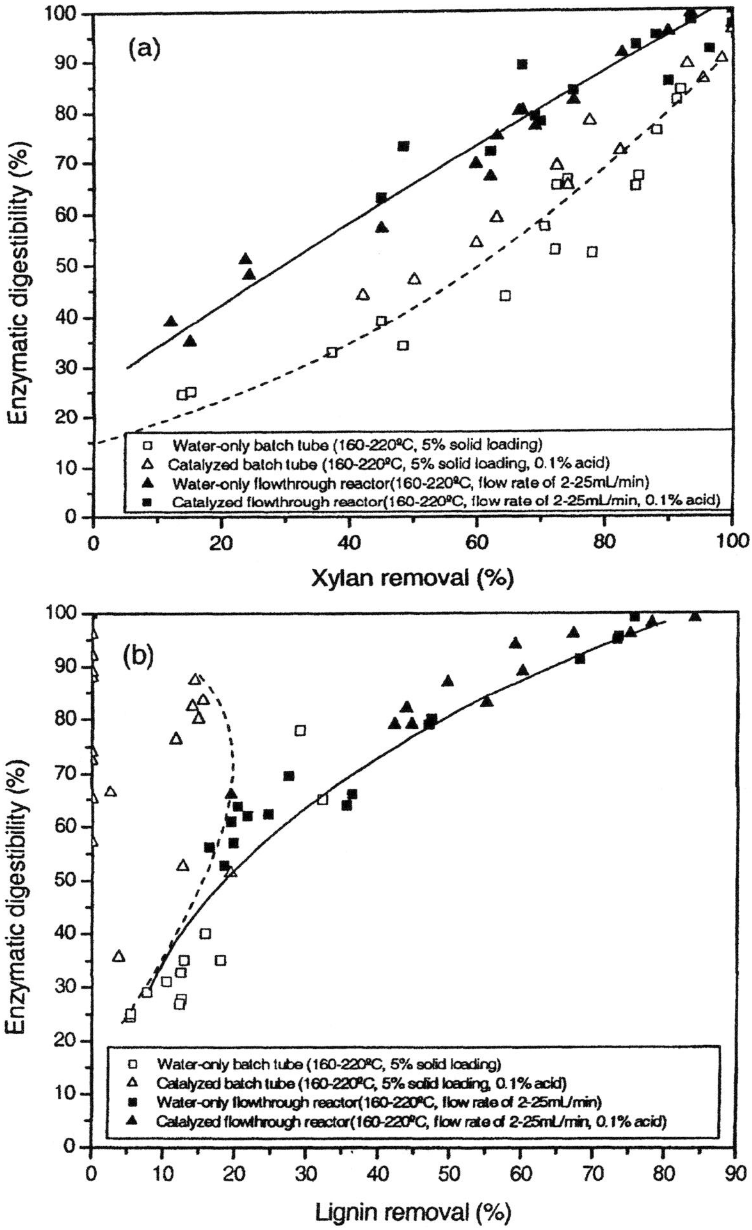

Figure 6. Effect of xylan removal (a) and lignin removal (b) on enzymatic digestibility of cellulose produced by hemicellulose hydrolysis of corn stover at the conditions noted.

considerably compared to batch operations, with as much as 70% of the total lignin being removed at high flow rates. In addition, the enzymatic digestibility of the cellulose in the solids from a flowthrough reactor was always greater than that from a batch system at the same degree of xylan removal. These flowthrough findings imply that lignin removal augments digestibility. Yet, adding acid to a batch system also increases cellulose digestibility even though lignin removal drops. The direct relationship observed between removal of lignin and xylan at high flow rates for flowthrough operations leads us to believe that lignin is solubilized during hemicellulose hydrolysis in both batch and flowthrough systems but that much of the lignin will react and precipitate back on the solid surface unless it is removed during pretreatment. Nonetheless, we believe that modification of lignin in this way improves the accessibility of enzymes to cellulose, increasing digestion yields. However, removal of lignin altogether before it can reform on the surface prevents it from adsorbing enzyme and from interfering with enzyme action, further enhancing performance. Although difficult to prove, lignin disruption or removal may actually be more important than xylan dissolution, and the latter may provide a simple marker of lignin alteration in batch systems. Our research continues to explore these relationships with the goal of improving the effectiveness of pretreatment in recovering hemicellulose sugars and enhancing cellulose digestibility.

Acknowledgments

We would like to recognize other members of our biomass hydrolysis research team for their contributions to these findings: Michael A. Brennan, Sigrid E. Jacobsen, and Xia Li. We are also grateful to each of the following sponsors for making it possible for us to develop these results: the Initiative for Future Agriculture and Food Systems Program of the US Department of Agriculture (award number 00-52104-9663), the National Institute of Standards and Technology (60NANB1D0064), the National Science Foundation Division of Bioengineering and Environmental Systems (BES-9985351), the National Research Initiative Competitive Grants Program of the US Department of Agriculture (2001-35504-10041), the Office of the Biomass Program of the US Department of Energy (DE-FC36-00GO10589 and DE-FC36-01GO11075), and the Thayer School of Engineering at Dartmouth College.

References

1. Lynd, L.R.; Cushman, J.H.; Nichols, R.J.; Wyman, C.E. *Science (Washington, DC)*. **1991**, 251(4999), 1318-23.

2. Wyman, C.E.; Goodman, B.J. *Appl. Biochem. Biotechnol.* **1993**, 39, 41-59.
3. Lynd, L.R. *Ann. Rev. Energy Environ.* **1996**, 21, 403-465.
4. Wright, J.D. *Chem. Eng. Prog.* **1988,** 84(8), 62-74.
5. Wyman, C.E. In *Handbook on Bioethanol: Production and Utilization;* Wyman, C.E. Ed.; Taylor & Francis, Washington, DC, 1996: pp 1-18.
6. Lynd, L.R.; Wyman, C.E.; Gerngross, T. U. *Biotechnol. Prog.* **1999,** 15, 777-793.
7. Hinman, N.D.; Schell, D.J.; Riley, C.J.; Bergeron, P.W.; Walter, P.J. *Appl. Biochem. Biotechnol.* **1992,** 34/35, 639-649.
8. Wright, J.D. In *Energy from Biomass and Wastes XII*; Klass, D.L.; Eds., : Institute of Gas Technology: Chicago, IL. 1989, pp1247-1296.
9. Wyman, C.E. *Ann. Rev. Energy Environ.* **1999,** 24, 189-226.
10. Wyman, C.E. *Biotechnol. Prog.* **2003,**19(2), 254-262.
11. Grethlein, H.E. US patent 4237226, 1980.
12. Knappert, D.; Grethlein, H.; Converse, A.O. *Biotechnol. Bioeng.* **1980,** 22(7), 1449-63.
13. Heitz, M.; Capekmenard, E.; Koeberle, P.G.; Gagne, J.; Chornet, E.; Overend, R.P.; Taylor, J.D.; Yu, E. *Bioresour. Technol.* **1991,** 35(1), 23-32.
14. Ramos, L.P.; Breuil, C.; Saddler, J.N. *Appl. Biochem. Biotechnol.* **1992,** 34-35, 37-48.
15. Wooley, R.; Ruth, M.; Glassner, D.; Sheehan, J. *Biotechnol. Prog.* **1999,** 15, 794-803.
16. Torget, R.; Werdene, P.; Himmel, M.; Grohmann, K. *Appl. Biochem. Biotechnol.* **1990,** 24-25, 115-26
17. Lee, Y.Y.; Wu, Z.; Torget; R.W. *Bioresour. Technol.* **1999**, 71(1), 29-39.
18. Torget, R.W.; Kim, J.S.; Lee, Y.Y. *Ind. Eng. Chem. Res.* **2000**, 39(8), 2817-2825.
19. Torget, R.; Werdene, P.; Himmel, M.; Grohmann, K. *Appl. Biochem. Biotechnol.* **1990,** 24-25, 115-26.
20. Grohmann. K.; Torget, R.; Himmel, M. *Biotechnol. Bioeng.* **1985,** 15, 59-80.
21. Hsu T-A. In *Handbook on Bioethanol: Production and Utilization;* Wyman, C.E. Ed.; Taylor & Francis, Washington, DC, 1996: pp 179-212.
22. McMillan, J.D. In *Enzymatic Conversion of Biomass for Fuels Production.* Himmel, M.E.; Baker, J.O.; Overend, R.P. Eds, ACS Symposium Series 566, American Chemical Society, Washington, DC, 1994, 566, 292-324.
23. von Sivers, M.; Zacchi, G. *Bioresour. Technol.***1995,** 51, 43-52.
24. Bobleter, O. *Prog. Polym. Sci.* **1994**, 19(5), 797-841.

25. Bonn, G.; Oefner, P.J.; Bobleter, O. *Fresenius' Z. Anal. Chem.* **1988**, 331(1), 46-50.
26. Torget, R.; Hatzis, C.; Hayward, T.K.; Hsu, T-A.; Philippidis, G.P. *Appl. Biochem. Biotechnol.* **1996**, 57/58, 85-101.
27. Hoermeyer, H.F.; Schwald, W.; Bonn, G.; Bobleter, O. *Holzforschung*. **1988**, 42(2), 95-98.
28. Laser, M.; Schulman, D.; Allen, S.G.; Lichwa, J.; Antal, M.J.; Lynd, L.R. *Bioresour. Technol.* **2001**, 81(1), 33-44.
29. van Walsum, G.P.; Allen; S.G.; Spencer, M.J.; Laser, M.S.; Antal, M.J. Jr.;, Lynd, L.R. *Appl. Biochem. Biotechnol.* **1996**, 57/58, 157-170.
30. Saeman, J.F.; Bubl, J.L.; Harris, E.E. *Ind. Eng. Chem. Res.* **1945**,17, 35-7.
31. Kobayashi, T.; Sakai, Y. *Mokuzai Toka Shingikai Hokoku.* **1956**, 5, 1-9.
32. Conner, A.H.; Lorenz, L.F. *Wood and Fiber Sci.* **1986**, 18(2), 248-263.
33. Garrote, G.; Dominguez, H.; Parajo, J.C. *Proc. Biochem.* **2001,** 36, 571-578.
34. Jacobsen, S.E.; Wyman, C.E. *Appl. Biochem. Biotechnol.* **2000**, 84-86, 81-96.
35. Lloyd, T.; Wyman, C.E. *Appl. Biochem. Biotechnol.* **2003**, 105-108, 53-67.
36. Stuhler, S.L.; Wyman, C. E. *Appl. Biochem. Biotechnol.* **2003**, 105-108, 101-114.
37. Jacobsen, S.E.; Wyman, C.E. *Ind. Eng. Chem. Res.* **2002**, 41(6), 1454-1461.
38. Allen, S.G.; Schulman, D.; Lichwa, J.; Antal, M.J. Jr.; Jennings, E.; Elander, R. *Ind. Eng. Chem. Res.* **2001**, 40(10), 2352-2361.
39. Allen, S.G.; Schulman, D.; Lichwa, J.; Antal, M.J. Jr.; Laser, M.; Lynd, L.R. *Ind. Eng. Chem. Res.* **2001**, 40(13), 2934-2941.
40. Chang, V.S.; Holtzapple, M. T. *Appl. Biochem. Biotechnol.* **2000**, 84-86, 5-37.
41. Converse, A.O. In *Bioconversion of Forest and Agricultural Plant Residues.* Saddler, J.N.; Eds., C.A.B. International: Wallingford, UK, **1993,** pp 93-106.
42. Gharpuray, M.M.; Lee,Y-H; Fan, L.T. *Biotechnol. Bioeng.* **1983**, 25, 157-172.
43. Gould, J.M. *Biotechnol. Bioeng.* **1984.** 26(1), 46-52.
44. Yang, B.; Boussaid, A.; Mansfield, S.D.; Gregg, D.J.; Saddler, J.N. *Biotechnol. Bioeng.* **2002,** 77(6), 678-684.
45. Yamaguchi, A.; Shimizu, K.; Fujii, T. *Mokuzai Gakkaishi.* **1994,** 40(11), 1208-13.
46. Nelson, R. *J. Polym. Sci.* **1971**, 36, 305-320.
47. Mooney, C.A.; Mansfield, S.D.; Touhy, M.G.; Saddler, J.N. *Biores. Technol.* **1998**. 64, 113-119.

48. Donaldson, L.A.; Wong, K.K.Y.; Mackie, K.L. *Wood Sci. Technol.* **1988**, 22, 103-114.
49. Tanahashi, M. *Wood Res.* **1990**, 77, 49-117.
50. Kawamoto, H.; Nakatsubo, F.; Murakami, K. *Mokuzai Gakkaishi.* **1992,** 38(1), 81-4.
51. Girard, D.J.; Converse, A.O. *Appl. Biochem. Biotechnol.* **1993**, 39/40, 521-533.
52. Grethlein, H.E.; Allen, D.C.; Converse, A.O. *Biotechnol. Bioeng.* **1984,** 26(12), 1498-505.
53. Lu, Y.P.; Yang, B.; Gregg, D.; Saddler, J.N.; Mansfield, S.D. *Appl. Biochem. Biotechnol.* **2002**, 98, 641-654.
54. Ooshima, H.; Burns, D.S.; Converse, A.O. *Biotechnol. Bioeng.* **1990**, 36, 446-452.
55. Grohmann, K.; Himmel, M.; Rivard, C.; Tucker, M.; Baker, J. Torget, R.; Graboski, M. *Biotechnology and Bioengineering Symposium 14(Symp. Biotechnol. Fuels Chem.)* **1984**, 6, 137-57.
56. Grohmann, K.; Torget, R.; Himmel, M. *Biotechnol. Bioeng.* **1985**,15, 59-80.
57. Fan, L.T.; Lee, Y.H.; Gharpuray, M.M. *Adv. Biochem. Eng.* **1982**, 23, 157-87.
58. Millett, M.A.; Baker, A.J.; Satter, L.D. *Biotechnology and Bioengineering Symposium 6(Enzym. Convers. Cellul. Mater.: Technol. Appl., Symp. Proc., 1975).* **1976**,125-53.
59. Ehrman, T. *Laboratory Analytical Procedure No.001*, National Renewable Energy Laboratory: Golgen, CO. 1994.
60. Ehrman, T. *Laboratory Analytical Procedure No.012*, National Renewable Energy Laboratory: Golden, CO. 1994.
61. Ruiz, R.; Ehrman, T. *Laboratory Analytical Procedure No.002*, National Renewable Energy Laboratory: Gloden, Co. 1996.
62. Brennan, M. A. M.S. thesis, Thayer School of Engineering of Dartmouth College, Hanover, NH, 2003.
63. Liu, C.G.; Wyman, C.E. *Ind. Eng. Chem. Res.* **2003**, 42(21), 5409-5416.
64. Yang, B.; Wyman, C. E. *Biotechnol. Bioeng*. (accepted Nov. 2003).
65. Gray, M.C.; Converse, A.O.; Wyman, C.E. *Appl. Biochem. Biotechnol.* **2003**, 105-108,179-193.
66. Adney, B.; Baker, J. *Laboratory Analytical Procedure No.006,* National Renewable Energy Laboratory: Gloden, CO.1996.
67. Ruiz, R.; Ehrman, T. *Laboratory Analytical Procedure No.014*, National Renewable Energy Energy Laboratory: Gloden, CO.1996.
68. Templeton, D.; Ehrman, T. *Laboratory Analytical Procedure No.003,* National Renewable Energy Laboratory: Gloden, CO. 1995.
69. Lora, J.H.; Wayman, M. *Tappi J.* **1978**. 61, 47-50.
70. Bobleter, O.; Grif, M.; Huber, C. De patent 4401992, 1994.

71. Stuhler, S.L. M.S. thesis, Thayer School of Engineering of Dartmouth College, Hanover, NH, 2002.
72. Jozwiakowski, M.J.; Connors, K. A. *Carb. Res.* **1985**, 143, 51-59.
73. Jennings, D.W.; Rousseau, R.W. *Carb. Res.* **1995**, 273(2), 243-248.
74. Li, X.; Converse, A.O.; Wyman, C.E. *Appl. Biochem. Biotechnol.* **2003,** 105-108, 515-522.
75. Simha, R. *J. Appl. Phys.* **1941,** 12, 569-78.

Chapter 7

Laccase Modification of the Physical Properties of Bark and Pulp of Loblolly Pine and Spruce Pulp

William Kenealy[1], John Klungness[1], Mandla Tshabalala[1], Eric Horn[2], Masood Akhtar[2], Roland Gleisner[1], and Gisela Buschle-Diller[3]

[1]Forest Service, Forest Products Laboratory, U.S. Department of Agriculture, One Gifford Pinchot Drive, Madison, WI 53726–2398
[2]Biopulping International, 2403 Industrial Drive, Madison, WI, 53713
[3]Textile Engineering Department, Auburn University, 115 Textile Building, Auburn, AL 36849

Pine bark, pine pulp, and spruce pulp were reacted with laccase in the presence of phenolic laccase substrates to modify the fiber surface properties. The acid–base and dispersive characteristics of these modified steam-treated thermomechanical loblolly pine pulps were determined by inverse gas chromatography. Different combinations of substrates with laccase modified the Lewis acid–base characteristics of the pulp. The binding of methylene blue was found unsuitable as an indicator of negative charges on pulps modified by laccase with phenolic substrates; however, its binding did indicate undefined changes in the pulp. Thermomechanical spruce pulp treated with manganese peroxidase or laccase with 4-hydroxyphenylacetic acid increased handsheet strength. Manganese peroxidase treatment also decreased the amount of refining needed to obtain the same freeness, which would indicate potential energy savings. Incremental addition of resorcinol in the presence of laccase altered spruce pulp to a greater extent than did bulk addition. Laccase with substrates also altered the binding of phosphate by modified pine bark.

Introduction

Many species of woody plants are used for particular purposes, with their end use being directed by the properties of the wood and economic concerns. Generally, the functional use of wood products is determined by the hydrophilic, hydrophobic, dimensional, and ionic properties of the cellulosic fiber. Small diameter and mixed species of wood have limited uses and are thus low in value. If we could understand how to change the cellulose fiber properties of such materials, alternative sources of fiber could be better exploited and more valuable products made from mixed woody species or recycled fiber.

The structure of wood and other lignocellulosic materials is on first inspection quite simple. It consists of very few polymeric materials: cellulose, hemicellulose, and lignin. Upon further consideration, three-dimensional aspects and species-specific variability make the structure of lignocellulose very complicated. Understanding and optimizing the surface properties of lignocellulosic materials for structural or paper applications will result in better use of these resources.

Our knowledge of lignocellulose structure and function is increasing, but we cannot as yet convert a source of lignocellulose to a desired product at will. Much of our knowledge is based on empirically separating the components of the material to create a product with the desired strength and flexibility. Processes following this approach, which are still being developed, will reduce costs and provide enhanced products. Thus to use lower value resources, the best approach might be to regard them as a mixture of polymeric materials and empirically derive ways of altering their properties with a target product in mind. The next stage in our understanding will come from developing ways to actively build strength and other properties into the materials.

The goal of this research is to investigate the feasibility of grafting compounds or polymers onto softwood materials to alter the surface characteristics. Laccase and peroxidases have been used to modify the properties of lignocellulosic materials (*1-7*). We have used laccase with substrates and manganese peroxidase to alter lignocellulose materials. Methylene blue binding (to determine carboxylic acid content), inverse gas chromatography (to determine Lewis acid–base characteristics), and pulp handsheet properties were used to explore the effectiveness of the treatments in altering surface properties.

Materials and Methods

Pulp Sources

Dried thermomechanical pulp (TMP) (mature loblolly pine, *Pinus taeda*) was obtained from the U.S. Department of Agriculture at Pineville, Mississippi. Steam pressures used during the TMP process were 4, 8, and 12 bar. Spruce (*Picea glauca*) chips were refined at 103 kPa, followed by atmospheric refining to ~600 mL Canadian standard freeness (CSF). Spruce pulps were further refined to 70 mL CSF after enzyme treatments using a PFI mill to reduce the freeness.

Laccase Treatment

Dry loblolly pine pulp samples (2 g) were weighed into Erlenmeyer flasks, 75 mL deionized water was added, and the mixture was stirred for at least 3 h at room temperature to ensure complete wetting. Sodium tartarate buffer (5 mL, 200 mM, pH 5) and one unit laccase (NovoSample 51002, 1 unit/5µL, unit = 1 µMole syringaldazine oxidized/min at 30°C, pH 5.5) were added and stirred for 20–25 min before 1 mL substrate (0.3 M) was added. After approximately 25 min of stirring, another 20 µL laccase was added. The samples were stirred for 1 to 2 h at room temperature, then loosely covered with foil and refrigerated for 15 to 17 h. After warming to room temperature, the supernatant solution was filtered off and the pulps were washed three times with 1.5 L distilled water. Finally the pulps were strained through a sieve, freeze-dried for at least 48 h, and stored in a desiccator until analyzed.

Spruce pulps were treated at 37°C for 24 h with manganese peroxidase at 12% consistency with 50 mM malonate buffer (pH 5), 0.1 mM $MnSO_4$, and 0.8 mM H_2O_2 (added slowly with time). Treatment of spruce pulp with 4-hydroxyphenylacetic acid was performed under the conditions described by Chandra and Ragauskas (*8*). For treatments with resorcinol, samples of 243 g spruce pulp (37% solids) were weighed into plastic bags and mixed with 64 mL 200-mM tartarate buffer pH 4.5 and 250 mL distilled water. Laccase (0.6 mL) was added in 43 mL distilled water. Resorcinol (2 g/50 mL water) was either added with the 250 mL distilled water or added in increments (1/3 added at 0, 60, and 105 min of incubation). The bags were sealed and incubated at 37°C for 2 h with the contents mixed by hand every 30 min, followed by storage overnight at 5°C. Pulps were rinsed with distilled water until the filtrate was colorless; the filter cake was then dewatered by pressing on adsorbent sheets and frozen until used. Where oxygen was used, the plastic bags were flushed with a stream of oxygen for 1 min, then sealed.

Methylene Blue Binding

Known weights of dry pulp (up to 150 mg) were placed in 50-mL disposable centrifuge tubes with 5 mL 1M borate buffer (pH 8.5) and five clean steel balls and mixed to break clumps. Aqueous methylene blue (35 mL of 1 mM) was added, and the tubes were tumbled for 1 h. Samples (1.5 mL) were removed and centrifuged at 20,000× g for 5 min. Duplicate dilutions were made of 0.5 ml supernate into 5 mL H_2O + 0.125 mL 1 N HCl. Two samples (0.2-mL) from each dilution were analyzed at 610 nm on a Molecular Devices Spectra Max Plus microtiter plate reader. Micromoles of dye adsorbed by pulp equals $35(1-A_{sample}/A_{blank})$. Methylene blue binding was determined by plotting amount of dye bound as a function of weight of pulp and determining the slope of the line (with at least three points). All methylene blue results in which less than half the methylene blue was adsorbed from the assay were combined for one calculation for each sample.

Inverse Gas Chromatography

Approximately 1 g pulp was packed into stainless steel chromatographic columns of 0.5 m length and 0.63 mm diameter. A Hewlett Packard 5790A gas chromatograph was used with helium as carrier gas at a flow rate of 15 mL/min. The flame ionization detector was set to 200°C and the injector at 150°C. Measurements were performed at five different oven temperatures, from 40°C to 80°C in 10°C increments. The reference line used five n-alkanes (n-pentane to n-nonane), and tetrahydrofuran, acetone, and chloroform were selected as the basic, amphoteric, and acidic probes, respectively. Results for the dispersive energy component were obtained according to Felix and Gatenholm (*9*) by plotting $RT \ln V_n$ versus $2Na(\gamma_l^d)^{1/2}$, with V_n the net specific retention volume, a the surface area occupied by the probe, γ_l^d the dispersive energy of the probe, R the gas constant, and T the temperature. The plot of $RT \ln V_n$ versus $2Na(\gamma_l^d)^{1/2}$ gives a straight line. From the slope of this line, the surface energy of the substrate can be determined at each temperature. Polar probes deviate from the straight relationship due to specific interactions, including Lewis acid–base interactions. If the vertical distance of each polar probe is measured to the straight line, the free energy of sorption can be determined and plotted as a function of temperature (*10*). The slope of that line is related to the free enthalpy of adsorption/desorption. Donor (DN) and acceptor (AN) numbers of the polar probes are available from the literature. A plot of the free enthalpy of sorption versus DN/AN allows the determination of acidity (K_A, slope) and basicity (K_D, intercept) of the substrate (*10*). Determinations were made for two independent samples and the results averaged.

Paper Testing

Handsheets (60-g/m^2) were made according to TAPPI standard T 205 sp-95 (*11*). Burst, tear, and tensile strengths were determined according to TAPPI standards T 403 om-97, T 414 om-98, and T 494 om-96, respectively (*11*). Results represent the average of 10 handsheets.

Bark Reactions

Bark of *P. taeda* obtained from a pulp mill in Mississippi was milled, extracted, activated, and reacted with bisaminopropylethylenediamine (BAPED). The modification was made to increase the capacity of the bark to adsorb anions. Modified bark (0.5 to 0.7 g) in 5 mL of 40 mM tartarate buffer pH 5.0 and 10 μL laccase was mixed for 30 min at room temperature. Phloroglucinol (1 mL of 0.3 M in ethanol) or catechol (1 mL of 0.3 M) were added and incubated for 3 h at room temperature, followed by 18 h at 5°C. The particles were rinsed and recovered by filtration and freeze dried. Controls included laccase addition without substrate and untreated modified bark. The bark was tested for the ability to remove phosphate from solution.

Results and Discussion

In initial testing, loblolly pine fiberboard furnishes produced at 4, 8, and 12 bar steam pressure were evaluated for laccase reactivity. Laccase was incubated with a sample of each furnish at 37°C, and the rate and extent of oxygen consumption were measured. Oxygen consumption by laccase was highest and fastest for the 12 bar furnish, with 12.0 nmoles oxygen consumed per milligram furnish. This would correspond to 48 nmoles of one-electron oxidative events per milligram of furnish. Assuming the reaction was with lignin and 28% lignin in loblolly pine, this corresponds to the reaction of approximately 1 syringaldehyde-like unit in 30. The 4 bar and 8 bar furnishes had 4.0 and 8.4

nmoles oxygen consumed per milligram furnish. The majority of subsequent screening work was performed with the 12 bar pulp.

Oxygen was consumed during laccase treatment, but the actual structures in lignocellulose that are substrates could not be determined. The affinity of laccase for its substrates depends upon the structure and its accessibility. The affinity of enzymes for substrates and concentrations of substrates will alter the rate of reaction. Although it is difficult to assess the enzyme affinity for substrates already present in the lignocellulose, the affinity of the enzyme for phenolic substrates was measured. The approximate K_m values for several substrates were determined by measuring the oxygen consumption rate at different substrate concentrations (data shown in Table I). The laccase substrates tested ranged from quite insensitive with phenol (18 mM) to sensitive with 2,5-dihydroxy benzoic acid (0.04 mM). Though there are some differences with various laccases and their affinity for the substrates, the values presented here are similar to those reported for *Polysporous pinsitus* laccase (*12*).

Table I. Approximate K_m for Laccase Substrates Tested

Substrate[a]	K_m (mM)
Phenol	18
4-hydroxyphenylacetic acid	7
Guaiacol	3
4-methylcatechol	2
Resorcinol	2
o-coumaric acid	2
Isovanillic acid	2
Vanillic acid	1
p-coumaric acid	1
Hydroquinone	<1
3,4-dihydroxybenzoic acid	<1
Catechol	<1
3,4-dihydroxyphenylacetic acid	<1
2,5-dihydroxybenzoic acid	<1

[a]Substrates were tested in 20 mM pH 5.0 tartarate buffer and 0.2 units laccase using a YSI oxygen meter with a temperature-controlled stir cell (2 mL). Substrates are listed in order of increasing reactivity with laccase.

Laccase creates free radicals that can further react with other compounds. Ideally, conditions for each substrate would be optimized for the grafting reaction, which would require investigation of concentration, pH, affinity of the enzyme, and other reactants in the pulp. Because optimizing the conditions for each substrate is beyond the scope of a screening test, the concentration of substrate (3.75 mM), enzyme (5 units), and fiber were held constant. We incubated 12 bar loblolly pine pulp in the presence of laccase with and without various laccase substrates, as described in Methods. These pulps were treated with laccase, washed, and freeze dried. High affinity laccase substrates were oxidized once before 1 h had elapsed. The enzyme reaction was continued for several hours to allow for lower affinity substrates to react.

Carboxylic acid groups are beneficial in the bonding of pulp fibers in paper and can increase the strength of the paper (*13*). Phenolic substrates containing carboxylic acid groups could be attached by the grafting of the substrate. Carboxylic acids could also be created by extensive oxidation of substrates or the fiber. Anionic groups from dissociated carboxylic acids can be estimated by conductometric titrations or by adsorption of cations such as metal atoms or dyes (e.g., methylene blue) (*14, 15*). Conductometric titrations are very time consuming, and methylene blue adsorption correlated well with the conductometric results on pulp (*14*). Thus, we modified a method of methylene blue adsorption (*15*) to the fibers to determine the amount of anionic groups bound to the fiber.

Single measurements of methylene blue adsorption on pulp fibers gave different results, probably as a result of non-specific binding. The conditions of the assay (volume, sample amount, and use of new glassware and plastic ware) were adjusted so that the adsorption of methylene blue to a pulp sample could be determined reproducibly. The final form of the assay is described in Methds and uses at least three different samples of pulp, all of which adsorb less than half the methylene blue in the assay. The adsorption of methylene blue was determined from the slope of methylene blue adsorbed as a function of weight of the pulp (Figure 1).

Methylene blue is positively charged at pH 8.5, and adsorption should be correlated to negative charges on the pulp. We expected that the laccase substrates containing carboxylic acid groups would interact more than those without acid groups. However, methylene blue adsorption did not directly correlate with acid content of the substrate. The highest values of methylene blue adsorption were observed for pulp treated with resorcinol, which by itself contains no carboxylic acid groups. The nature of the attachment of resorcinol

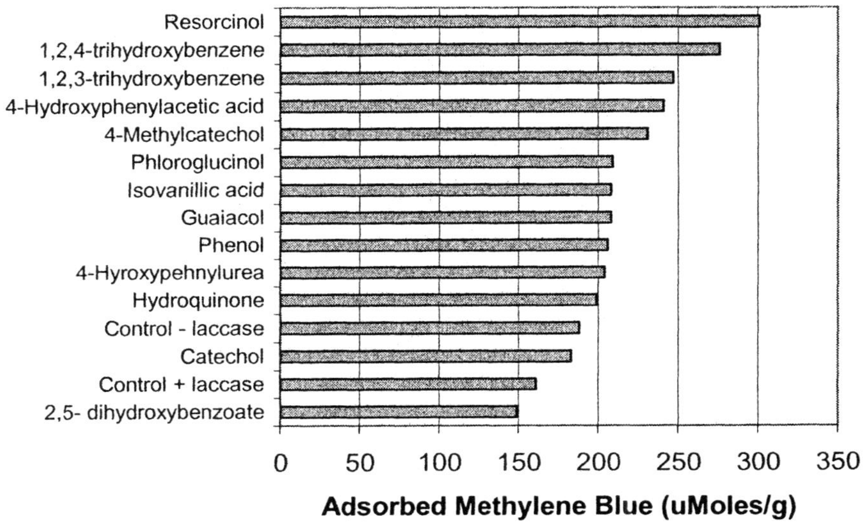

Figure 1. Methylene blue adsorption to 12 bar loblolly pine pulp after treatment with laccase and indicated substrates.

(or other substrates) to the pulp is unknown. A resorcinol free radical could also have reacted with oxygen, possibly increasing the acidity of the fiber.

We analyzed many of these pulps by IGC to determine if there was an increase in donor or acceptor character for the pulp or if the dispersive energy characteristics of the pulp were changed in a manner that correlates to the methylene blue binding results. Table II shows the surface acid–base characteristics and dispersive component for these pulps. The ability to act as a Lewis acid was the only statistically significant correlation when compared with methylene blue binding, with $r_s = -0.56$ (Spearman rank-order correlation coefficient) and $P = 0.03$. The ability to act as a Lewis base, the dispersive component, and the ratio of acid to base had no relationship with methylene blue binding.

Because methylene blue is a cationic dye, it would be expected to have activity as a Lewis acid. The correlation with methylene blue was weak but showed the expected trend, with increased methylene blue binding being negatively correlated to increased K_A. Phloroglucinol and isovanillic acid were the only laccase substrates that increased the K_D of the fiber. All substrates other than resorcinol, 4-hydroxyphenylacetic acid, 1,2,3-trihydroxybenzene, and 1,2,4-

trihydroxybenzene increased the K_A of the fiber. The ratio of K_A/K_D was increased for many samples after these treatments. We expected that the methylene blue adsorption would decrease for those treated fibers with K_A/K_D ratios greater than the control, but for many substrates the opposite effect was observed. This indicates that methylene blue does not interact with the fibers solely by Lewis acid–base interactions.

Table II. Analysis of Laccase Treated Pulps by IGC

Substrate for 12 bar pulp or pulp controls[a]	K_A	K_D	K_A/K_D	*Dispersive component*
2,5- Dihydroxybenzoate	0.261	0.019	13	29
12 bar control + laccase	0.223	0.037	6	28
Catechol	0.276	0.010	28	31
12 bar control – laccase	0.199	0.052	4	30
Hydroquinone	0.245	0.049	5	27
4-Hyroxyphenylurea	0.224	0.011	20	30
Phenol	0.268	0.017	16	27
Guaiacol	0.271	0.010	26	27
Isovanillic acid	0.277	0.074	4	28
Phloroglucinol	0.244	0.119	2	27
4-Methylcatechol	0.207	0.023	9	28
4-Hydroxyphenylacetic acid	0.180	0.035	5	29
1,2,3-Trihydroxybenzene	0.182	0.043	4	29
1,2,4-Trihydroxybenzene	0.186	0.011	16	26
Resorcinol	0.172	0.040	4	30
4 bar control – laccase	0.208	0.065	3	36
8 bar control – laccase	0.214	0.041	5	32

[a] The substrates are listed for the 12 bar pulp reacted with the substrate and laccase. The 4 bar, 8 bar, and 12 bar controls were also included.

Methylene blue binding does not predict the Lewis acid–base characteristics of the pulp but might be useful as an indicator of other properties in a product derived from the pulp. It does indicate that some change was made in the binding properties of the pulp.

Several conditions were selected to test the effects of oxidative enzyme treatment on spruce pulps and determine if handsheet properties were altered by the treatment. Spruce pulp of high freeness (CSF 600 mL) was treated with manganese peroxidase or laccase in the presence of 4-hydroxyphenylacetic acid, followed by refining in a PFI mill. Handsheets were made from the treated pulps. The results of methylene blue binding by the pulps and strength testing on handsheets made from these pulps are shown in Table III.

Table III. Properties of Treated Spruce TMP After Refining by PFI Mill and Handsheet Preparation

Pulp treatment	*Methylene blue adsorp.* (μMol/g)	*Tear index* (mN m^2/g)	*Burst index* (kN/g)	*Tensile index* (Nm/g)	*CSF* (mL)
Control	ND	3.52	1.54	33.5	65
MnP complete	54	3.80	1.64	34.1	69
H_2O_2 + Mn	50	3.51	1.52	33.5	70
Buffer	42	3.56	1.50	33.4	71
Laccase + 4-HPA	101	3.61	1.59	32.3	69

ND= not determined; 4-HPA is 4-hydroxyphenylacetic acid; and the complete reaction mix for MnP contained enzyme, buffer, H_2O_2, and Mn.

Handsheets from manganese peroxidase (MnP) treated pulp showed increases in burst, tear, and tensile indexes. The manganese peroxidase treated pulp also required less refining in the PFI mill to obtain the same degree of freeness (requiring only 8,500 revolutions compared with 10,000 for the controls and laccase treated material). This may indicate a savings in refining energy (*16*). The buffer, Mn, and H_2O_2 had no effect on the strength of the handsheets. The laccase treatment with 4-hydroxyphenylacetic acid brought about small increases in the burst and tear indexes and a decrease in the tensile index. Chandra and Ragauskas (*8*) reported an 18% increase in the burst index, an 11% increase in the tear index, and about an 8% increase in the tensile index for softwood linerboard pulp treated with laccase and 4-hydroxyphenylacetic acid.

The MnP treatment did not increase the binding of methylene blue above that of the controls. The laccase together with 4-hydroxyphenylacetic acid did increase the adsorption of methylene blue. Both the MnP treatment and laccase treatment together with 4-hydroxyphenylacetic acid increased handsheet strength. Methylene blue adsorption does not appear to correlate with handsheet strength.

The conditions under which the pulp is treated can influence the extent of the modification. Using resorcinol as the modifying substrate on spruce TMP, we investigated the method of substrate addition using the adsorption of methylene blue to assay the treated pulp. Table IV shows the results of the methylene blue adsorption on pulps treated with the same amount of resorcinol but under different conditions. When the entire amount of resorcinol was added at the start of the experiment, there was an increase in methylene blue adsorption above that of the laccase treated control. If the atmosphere of the incubation bag was replaced with oxygen, then no discernable changes in methylene blue adsorption were observed. Methylene blue adsorption was highest when resorcinol was added in three aliquots during incubation. The color of all the resorcinol treated pulps changed from brown to deep red and could not be removed by washing with distilled water. Samples reacted in an oxygen atmosphere turned red much more rapidly those reacted in air.

The conditions used for these experiments indicate that a greater effect can be created by metering in the substrate. Oxygen is required in these incubations to reoxidize the laccase. It might also react with laccase generated radicals and propagate oxygen free radicals that could be additional reactants.

Table IV. Effect of substrate addition on methylene blue binding to treated spruce TMP

Pulp treatment	Methylene blue adsorption (μMol/g)
No substrate + Laccase	66
Resorcinol–all at start	99
Resorcinol–all at start–Oxygen	60
Resorcinol–3 aliquots	126

Southern pine bark treated with BAPED is effective at removing phosphorous from aqueous solutions. The bark was treated with laccase and also with phloroglucinol and catechol to generate additional sites where more BAPED could be bound. The laccase treated material decreased the capacity of the bark to remove phosphorous by 31%. The catechol and phloroglucinol treatments decreased this capacity by 38% and 45%, respectively. It appears the reaction with laccase and substrate competes for sites already occupied by BAPED, diminishing the capacity of the bark. It is also possible the phenolic substrates attached to the nitrogen of BAPED and reduced the phosphorous binding capacity. These reactions will be explored further.

Conclusions

Laccase and manganese peroxidase are able to generate free radicals in pulps and substrates. In this paper, we showed that oxygen was consumed but we did not directly demonstrate free radical formation. The results of fiber treatment are influenced by treatment of the pulp prior to enzyme treatment, the conditions under which the pulp is treated with enzymes, and the types of substrates used for the treatment. Laccase reactions changed the surface characteristics of the fiber, and the changes were different with different substrates. Results obtained with IGC indicated that both acidic and basic modifications could be made. If these changes can be made economically, there may be some use in increasing paper strength. Methylene blue adsorption did correlate with the K_A results from IGC but is not suitable as a primary indicator of carboxyl group attachment on these treated pulps. Incremental additions of laccase substrate appeared to alter the fiber to a greater extent than did bulk addition. Some oxidative modifications by enzymes can result in increased strength properties or lower energy needs during refining. These enzyme studies need to be explored further to determine if there is a possibility of applications in industry.

Acknowledgments

We thank Carl Houtman for valuable suggestions in the presentation of this material.

References

1. Lund, M.; Ragauskas, A. J. *Applied Microbiology and Biotechnology* **2001,** 55, 699-703.
2. Huttermann, A.; Mai, C.; Kharazipour, A. *Applied Microbiology and Biotechnology* **2001,** 55, 387-394.
3. Huttermann, A.; Majcherczyk, A.; Braun, L. A.; Mai, C.; Fastenrath, M.; Kharazipour, A.; Huttermann, J.; Huttermann, A. H. *Naturwissenschaften* **2000,** 87, 539-541.
4. Mai, C.; Milstein, O.; Huttermann, A. *Applied Microbiology and Biotechnology* **1999,** 51, 527-531.
5. Kharazipour, A.; Schindel, K.; Huttermann, A. In *Enzyme applications in fiber processing*, ed.; Paulo, A. C., American Chemical Society: Washington, 1998; pp 99-115.

6. Felby, C.; Olesen, P. O.; Hansen, T. T., In *Enzyme applications in fiber processing*, ed.; Paulo, A. C., American Chemical Society: Washington, 1998; pp 88-98.
7. Felby, C.; Hassingboe, J.; Lund, M. *Enzyme and Microbial Technology* **2002,** 31, 736-741.
8. Chandra, R. P.; Ragauskas, A. J. In *Laccase: The renegade of fiber modification*, Tappi Pulping Conference, 2001. Tappi Press, Atlanta: 2001.
9. Felix, J. M.; Gatenholm, P. *Nordic Pulp and Paper Research Journal* **1993,** 1, 200.
10. Tshabalala, M. A. *Journal of Applied Polymer Science* **1997,** 65, 1013-1020.
11. *Tappi test methods.* ed.; TAPPI Press: Atlanta, GA, 2000..
12. Xu, F., *Biochemistry* **1996,** 35, 7608-7614.
13. Waleka, J. A., *TAPPI* **1956,** 39, (7), 458-463.
14. Fardim, P.; Holmbom, B.; Ivaska, A.; Karhu, J. *Nordic Pulp and Paper Research Journal* **2002,** 17, 346-350.
15. Philipp, B.; Rehder, W.; Lang, H., *Papier* **1965,** 19, 1-10.
16. Leatham, G.; Myers, G. C. *Tappi Journal* **1990**, (April), 192-197.

Chapter 8

Degradation of Wood and Pulp by Three Fungi, *Pycnoporus cinnabarinus, Trichophyton rubrum* LKY-7 and *Trichophyton rubrum* LSK-27

Xinglian Geng[1], Hyunchae Jung[2], and Kaichang Li[1*]

[1]**Department of Wood Science and Engineering. Oregon State University, Corvallis, OR 97331**
[2]**Department of Forest Resources, Sunchon National University, Sunchon 540-742, Korea**

Degradation of pine, yellow poplar and sweet gum by three fungi *Pycnoporus cinnabarinus, Trichophyton rubrum* LKY-7 and *Trichophyton rubrum* LSK-27 was investigated. The two *Trichophyton* species have not been identified, i.e., they are unknown. However, they have been putatively characterized with a chemical signature method. Their chemical signatures were compared to known fungi in a database of Microbe Inotech Labs, Inc. (St. Louis, MO). *P. cinnabarinus* degraded pine block samples much faster than *T. rubrum* LKY-7 and *T. rubrum* LSK-27, whereas *P. cinnabarinus* and *T. rubrum* LSK-27 degraded yellow poplar and sweet gum at almost the same rate. Among three fungi, LKY-7 had the lowest ability to degrade all three types of wood. In an effort to get a better understanding of how fungi degrade lignin in wood, contents of various functional groups were analyzed. After three-months of degradation of pine flour by the three fungi, the following changes were observed: an increase in condensed phenolic OH group content and carboxylic acid group content, a decrease in the guaiacyl phenolic OH content, and little change of aliphatic OH group content. Further studies in the degradation of pine flour by *P. cinnabarinus* revealed that the increase in condensed phenolic OH group content and the decrease in guaiacyl phenolic OH group content occurred in the first month of the degradation. The changes of functional group contents in the degradation of unbleached softwood kraft pulp by *P. cinnabarinus* had the same trends as those in the degradation of pine flour. In other words, structural alterations of lignin due to the kraft pulping process had little effect on how *P. cinnabarinus* degraded lignin. Action modes in the fungal degradation of lignin are discussed.

Lignin is the second most abundant renewable biopolymer on earth and its biodegradation is of great importance for the global carbon cycle. Selective biodegradation of lignin has drawn particular attention because lignin is one of the major obstacles for the efficient utilization of lignocellulosic materials in many industries such as the pulp and paper industry and the animal feed industry. White-rot fungi are the only microorganisms able to selectively and efficiently degrade lignin in natural environments. However, the mechanisms by which white-rot fungi degrade lignin are still poorly understood.

It is believed that three types of potentially ligninolytic enzymes, lignin peroxidase (LiP), manganese peroxidase (MnP) and laccase are involved in the fungal degradation of lignin. Extensive studies reveal that these enzymes degrade lignin in different ways. For instance, LiP is able to degrade non-phenolic lignin substructures that account for 90% of all lignin substructures in wood, whereas MnP and laccase are only able to degrade phenolic lignin substructures that accounts for only about 10% of all lignin substructures in wood. White-rot fungi secrete one or more of these three enzymes for lignin degradation, which suggests that there is likely more than one mechanism for fungal degradation of lignin.

We have been using the white rot fungus *Pycnoporus cinnabarinus* as a model white-rot fungus for studying mechanisms of fungal degradation of lignin because of its simple ligninolytic enzyme system. Only one of the three ligninolytic enzymes, laccase, is produced in the degradation of lignocellulosic materials by *P. cinnabarinus* although the fungus can degrade lignin efficiently (*1-4*). It is well established that laccase can only oxidize non-phenolic lignin substructures in the presence of a redox mediator (*5,6*). Our recent studies revealed that *P. cinnabarinus* was able to degrade non-phenolic lignin substructures (*7*). However, identification of a natural mediator compound that enables laccase to degrade non-phenolic lignin substructures remains elusive despite extensive efforts (*8*).

We are also investigating two newly isolated wood-degrading fungi, *Trichophyton rubrum* LKY-7 and *Trichophyton rubrum* LSK-27, for their ability to degrade lignin. In this study, we investigated the abilities of *P. cinnabarinus, T. rubrum* LKY-7 and *T. rubrum* LSK-27 to degrade wood and unbleached kraft pulp. We analyzed the changes of various functional group contents in pine lignin and kraft lignin during the degradation of wood and unbleached kraft pulp by these fungi in an effort to gain a better understanding of how white-rot fungi degrade lignin.

Materials and Methods

Microorganisms

Two fungal strains, *Trichophyton rubrum* LKY-7 and *Trichophyton rubrum* LSK-27 were maintained on yellow poplar blocks at –80 °C before use. *P. cinnabarinus,* strain PB (ATCC 204166) was maintained on 2% (w/v) potato dextrose agar (PDA) plates at 4 °C. *P. cinnabarinus, T. rubrum* LKY-7 and *T. rubrum* LSK-27 were separately inoculated on PDA plates and incubated at 30 °C for 5 days before inoculated on wood blocks. *P. cinnabarinus* on a PDA plate was inoculated on autoclaved long-grain polished rice (60 g/l-L flask) for production of fungal spores. After cultivation at 30 °C for 8 days, the fungal spores were extracted with sterile 0.9% (w/v) NaCl solution, filtered with cheesecloth to remove the rice, and extensively washed with sterile 0.9% (w/v) NaCl solution (*4*). The spore suspension contained about 3 x 10^7 spores/mL and was used for the degradation of pine wood flour.

Degradation of Wood Blocks by Three Fungi

A scaled-down version of the soil block test developed by Scheffer et al was used to investigate the fungal degradation of wood blocks (*9,10*). More specifically, glass chambers (300 mL) were half-filled with moist garden loam followed by addition of 50 mL tap water. An Alder feeder block (3 cm x 3 cm x 2 cm) was put on the surface of the soil and autoclaved at 121 °C for 45 min. After it was cooled down to room temperature, the feeder block in the soil-containing chamber was inoculated with a PDA disc (Φ 6 mm) taken from the hyphal tip in a fungus-growing PDA plate, and incubated at 30 °C until the fungus had thoroughly colonized the feeder. A wood block specimen (1 cm x 1 cm x 1cm) cut from pine sapwood, sweet gum or yellow poplar was dried at 104 °C overnight, cooled in a desiccator, and then weighed. All wood block specimens were autoclaved at 121 °C for 45 min, cooled in a desiccator and placed on the corner of the feeder block in the chamber. The yellow poplar and sweet gum block specimens were incubated for 2 months at 30 °C, while the pine block specimens were incubated for 3 months at 30 °C. After the mycelia were carefully removed, the decayed wood block specimens were rinsed with water, dried at 105 °C, and then weighed to determine the weight loss. Nine replicates were investigated for each fungus on each wood species.

Degradation of Pine Flour by Three Fungi

Experiments for fungal degradation of pine flour were carried out by following an established procedure for fungal degradation of pulp (*3*). *Pondarosa* pine sapwood flour (30 mesh) was extracted with acetone/water (v/v, 9/1) and dried under vacuum. The moisture content was determined as 9.8 %. A mixture of wood flour (9.07 g oven-dried) and de-ionized water (10 mL) was autoclaved in 250 mL Erlenmeyer flask at 121 °C for 1 h. The autoclaved wood flour was inoculated with the *P. cinnabarinus* spore suspension (5 mL), and the moisture content of the wood flour adjusted to 80% with sterilized water. The culture was then incubated at 30 °C for pre-determined times. Twelve replicates were carried out for each cultivation time.

Twelve PDA discs (Φ 6 mm) of the pre-incubated *T. rubrum* LKY-7 and *T. rubrum* LSK-27 were inoculated with 250 mLof a cultivation medium in a 1-L flask and incubated at 30 °C in a shaker (120-150 rpm, Φ 18 mm). The culture medium (per liter) for *T. rubrum* LSK-27 contains 30 g glucose, 10 g Bacto-peptone, 1.5 g KH_2PO_4, 0.5 g $MgSO_4$, 20 mg $CuSO_4$, 50 mg $MnSO_4$ and 10 mg thiamine-HCl. The culture medium (per liter) for *T. rubrum* LKY-7 contains 30 g glucose, 10 g Bacto-peptone, 1.5 g KH_2PO_4, 0.5 g $MgSO_4$, 20 mg $CuSO_4$, 50 mg $MnSO_4$ and 2 mg thiamine-HCl. The fungal mycelia were filtered with filter paper, washed with sterilized water and then homogenized in 40 mL water. Mycelial suspension (5 mL) was inoculated on pine flour (3.61 g, oven-dried) in a 125 mL flask and the moisture content adjusted to 83%. Six replicates were inoculated for each fungus.

After a predetermined incubation time at 30 °C, fungal mycelia were carefully removed from wood flour, and wood flour was dried under vacuum at 40 °C for further analyses.

Isolation of Lignin from Wood

Lignin in pine flour was isolated before and after the fungal degradation with an acidic dioxane extraction method, i.e., the pine flour samples (10 g) were refluxed under nitrogen for 2 h with 200 mL 0.1 M HCl in dioxane/water (85/15, v/v) at solvent/wood weight ratio 20:1 (*11,12*). The dioxane-water extract was neutralized with solid sodium bicarbonate. Dioxane in the extract was removed through a rotary evaporator, and lignin left in the solution was precipitated in water (500 mL, pH 2-3). The precipitated lignin was collected by centrifugation, washed with acidified water (pH 2-3) and then freeze-dried. lignin extracted from pine flour before the fungal degradation was used as a control.

Degradation of Unbleached Softwood Kraft Pulp by *P. cinnabarinus*

This experiment was carried out in accord with an established procedure (*3*). Unbleached softwood kraft pulp (USKP) from Douglas fir was obtained from Pope and Talbot (Halsey, Oregon) and washed with deionized water until the effluent was neutral and colorless. USKP (8.0 g, oven-dried) was autoclaved at 121 °C for 40 min, cooled down at room temperature and then inoculated with 3 mL of *P. cinnabarinus* spore suspension (ca. 9×10^7 spores/mL). The moisture content of the culture was adjusted to 80% and the culture was incubated statically at 30 °C. After a pre-determined incubation time, mycelia were carefully removed from pulp. The treated pulp was analyzed for kappa number.

Measurement of Kappa Number

The kappa number was determined in accordance with TAPPI Test Methods (1994-1995) T236 om-85. At least two replicates were performed for each sample.

Isolation of Lignin from Pulp

The pulp samples before and after the fungal degradation were air-dried and milled with Wiley Mill to flour (30 mesh). Lignin in pulp was isolated with the acidic dioxane extraction method described previously. Lignin extracted from the milled pulp before fungal degradation was used as a control.

Quantitative Analysis of Lignin Functional Groups with ^{31}P NMR

Lignin structures were characterized with ^{31}P NMR according to established methods (*13-15*). Accurately weighed lignin (40 mg/mL) together with cholesterol (3.5 mg/mL, an internal standard) and chromium acetylacetonate (5.0 mg/mL, relaxation reagent) were dissolved in a mixture of pyridine/$CDCl_3$ (1.6/1, v/v) in a 4-mL vial with magnetic stirring. The phosphorylation agent 2-chloro-4, 4, 5, 5-tetramethyl-1, 3, 2-dioxaphospholane (100 μl) was added and stirred for 20 min. The resulting solution was then transferred to a 5 mm NMR tube for the acquisition of ^{31}P NMR spectra. The ^{31}P NMR spectra were acquired by the inverse gated decoupling sequence on a Bruker 300 MHz NMR spectrometer (*13,16*). All chemical shifts in this paper were referenced by the reaction product of water with the phosphorylation agent, which gives a sharp ^{31}P NMR signal at 132.2 ppm in pyridine/$CDCl_3$ (*14*). The integral value of the internal standard (^{31}P NMR signals at 146.1 ppm) was used to calculate the amount of various hydroxyl groups. The peaks of quantitative

^{31}P NMR spectra were assigned as follows: aliphatic alcoholic units 149.6-146.4 ppm; condensed phenolic units 145.7-141.6 ppm; guaiacyl phenolic units 141.6-140.1 ppm; and carboxylic acids 136.8-135.8 ppm.

FTIR Spectroscopy

Fourier Transfer Infrared (FTIR) spectra of lignin were acquired on a Thermo Nicolet Nexus™ 470 FTIR spectrometer with a Golden Gate Heatable Diamond ATR (attenuated total reflectance) accessory at room temperature. The number of scans for each sample was 32. The assignments of absorbance peaks were based on published FTIR database for lignin (*17*).

Results and Discussion

Weight Loss of Decayed Wood Blocks

After three months of degradation, the weight losses of pine blocks caused by *P. cinnabarinus, T. rubrum* LSK-27 and *T. rubrum* LKY-7 were 87.6%, 39.2%, and 3.1%, respectively, for these three fungi (Table I). Because lignin content in pine is at least 30%, a weight loss of 87.6% implies that *P. cinnabarinus* efficiently degrade all wood components including lignin. The degradation ability of *T. rubrum* LKY-7 on pine was much lower than that of *P. cinnabarinus* and *T. rubrum* LSK-27. As a matter of fact, *T. rubrum* LKY-7 only caused a slight decrease in weight loss as compared to the control.

The ability of these three fungi to degrade the hardwoods yellow poplar and sweet gum was also investigated (Table II). The weight losses of yellow poplar blocks after two months of degradation by *P. cinnabarinus, T. rubrum* LSK-27 and *T. rubrum* LKY-7 were 30.1%. 39.2%, and 7.5%, respectively (Table II). Statistically, *P. cinnabarinus* and *T. rubrum* LSK-27 degraded yellow poplar blocks at the same rate. However, *P. cinnabarinus* and *T. rubrum* LSK-27 degraded yellow poplar blocks much more efficiently than *T. rubrum* LKY-7. For the degradation of sweet gum blocks, there were no statistically significant difference of the weight losses caused by *P. cinnabarinus* and *T. rubrum* LSK-27. *T. rubrum* LKY-7 degraded sweet gum blocks much slower than *P. cinnabarinus* and *T. rubrum* LSK-27.

It is perplexing that *P. cinnabarinus* degraded pine blocks much more efficiently than *T. rubrum* LSK- 27 and *T. rubrum* LKY-7, whereas *P. cinnabarinus* and *T. rubrum* LSK-27 degraded hardwood blocks (yellow poplar and sweet gum) at the same rate. Both *T. rubrum* LSK-27 and *T. rubrum* LKY-7 were recently isolated from a decayed hardwood chip pile in the vicinity of

Table I. Weight Loss of Pine Blocks Caused by Three Fungi after a Three-month Incubation [a]

	P. cinnabarinus	*T. rubrum LSK-27*	*T. rubrum LKY-7*	*Control*
Weight loss (%)	87.6 ± 2.7	20.8 ± 1.5	3.1 ± 0.7	2.3 ± 0.5

[a] Data were the mean of results from nine wood samples

Table II. Weight Loss of Hardwood Blocks Caused by Three White-rot Fungi after a Two-month Incubation [a]

	P. cinnabarinus	*T. rubrum LSK-27*	*T. rubrum LKY-7*	*Control*
Yellow Poplar	30.1 ± 9.5	39.2 ± 2.3	7.5 ± 0.7	0.9 ± 0.3
Sweet Gum	24.8 ± 14.6	28.4 ± 4.4	3.2 ± 2.5	0.23 ± 0.10

[a] Data were the mean of results from nine wood samples

Chonnam, Korea. The species of these two fungi were characterized through a GC-FAME (gas chromatography-fatty acid methyl ester) method (Microbe Inotech Labs, Inc., St. Louis, MO). When searched against the database of Microbe Inotech Labs, Inc., the most similar fungal species for these two fungi was *Trichophyton rubrum.* The Similarity and Distance Coefficient of these two fungi to *Trichophyton rubrum* were 0.442 and 4.426 for *T. rubrum* LKY-7 and 0.688 and 2.995 for *T. rubrum* LSK-27. Two identical fungal strains in the database would have similarity coefficient of one and a distance of zero. Therefore, these two fungi may not truly belong to *Trichophyton rubrum.* The great difference of these two fungi's ability to degrade both softwood and hardwood blocks suggests that they might belong to different species.

Activities of three ligninolytic enzymes, lignin peroxidase, manganese peroxidase, and laccase were detected during the incubation of pine flour with *T. rubrum* LSK-27 and *T. rubrum* LKY-7, whereas only laccase activity was detected with *P. cinnabarinus* (data not shown). It appeared that these three fungi used different ligninolytic enzyme systems for lignin degradation. To gain a better understanding of the mechanisms by which these three fungi degraded lignin, we investigated the structural changes of lignin during the course of the fungal degradation of wood. Pine flour samples rather than pine block samples were used for the fungal degradation experiments so that lignin could be directly and efficiently extracted with dioxane/water from the decayed wood residue without a post-degradation grinding because the grinding may cause structural changes in the lignin. There were large variations in the weight loss results because fungal mycelia were difficult to remove from the decayed pine flour samples. Therefore the weight loss results are not reported here. It is worthy noting that *T. rubrum* LKY-7 degraded pine flour much faster than pine blocks. As much as a 20% weight loss of pine flour was detected after a three-month incubation with *T. rubrum* LKY-7 even though the fungal mycelia could not be completely removed.

Fungal Modification of Lignin Structures in Wood

Extracted lignin samples were analyzed with ^{31}P NMR for the content of various functional groups (Table III). Aliphatic hydroxyl groups in lignin mainly refer to those hydroxyl groups of compound **I** (Figure 1). Compound **II** in Figure 1 is a representative lignin substructure with condensed phenolic hydroxyl groups. Compound **III** is a representative lignin substructure containing guaiacyl phenolic hydroxyl groups (Figure 1). When compared with the control, the contents of condensed phenolic OH groups and carboxylic acid groups increased, the content of guaiacyl phenolic OH groups decreased, and the contents of aliphatic OH groups remained statistically the same after a three-month's degradation of pine flour by *P. cinnabarinus.* For *T. rubrum* LSK-27, the trends in functional group changes were the same as those for *P. cinnabarinus,* i.e., the contents of the aliphatic OH groups were basically the

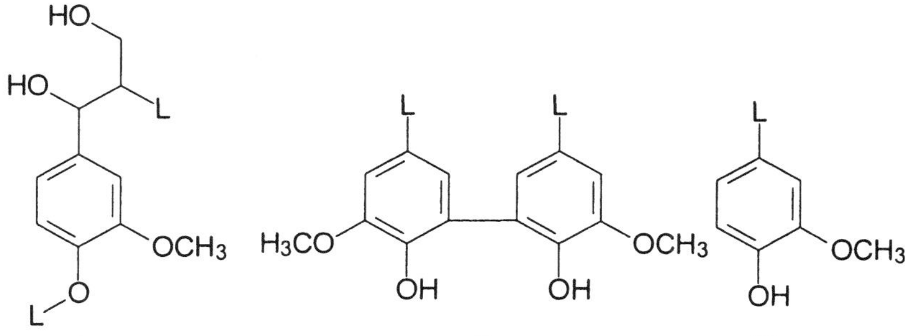

Figure 1. Representative lignin substructure (L = lignin matrix)

same, the contents of condensed phenolic OH groups and carboxylic acid groups slightly increased; and the content of the guaiacyl phenolic hydroxyl groups decreased. For *T. rubrum* LKY-7, the only difference for the trends of the functional groups change was that the carboxylic acid contents remained virtually the same after three months of fungal degradation.

Pine lignin is predominately derived from conifer alcohol. In pine lignin, about 10% of lignin substructures are phenolic, i.e., those substructures with a guaiacyl phenolic hydroxyl group, whereas about 90% of lignin substructures are non-phenolic, i.e., the guaiacyl phenolic hydroxyl group is covalently linked to the lignin matrix. It is well established that a phenolic hydroxyl group is easy to oxidize. In other words, a phenolic lignin substructure is easier to oxidize than a non-phenolic lignin substructure with the exception that condensed lignin structures are especially difficult to degrade because of the carbon-carbon linkages between two phenylpropane units. The ^{31}P NMR results indicated that three fungi used in this study all caused decreases in the guaiacyl phenolic OH group content and increases in the condensed phenolic hydroxyl group content, which implied that these fungi first degraded the easily oxidizable lignin substructures with guaiacyl phenolic OH group and left the condensed lignin substructures behind regardless of the ligninolytic enzymes they secreted for lignin degradation. Little change of the aliphatic OH groups suggested that the oxidization of the phenylpropane side chain in the pine lignin was not the important action mode in the degradation of lignin by these three fungi. The increase in the carboxylic acid content probably resulted from the ring-opening reactions in the oxidation of the guaiacyl phenolic substructure.

The previously proposed action modes for the fungal degradation of lignin were derived from a three-month fugal degradation of pine flour. Additional studies were performed to determine the effect of incubation time, i.e., the degree of wood degradation. Results in Table IV revealed that the incubation time had little effect on the change of the aliphatic OH content. The condensed phenolic OH content significantly increased in the first month of the incubation,

Table III. The Contents of Various Functional Groups (mmol/g Lignin) after a Three-month Degradation of Pine Flour by Three Fungi as Determined by ^{31}P NMR Techniques [a]

		Phenolic OH		
	Aliphatic OH	*Condensed*	*Guaiacyl*	*R-COOH*
Control	3.80 ± 0.37	0.35 ± 0.20	1.19 ± 0.04	0.07 ± 0.04
Pc [b]	4.31 ± 0.34	0.54 ± 0.10	0.78 ± 0.03	0.20 ± 0.01
Tr LSK-27 [c]	3.78 ± 0.34	0.39 ± 0.02	0.80 ± 0.01	0.16 ± 0.01
Tr LKY-7 [c]	3.43 ± 0.51	0.45 ± 0.05	0.79 ± 0.08	0.09 ± 0.02

[a] Data were the mean of results from two wood samples

[b] Pc refers to *Pycnoporus cinnabarinus*

[c] Tr refers to *Trichophyton rubrum*

Table IV. Effects of the Incubation Time on the Contents of Various Functional Groups (mmol/g Lignin) in the Degradation of Pine Flour by *P. cinnabarinus* [a]

Incubation time		*Phenolic OH*		
(month)	*Aliphatic OH*	*Condensed*	*Guaiacyl*	*R-COOH*
Control	3.80 ± 0.37	0.35 ± 0.20	1.19 ± 0.04	0.07 ± 0.04
1	3.74 ± 0.33	0.54 ± 0.10	0.81 ± 0.04	0.17 ± 0.04
2	3.75 ± 0.33	0.58 ± 0.00	0.80 ± 0.00	0.21 ± 0.03
3	4.31 ± 0.34	0.54 ± 0.10	0.78 ± 0.03	0.20 ± 0.01
5	3.71 ± 0.14	0.38 ± 0.01	0.67 ± 0.00	0.20 ± 0.01

[a] Data were the mean of results from two wood samples

remained at the increased level during the second and the third months of the incubation, and then decreased after the fifth month of the incubation (Table IV). In contrast with the significant increase of the condensed phenolic structures, there was a significant reduction of the guaiacyl phenolic OH content after one month of degradation. The guaiacyl phenolic OH content remained unchanged between one month and three months of incubation. However, the guaiacyl phenolic OH content further decreased when the pine flour samples were incubated with *P. cinnabarinus* for five months. Carboxylic acids content gradually increased in the first two months of the fungal degradation and remained constant thereafter.

It appeared that some easily accessible guaiacyl phenolic substructures were preferentially degraded in the first month, which led to the increase in the relative content of the condensed phenolic substructures. Lignin is a complex polymeric material. Guaiacyl phenolic substructures are widely distributed in the lignin matrix. Therefore, some guaiacyl phenolic substructures were not readily accessed by the fungus *P. cinnabarinus,* which accounts for the detection of a significant amount of the guaiacyl phenolic OH group even after five months of degradation. The ^{31}P NMR data indicated a rapid and preferential degradation of guaiacyl phenolic substructures. This appeared to be in accord with the FTIR results: the intensity of the characteristic absorbance of guaiacyl substructures at 1265 cm^{-1} significantly decreased after one-month of degradation, and the intensity ratio of the absorbance at 1510 cm^{-1} vs. 1452 cm^{-1}, another characteristic for guaiacyl phenolic substructures, decreased from 1.38 in pine lignin before decay to 1.21 in pine lignin after one month of degradation (*17*).

It is not clear whether all lignin substructures including the predominant non-phenolic lignin substructures and the condensed phenolic substructures were degraded at approximately the same rate after the readily accessible guaiacyl phenolic substructures were degraded in the first month of the incubation. The small changes in the guaiacyl phenolic OH group content during the one to three months of the fungal degradation appeared to suggest that the fungal degradation of guaiacyl phenolic substructures were controlled by the accessibility of these substructures to the fungus. The relatively constant level of the condensed phenolic substructures appeared to suggest that the condensed phenolic substructures were also degraded at the time when non-phenolic lignin substructures were degraded.

The structures of lignin in unbleached softwood kraft pulp (USKP) were significantly altered during the Kraft pulping process. We investigated whether the alteration of lignin structures had effects on the action modes of degradation for *P. cinnabarinus* (Table V). In contrast to the initial increase in the condensed phenolic OH group content in the degradation of pine flour, the condensed phenolic OH group content gradually decreased in the degradation of USKP. The aliphatic OH group content first increased and then decreased during the

Table V. The Contents of Various Functional Groups (mmol/g Lignin) in the Degradation of Unbleached Softwood Kraft Pulp by *P. cinnabarinus* [a]

Incubation time (day)	*Aliphatic OH*	*Phenolic OH*		*R-COOH*
		Condensed	*Guaiacyl*	
Control	2.18 ± 0.04	0.76 ± 0.02	0.68 ± 0.01	0.21 ± 0.04
25	2.72 ± 0.04	0.72 ± 0.01	0.48 ± 0.00	0.28 ± 0.01
40	2.37 ± 0.26	0.60 ± 0.15	0.33 ± 0.04	0.46 ± 0.19

[a] Data were the mean of results from three pulp samples

fungal degradation of USKP, which is also contrary to the small changes observed in the aliphatic OH group content in the fungal degradation of pine flour. The trends of the content change for guaiacyl phenolic OH group and carboxylic acids group during fungal degradation were the same for both pine flour and USKP. It is worth noting that both the increased rate of loss of the guaiacyl phenolic OH group content and the decreased rate of loss of the carboxylic acid group content in the fungal degradation of USKP were higher than those in the fungal degradation of pine flour. This is probably due to the fact that the guaiacyl phenolic substructures in USKP were more accessible than those in pine flour.

Treatment of USKP with the fungus *P. cinnabarinus* led to decrease in Kappa number, which implied that the fungus selectively degraded lignin (Table VI). The Kappa number decreased from 23.34 to 5.46 after a 25-day-incubation. However, an increase of the incubation time from 25 to 40 days resulted in little change of the Kappa number. The fungus was also able to degrade cellulose and hemicelluloses although it preferentially degraded lignin. Where the lignin content was low, the fungal degradation rates for carbohydrates increased, which might account for the small change in Kappa number when the incubation time increased from 25 days to 40 days.

Table VI. Kappa Number of Pulp before and after Treatment of Unbleached Softwood Kraft Pulp by *P. cinnabarinus* [a]

Incubation time (day)	*Kappa number*
Control	23.34 ± 0.10
25	5.46 ± 0.30
40	5.67 ± 0.39

[a] Data were the mean of results from three pulp samples

Acknowledgments

This research was supported by a grant from the National Research Initiative Competitive Grants Program of the USDA (award number: 2001-35103-10149).

References

1. Hatakka, A.; Uusi-Rauva, A. K. *Eur. J. Appl. Microbiol. Biotechnol.* 1983, *17*, 235-242.
2. Eggert, C.; Temp, U.; Eriksson, K.-E. L. *FEBS Lett.* 1997, *407,* 89-92.
3. Bermek, H.; Li, K.; Eriksson, K.-E. L. *J. Biotechnol.* 1998, *66*, 117-124.
4. Eggert, C.; Temp, U.: Eriksson K.-E. L. *Appl. Environ. Microbiol.* 1996, *62,* 1151-1158.
5. Bourbonnais, R.; Paice, M. G. *FEBS Lett.* 1990, *267*, 99-102.
6. Bourbonnais, R.; Paice, M. G. *Appl. Microbiol. Biotechnol.* 1992, *36,* 823-827.
7. Geng, X.; Li, K. *Appl. Microbiol. Biotechnol.* 2002, *60,* 342-346.
8. Li, K.; Horanyi, P. S.; Collins, R.; Phillips, R. S.; Eriksson, K. L. *Enzyme Microb. Technol.* 2001, *28,* 301-307.
9. Scheffer, T. C.; Morrell, J. J.; Newbill, M. A. *Forest Prod. J.* 1987, *37,* 51-54
10. Freitag, C. M.; Li, K.; Morrell, J. J. *Forest Prod. J.* 2003, *53,* 77-79.
11. Pepper, J. M.; Baylis, P. E. T.; Adler, E. *Can. J. Chem.* 1959, *37,* 1241-1248.
12. Jiang, Z. H.; Argyropoulos, D. S. *J. Pulp Paper Sci.* 1994, *20,* 183-188.
13. Argyropoulos, D. S. *J. Wood Chem. Technol.* 1994, *14,* 45-63.
14. Granata, A. A.; Argyropoulos, D. S. *J. Agric. Food Chem.* 1995, *43,* 1538-1544.
15. Tohmura, S.; Argyropoulos. D. S. *J. Agric. Food Chem.* 2001, *49*, 536-542.
16. Jiang. Z.-H.; Argyropoulos, D. S.; Granata, A. A. *Mag. Resonan. Chem.* 1995, *33,* 375-382.
17. Buta, J. G.; Zadrazil. F.; Galletti. G. *C. J. Agric. Food Chem.* 1989, *37,* 1382-1384.

Enzyme Characterization

Chapter 9

Productive Cellulase Adsorption on Cellulose

Hanshu Ding and Feng Xu*

Novozymes Biotech, 1445 Drew Avenue, Davis, CA 95616

Hydrolysis of insoluble polymeric cellulose by cellulases requires enzyme adsorption onto the substrate. Being heterogeneous on its surface, cellulose in general has different sites for different cellulases to adsorb. Only a portion of the adsorbed cellulase is productive, able to lead to subsequent glycosidic bond cleavage. We applied the dependence of initial hydrolysis rate on the concentration of enzyme and substrate to probe the "productive" adsorption of four different cellulases on two types of cellulose. We found that different celluloses could have significantly different cellulase accessibility; that different cellulases, even from the same family, could have significantly different "productive" adsorption on cellulose; and that functionally analogous cellulases could have overlapping adsorption sites, while functionally dissimilar cellulases could have separate adsorption sites. The usefulness of the method for mechanistic study and industrial application of cellulases was discussed.

To efficiently hydrolyze cellulose, cellulases are often required to tightly bind onto their insoluble substrate (*1-6*). Many cellulases are modularly structured, possessing a catalytic core and one or more carbohydrate-binding modules (CBM) (*2*, *3*, *7-9*). The function of CBM, which has high affinity for crystalline cellulose, is believed to include anchoring the catalytic core onto the surface of a cellulose microfibril, disrupting hydrogen bonds in cellulose, and facilitating the binding of the cellulose chain at core's active site. For cellobiohydrolases (CBH), the typical catalytic cycle involves binding at the end of a cellulose chain, cleavage of a β-1,4 glycosidic bond to yield cellobiose, and finally translocation along the chain for the next cycle ("processivity") or desorption into solution when the chain end is reached. *Trichoderma reesei* and *Humicola insolens* CBH-I (Cel7A) act from the reducing end of a cellulose chain, while their CBH-II (Cel6A) prefer the non-reducing end (*10*, *11*). For endo-β-1,4-glucanases (EG), the typical catalytic cycle involves random binding along a cellulose chain, cleavage of a β-1,4 bond to fragment the chain or release soluble cellooligomers, and desorption from the chain before the next cycle. It has been proposed that it is the core, rather than CBM, that specifies where on the cellulose the cellulase adsorbs (*12*).

In the past two decades, tremendous progress has been made in determining the structure of cellulases (*2*, *13-16*). Many cellulase families have been discovered based on sequence homology (*17*). Available crystallographic data indicate that cellulases belonging to the same family share highly homologous 3-D structures. Substrate specificity, reactivity and mechanism of cellulases have been probed by various techniques including X-ray crystallography and site-directed mutagenesis. For CBH, their processivity has been related to the tunnel morphology of their catalytic site where long loops can open/close to facilitate CBH to adsorb to, slide along, or desorb from a cellulose chain (*18*, *19*). Principles of cellulase action on cellulose have been established (*1*, *2*, *5*, *20*). Several mathematic models have been developed to analyze the multiphasic kinetics of the heterogeneous cellulose hydrolysis by cellulase (*6*, *10*, *21-28*).

Despite the significant progress achieved so far, several mechanistic details governing CBM-cellulose interaction and its effect on catalysis remain unclear (*3*, *12*, *16*, *29-31*). All fungal CBM belong to CBM family 1 because of their sequence homology (*16*). Yet different fungal cellulases can have significantly different cellulose adsorption in terms of adsorption rate, affinity constant (K_a) and reversibility. For instance, *T. reesei* CBH-I binds cellulose reversibly while *T. reesei* CBH-II binds irreversibly (*32*, *33*). It is not known whether CBH-I and CBH-II, or even CBH-I analogs from different fungi, adsorb onto the same or different regions of a cellulose microfibril. A general correlation between cellulose affinity and catalytic reactivity for cellulases has yet to be established.

Conventionally, cellulase adsorption onto (or accessibility for) cellulose, such as that of *T. reesei* cellulases onto amorphous/crystalline/lingo-cellulose, has been investigated by Langmuir-type isotherms (*20*, *34*). They measure overall cellulase protein adsorption, assuming a uniform cellulose surface, adsorption enthalpy independent on surface coverage, and negligible interaction among adsorbed molecules. The main shortcoming of the methods is their inability of differentiating "productive" adsorption, which leads to subsequent β-1,4 glycosidic bond cleavage, from "nonproductive" adsorption, which leads to either desorption or enzyme-inactivating immobilization (*10*, *35*). We developed a new method capable of measuring the "productive" cellulase adsorption, based on the dependence of initial hydrolysis rate on the concentration of cellulase and cellulose. In contrast to the "static" cellulose accessibility revealed by Langmuir isotherm, the accessibility detected by the new method is "kinetic" since it relates to the surface region hydrolysable by adsorbed cellulase. Four representative cellulases (CBH-I from *T. reesei* and *H. insolens*, CBH-II from *H. insolens* and EG-I (Cel7B) from *T. reesei*) were studied for their "kinetic" accessibility (related to their "productive" adsorption) towards two representative celluloses (amorphous phosphoric acid-swollen cellulose (PASC) and microcrystalline Avicel), in comparison with their "static" Langmuir accessibility. In addition to the cellulose accessibility, we also probed whether these cellulases bind onto the same or different regions of the celluloses.

Mathematic modeling for cellulase:cellulose interaction

When CBH encounters cellulose, it first adsorbs/binds onto cellulose, complexes a cellooligomeric chain at its active site, hydrolyzes a β-1,4 bond to release a cellobiose product, then slides along the chain processively for the next hydrolysis (*1*, *2*, *13-15*, *36*). At the initial stage of hydrolysis, when product inhibition is negligible, the reaction can be represented by the simplified mechanism shown in Figure 1, with the following symbols: E, free enzyme; S, enzyme-accessible/hydrolysable cellobiosyl unit in cellulose; β, number of the cellobiosyl units covered by a bound CBH molecule; $E_a{\cdot}\beta S$, adsorbed enzyme; "E·βS", pseudo-Michaelis intermediate ("pseudo" because of the insoluble nature of substrate); P, released hydrolysis product (cellobiose); $E_a{\cdot}(\beta\text{-}1)S$, enzyme:cellulose complex right after cellobiose release. Four major steps are depicted in Figure 1: Enzyme adsorption onto cellulose ($K_a = k_1/k_{-1}$), substrate activation (pseudo-Michaelis constant $K_m = (k_{-2}+k_3)/k_2$), hydrolysis (rate constant k_3), and processive translocation (procession constant $K_p = k_4/k_{-4}$).

$$E + \beta S \underset{k_{-1}}{\overset{k_1}{\rightleftharpoons}} E_a{\cdot}\beta S \underset{k_{-2}}{\overset{k_2}{\rightleftharpoons}} E{\cdot}\beta S \xrightarrow[\;-P\;]{k_3} E_a{\cdot}(\beta\text{-}1)S \underset{k_{-4}}{\overset{k_4}{\rightleftharpoons}} E_a{\cdot}\beta S$$

Figure 1. Simplified mechanism of cellulase-catalyzed cellulose hydrolysis.

Under steady-state conditions, the hydrolysis step becomes rate-limiting, while other steps reach equilibrium, yielding the following kinetic equations:

$$d[E\beta S]/dt=k_2[E_a\beta S]-(k_{-2}+k_3)[E\beta S]=0, \text{ or } [E_a\beta S]=(k_{-2}+k_3)[E\beta S]/k_2=K_m[E\beta S] \quad (1)$$

$$[E][\beta S]/[E_a\beta S]=\beta[E][S]/[E_a\beta S]=\beta/K_a, \text{ or } [E]=[E_a\beta S]/(K_a[S])=K_m[E\beta S]/(K_a[S]) \quad (2)$$

$$[E_a\beta S]/[E_a(\beta\text{-}1)S] = K_p, \text{ or } [E_a(\beta\text{-}1)S] = [E_a\beta S]/K_p = K_m[E\beta S]/K_p \quad (3)$$

From eqs 1 to 3 and the mass balance $[E]_0 = [E]+[E_a\beta S]+[E\beta S]+[E_a(\beta\text{-}1)S]$, in which $[E]_0$ is the total enzyme concentration, eq 4 is obtained:

$$[E\beta S] = [E]_0[S]/\{(1 + K^*)(\alpha + [S]\} \quad (4)$$

in which $K^* = K_m(1 + 1/K_p)$, $\alpha = K_m/\{K_a(1 + K^*)\}$.

The mass balance $[S]_0 = [S] + \beta[E_a\beta S] + \beta[E\beta S] + (\beta\text{-}1)[E_a(\beta\text{-}1)S] + [P]$, in which $[S]_0$ is the total enzyme-accessible/hydrolysable cellobiosyl unit concentration, applies to the substrate. Because $\beta \approx 39$ (*22*, *37*, *38*), $1/\beta \ll 1$. When hydrolysis extent is low, [P] is negligible, thus $[S]_0 \approx [S]+\beta\{[E_a\beta S]+[E\beta S]+[E_a(\beta\text{-}1)S]\}=[S]+\beta[E\beta S]\{K_m+1+K_m/K_p\}$. Considering eq 4, this leads to:

$$[S]_0=[S](\alpha+\beta[E]_0+[S])/(\alpha+[S]), \text{ or } [S]^2 + (\alpha + \beta[E]_0 - [S]_0)[S] - \alpha[S]_0 = 0 \quad (5)$$

Because the hydrolysis rate $v = k_3[E\beta S]$, eqs 4 and 5 can lead to eq 6:

$$v = K[E]_0[S]_0/(\alpha + \beta[E]_0 + [S]) \quad (6)$$

in which $K = k_3/(1 + 1/K^*)$. Since cellulase adsorption on cellulose is generally very strong ($1/K_a \to 0$), α becomes negligible. Assuming $[S]_0=\phi[S]_t$, in which $[S]_t$ is the total cellobiosyl unit concentration of substrate, then the initial rate can be expressed as:

$$v = K\phi[E]_0[S]_t/(\beta[E]_0 + [S]) \quad (7)$$

in which [S] is a function of $[S]_t$ according to $[S]_t = [S]_0/\phi$ and eq 5.

For non-processive EG, the k_4, k_{-4} reactions in Figure 1 can be omitted, and the $E_a\cdot(\beta\text{-}1)S$ can be replaced by E + P. Eqs 6 and 7 can be derived with $\alpha = K_m/\{K_a(1 + K_m)\}$ and $K = k_3/(1 + 1/K_m)$.

Dependence of initial rate on concentration

When initial substrate concentration $[S]_t$ is given, the v dependence on initial enzyme concentration $[E]_0$ can be characterized by an initial linear phase and a final saturation phase: When $[E]_0 \ll [S]_t$, $[S] \approx \phi[S]_t$ and eq 7 becomes $v \approx K[E]_0$; when $[E]_0 \gg [S]_t$, eq 7 becomes $v \approx K\phi[S]_t/\beta$. The two lines intersect at $[E]_0^s$, leading to:

$$[E]_0^s/[S]_t = \phi/\beta \tag{8}$$

When initial enzyme concentration $[E]_0$ is given, the v dependence on initial substrate concentration $[S]_t$ can also be characterized by an initial linear phase and a final saturation phase: When $[S]_t \ll [E]_0$, eq 7 becomes $v \approx K\phi[S]_t/\beta$; when $[S]_t \gg [E]_0$, $[S] \approx \phi[S]_t$ and eq 7 becomes $v \approx K[E]_0$. The two lines intersect at $[S]_t^s$, leading to

$$[E]_0/[S]_t^s = \phi/\beta \tag{8'}$$

Based on crystallographic data, the number of cellobiose unit covered by cellulose-absorbed CBH, β, can be estimated as ~39 (*22*, *37*, *38*). Thus both rate-concentration dependences can reveal ϕ, a parameter reflecting the initial "kinetic" accessibility of cellulose for cellulase.

Difference between ϕ and Langmuir-type binding capability

Langmuir-type isotherms, governed by an adsorption constant and a capability N_0, are widely used to measure the adsorption of cellulase onto cellulose. The parameter N_0 reflects the portion of surface cellobiosyl/lattice units accessible for cellulase adsorption, regardless whether or not the adsorption is followed by hydrolysis. In contrast to the "static" N_0, ϕ is "kinetic" and measures those "productive" cellulase adsorptions that can lead to subsequent β-1,4 bond cleavage.

It is generally believed that EG cleave β-1,4 bonds randomly on the surface of cellulose, and CBH cleave β-1,4 bonds in a processive manner from either the reducing or non-reducing end of a cellulose chain (*2*, *11*, *13*, *15*). It is not clear, however, whether different cellulases, belonging to either the same or different functional group/family, adsorb initially at the same cellulose surface region. Measuring ϕ for individual and mixed cellulases would address this question.

Accessibility measurement

H. insolens CBH-I and CBH-II were purified as reported (*39*). *T. reesei* CBH-I and EG-I were purified to apparent electrophoretic homogeneity by chromatographical methods similar to those published (*40*). Briefly, a commercial *T. reesei* cellulase mix, Celluclast (activity: ~60 Filter-Paper-Unit/mL, protein content: ~100 mg/mL, Major component: Cel7A, 6A, 7B and 5A, Novozymes), was first buffer-exchanged by ultrafiltration (Pall Filtron ultrafiltration apparatus and polysulfone membrane of 10K MW cutoff), then subjected to Q-Sepharose (pH 7) and Mono-S (pH 4) chromatography (on a Pharmacia FPLC). Active fractions were monitored by carboxymethyl cellulose (CMC) and PASC-hydrolyzing activity at pH 5 and 30 °C. Purity was measured by SDS-PAGE and IEF, and identity was verified by mass spectrometry-based protein sequencing. Purified *T. reesei* CBH-I, *T. reesei* EG-I, *H. insolens* CBH-I and *H. insolens* CBH-II showed respectively a molecular mass of 65000, 60000, 72000 and 65000, isoelectric point of 4.1, 4.8, 4.5 and 4.9, initial PASC-hydrolyzing activity of 0.86, 36, 2.0 and 5.5 IU/mg, Avicel-hydrolyzing activity of 0.12, 0.3, 0.12 and 0.2 IU/mg, and CMC-hydrolyzing activity of 0.6, 84, 0.4 and 0.1 IU/mg. Table I summarizes the extended CBH-I activity on PASC and Avicel in the presence of *Aspergillus oryzae* β-glucosidase (to prevent cellobiose inhibition).

Table I. Extent og long-term cellulose hydrolysis by CBH-I

CBH-I	*PASC*		*Avicel*	
	2 h	*24 h*	*2 h*	*24 h*
T. reesei	49%	99%	3%	19%
H. insolens	54%	83%	1%	5%

Buffer: 50 mM Na-acetate, pH 5. Temperature: 40 °C. Concentration: 1.8 g/L PASC, 10 g/L Avicel, 50 nM *A. oryzae* β-glucosidase, 3 μM CBH-I.

Chemicals used as substrate, chromogenic agent, or buffer were commercial products of at least reagent grade. PASC was prepared from Avicel (PH101 from FMC) (*39*). Acid pretreated corn stover (PCS) was obtained from National Renewable Energy Laboratory, with a composition of ~57% cellulose, 28% lignin, 5% hemicellulose and 4% protein.

Cellulose-cellulase incubation was done in 50 mM Na-acetate of pH 5 with 0.5 g/L bovine serum albumin (BSA; to reduce nonspecific binding of cellulase on inner surface of reaction vessel) at 40 °C. To study v dependence on [cellulase], 0.3 – 15 μM cellulase was reacted with 1.8 g/L cellulose. To study v dependence on cellulose dose, denoted as "[cellulose]" (amount divided by suspension volume), 0.2 – 10 g/L cellulose was reacted with 5 μM cellulase.

Reducing sugar production was monitored with *p*-hydroxybenzoic acid hydrazide (PHBAH) (*41*). First, 100 μL sample was mixed with 50 μL 1.5 % PHBAH in 0.5 M NaOH in a 96-well plate. After 10 min at 95 °C, 50 μL H_2O was added and 100 μL solution was read at 410 nm in another 96-well plate. D-Glucose served as quantification standard. Initial v was derived from five samplings taken during the first hour of hydrolysis.

Because of its endocellulase nature, *T. reesei* EG-I could hydrolyze a cellulose chain without releasing soluble sugar. To detect such endo- reactivity, insoluble cellulose, after being incubated with EG-I, decanted and washed, was subjected to PHBAH reaction. Reduced PHBAH in solution was measured colorimetrically. Under our conditions, the majority (~90%) products of EG-I's initial reaction was found in solution, likely as glucose (*36*). As detected by HPLC, the main initial product from < 1 and > 4 μM EG-I was cellobiose and glucose, respectively; with > 15 μM EG-I, glucose accounted for more than 85% of the initial products. When initial rate v was plotted against initial enzyme concentration $[EG\text{-}I]_0$, two v "saturations" were observed: The first one took place at ~2 μM $[EG\text{-}I]_0$ and the second at ~11 μM $[EG\text{-}I]_0$. Likely the first and second saturation corresponded to cellobiose and glucose production by EG-I, respectively. Only the first $[EG\text{-}I]_0^s$ was used to derive ϕ for EG-I.

Langmuir adsorption measurement

Cellulose-cellulase incubation was made at 25 °C with 1 mL of 50 mM Na-acetate of pH 5, 0.8 – 62 μM cellulase, 1.8 g/L cellulose and 8 μM BSA in 1.2-mL Pierce ImmunoWare tubes. After 30 min, the supernatant was decanted, filtered (by 0.45 μ Millipore Multi Screen-HV), and assayed by PASC hydrolysis activity for free cellulase. Langmuir adsorption equation, $1/[E]_{adsorbed} = 1/([E]_0 - [E]) = 1/(N_0 K'[E]) + \beta/N_0$ (*38*), was then applied to calculate apparent capacity N_0 (in mole/mole) and adsorption constant K'.

For PASC hydrolysis assay, 10 μL sample was mixed with 190 μL solution containing 2.1 g/L PASC and 8 μM BSA in 50 mM Na-acetate of pH 5 in a 96-well plate. After 30 min at 50 °C, 50 μL 0.5 M NaOH was added to stop hydrolysis. After 5 min centrifugation at 2000 rpm, 100 μL supernatant was subjected to PHBAH reducing sugar assay.

Accessibility of individual cellulase

Under our conditions, mixing cellulase with cellulose led to immediate release of reducing sugar. Figure 2A shows the dependence of initial rate v on initial enzyme concentration $[CBH\text{-}I]_0$ for the PASC hydrolysis by *T. reesei* CBH-I. At low $[CBH\text{-}I]_0$, v was proportional to $[CBH\text{-}I]_0$. At high $[CBH\text{-}I]_0$, however, a "saturating" phase was achieved where v became independent on $[CBH\text{-}I]_0$. Crossing the $v \propto [CBH\text{-}I]_0$ and $v = constant$ lines, a $[CBH\text{-}I]_0^s$ was obtained and ϕ was deduced according to eq 8. A similar situation was observed when $[PASC]_t$ varied (Figure 2B). The ϕ obtained from $[PASC]_t^s$ and eq 8' was close to that obtained from $[CBH\text{-}I]_0^s$ and eq 8. Table II summarizes ϕ measured for the four cellulases and two celluloses.

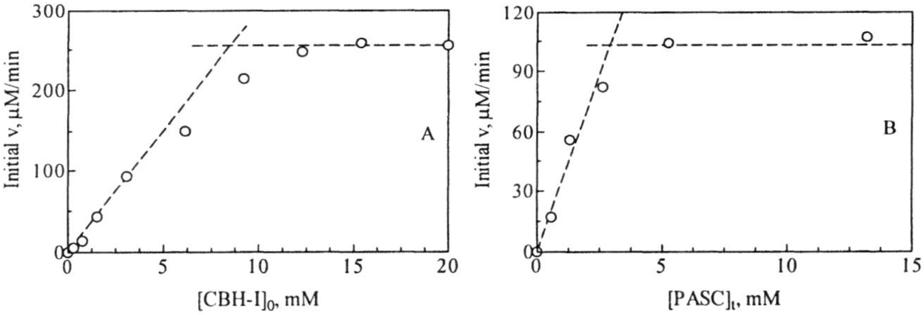

Figure 2. PASC hydrolysis by T. reesei CBH-I. Initial rate v dependence on (A) $[CBH\text{-}I]_0$ or (B) $[PASC]_t$. (A): $[PASC]_t$: 5 mM cellobiose equivalent (1.8 g/L). Lines $v = 30[CBH\text{-}I]_0$ and $v = 255$ intercept at $[CBH\text{-}I]_0^s = 8.5\ \mu M$ (0.55 g/L). (B): $[CBH\text{-}I]_0$: 4.6 μM (0.3 g/L). Lines $v = 35.5[PASC]_t$ and $v = 103$ intercept at $[PASC]_t^s = 2.8$ mM cellobiose equivalent (1 g/L). For clarity purpose, "saturating" v data beyond the depicted concentration range are not included.

For a given cellulase, Avicel's ϕ was less than PASC's ϕ, suggesting less "productive" cellulase adsorption on more crystalline cellulose surface. With crystalline cellulose substrate, it could be more difficult for the cellulase to thread a cellulose chain into its active site cleft/tunnel, since more inter-chain H bonds would need to be broken in a highly crystalline region than in an amorphous/chain end region.

Table II. Cellulase accessibility towards cellulose

Cellulase	*Cellulose*	$[E]_0/[S]_t^s$	$[E]_0^s/[S]_t$	ϕ
T. reesei CBH-I	PASC	0.0016	0.0016	0.062
T. reesei EG-I	PASC	0.00011	0.00010	0.0043
H. insolens CBH-I	PASC	0.0011	0.0011	0.043
H. insolens CBH-II	PASC	0.00019	0.00017	0.0070
T. reesei CBH-I	Avicel	0.00035	0.00035	0.014
T. reesei EG-I	Avicel	-	0.000032	0.0012
H. insolens CBH-I	Avicel	0.000036	0.000048	0.0016
H. insolens CBH-II	Avicel	0.000029	0.000034	0.0012
T. reesei CBH-I	PCS	-	0.00023	0.0090

$[E]_0$: initial enzyme concentration. $[S]_t$: initial cellulose concentration. $[E]_0^s/[S]_t \approx [E]_0/[S]_t^s \approx \phi/\beta$, $\beta \approx 39$.

The ϕ_{PASC} - ϕ_{Avicel} difference varied among cellulases, indicating differential cellulose accessibility of these cellulases, even when they were close structural/functional analogs. For instance, *T. reesei* CBH-I had a ϕ on PASC ~3 times larger than that on Avicel, but *H. insolens* CBH-I had a ϕ on PASC ~26 times larger than that on Avicel.

For a given cellulose, different cellulases (even from the same family) showed different ϕ. For instance, *T. reesei* and *H. insolens* CBH-I showed a ϕ of 0.014 and 0.0016, respectively, on Avicel, although they both are family 7 glycoside hydrolase and possess family 1 CBM.

Accessibility of cellulase mixtures

Table III summarizes the ϕ measured with 1:1 binary cellulase mixtures. For PASC, *T. reesei* CBH-I and *H. insolens* CBH-I mixture showed a ϕ (0.062) similar to that of *T. reesei* CBH-I (0.062), rather than the ϕ sum (0.062 + 0.043 ≈ 0.11) of the two cellulases (Table II), indicating that the two CBH-I were adsorbed at the same region (likely rich in reducing chain ends) on PASC. Similar result was seen when *T. reesei* CBH-I and *H. insolens* CBH-I were mixed at 1.4:1 ratio ($\phi \approx 0.07$). For Avicel, *T. reesei* CBH-I and *H. insolens* CBH-I mix showed a ϕ (0.013) similar to that of *T. reesei* CBH-I (0.014).

For PASC, *H. insolens* CBH-I and CBH-II mix showed a ϕ (0.055) similar to the ϕ sum (0.043 + 0.007 = 0.05) of the two cellulases, rather than the individual ϕ of either cellulase (Table II), suggesting that the CBH-I and CBH-II were adsorbed at different regions on PASC. A similar result was seen when CBH-I and CBH-II were mixed at 6.7:1 ratio ($\phi \approx 0.06$). For Avicel, *H. insolens*

CBH-I and CBH-II mix showed a ϕ (0.0023) again similar to the ϕ sum (0.0016 + 0.00012 = 0.0028) of the two cellulases. The ϕ observation indicated that the region CBH-I could "productively" adsorb (likely enriched in reducing ends) was separated from that of CBH-II (enriched in non-reducing ends) at the initial hydrolysis stage.

Unlike processive *T. reesei* CBH-I's preference for non-reducing cellulose chain ends, non-processive *T. reesei* EG-I is believed capable of randomly act along the cellulose chain (*2*, *13*, *15*). Under our conditions, 1:1 *T. reesei* CBH-I and EG-I mix had a ϕ of 0.082 for PASC, while 15:1 or 17:1 mix had a ϕ of 0.067. For Avicel, 1:1 *T. reesei* CBH-I and EG-I mix had a ϕ of 0.015. These ϕ seemed to correspond to the ϕ sums of the two cellulases, indicating different "productive" adsorption regions for them. However, because the ϕ of *T. reesei* EG-I was ~15 times smaller than that of *T. reesei* CBH-I, our experiment could not conclusively differentiate a true ϕ sum that was ~15% higher than ϕ(CBH-I) from a ϕ(CBH-I) with ~10% experimental error.

Table III. Accessibility of 1:1 cellulase mix.

Cellulases	*Cellulose*	$[E]_0/[S]_t^s$	$[E]_0^s/[S]_t$	ϕ
T. r. CBH-I/*T. r.* EG-I	PASC	0.0023	0.0020	0.082
T. r. CBH-I/*H. i.* CBH-I	PASC	0.0017	0.0015	0.062
H. i. CBH-I/*H. i.* CBH-II	PASC	0.0014	0.0014	0.055
T. r. CBH-I/*T. r.* EG-I	Avicel	0.00038	0.00040	0.015
T. r. CBH-I/*H. i.* CBH-I	Avicel	0.00044	0.00021	0.013
H. i. CBH-I/*H. i.* CBH-II	Avicel	0.000053	0.000065	0.0023
Celluclast	PCS	0.00032	0.00030	0.012

For the definition of $[E]_0$, $[S]_t$ and ϕ, see Table II.

Comparison between ϕ and N_0

With 1.8-2.3 g/L or 5.3-6.6 mM cellobiose-equivalent PASC, our Langmuir adsorption measurement yielded an N_0 of 0.23, 0.43 and 0.07, as well as a K' of 0.01, 0.004 and 0.002 μM^{-1} for *T. reesei* CBH-I, *T. reesei* EG-I and *H. insolens* CBH-I, respectively. The N_0 was ~4, 100 and 1.6 times the ϕ for these enzymes. Thus, only ~25, 1 or 60% of adsorbed *T. reesei* CBH-I, *T. reesei* EG-I or *H. insolens* CBH-I was "productive" at the initial hydrolysis stage, leading to subsequent β-1,4 bond cleavage.

Assuming $\beta \approx 39$, N_0 of ~0.03 - 0.05 can be derived for *T. reesei* EG-I from published Langmuir isotherms on Avicel (*42*, *43*). The N_0 thus derived were

~30 times the ϕ for *T. reesei* EG-I, indicating that only ~3% of adsorbed enzyme was "productive" at the initial hydrolysis stage. N_0 of ~0.006 - 0.013 (along with *K*' of 0.1 - 0.3 μM^{-1}) have been reported for Langmuir adsorption of *T. reesei* CBH-I onto Avicel (*20, 44, 45*). Thus, *T. reesei* CBH-I's N_0 and ϕ were comparable, suggesting that the enzyme's initial adsorption onto Avicel was ~100% "productive".

It is known that for crystalline cellulose, *T. reesei* EG-I has a stronger overall adsorption than *T. reesei* CBH-I, although the former's activity is much weaker than the latter (*12*). The significant difference between the two cellulases' "productive" adsorption ("kinetic" accessibility) may contribute to the activity difference.

Being close functional (reducing-end specific, processive) and structural (family 7 cellulases) analogs, *T. reesei* and *H. insolens* CBH-I showed comparable activity on amorphous PASC; however, the former was significantly more active on microcrystalline Avicel than the latter (Table I). This difference might be related to the high "kinetic" Avicel accessibility of *T. reesei* CBH-I.

For bacterial microcrystalline cellulose, N_0 of ~0.03 - 0.06 and *K*' of 0.1 - 0.5 μM^{-1} have been reported with *T. reesei* CBH-I (*38, 45*). For α-cellulose, an N_0 of ~0.02 has been reported (*6*). For acid pretreated filter paper and wood, N_0 of ~0.003 - 0.019 have been reported, respectively (*25, 34, 37*). They are comparable to the ϕ of *T. reesei* CBH-I and Celluclast, in which *T. reesei* CBH-I accounted for ~60% of total protein, towards PCS (Tables II, III).

Structural basis for different ϕ among cellulases

All of the fungal CBM found so far are grouped to CBM family 1 due to their sequence similarity (*13, 15*). Figure 3 compares the CBM of the cellulases tested in this study.

```
T.r.CBH-I    HYGQCGGIGYSGPTVCASGTTCQVLNPYYSQCL
T.r.EG-I     HWGQCGGIGYSGCKTCTSGTTCQYSNDYYSQCL
H.i.CBH-I    RWQQCGGIGFTGPTQCEEPYTCTKLNDWYSQCL
H.i.CBH-II   TWGQCGGIGFNGPTCCQSGSTCVKQNDWYSQCL
              * *                     ** ** *
```

Figure 3. Sequence alignment of the CBM. "" denotes the key amino acids on the putative cellulose-binding surface (33).*

The CBM from *H. insolens* CBH-I is 58% homologous to that of *T. reesei* CBH-I. Between the two CBM, all the key amino acid residues, located at the flat face of the wedge-shaped CBM (*12*, *46*, *47*), are conserved, except two changes from tyrosine to tryptophan. On PASC, *H. insolens* CBH-I showed a ϕ (0.043) being 69% of that for *T. reesei* Cel7A (0.062). On Avicel, however, the ϕ for *H. insolens* CBH-I (0.0016) was only 11% of that for *T. reesei* CBH-I (0.014). Such accessibility difference towards highly crystalline and amorphous celluloses might be attributed to the tyrosine/tryptophan changes, or the variable amino acid residues adjacent to the key residues, which might affect CBM interaction with cellulose. Since protein adsorption onto carbohydrate involves hydrogen bonds, van der Waals interactions, polarization or hydrophobic interactions, variation of other amino acid residues might impact CBM adsorption as well (*12*, *16*, *30*, *37*, *38*, *48*). How the amino acid changes from *T. reesei* CBM to *H. insolens* CBM affect their cellulose adsorption needs to be further elucidated by techniques such as site-directed mutagenesis.

Summary

Being a heterogenous reaction, the hydrolysis of cellulose by cellulase requires adsorption of the enzyme onto the substrate. In general, insoluble cellulose has a non-uniform surface due to factors such as shape, crystallinity, lattice defect, and non-cellulosic moieties (*i.e.* lignin, hemicellulose). PASC is almost completely amorphous. Avicel, having a crystallinity of 0.47, is partly amorphous and partly crystalline (*49*). In the crystalline region of Avicel, different faces and corners exist. In PCS, lignin and hemicellulose entangle/overlap with cellulose. Thus a cellulose substrate can have different binding sites with intrinsically different accessibility for different cellulases (*14*, *36*). Such heterogeneity may force an enzymatic cellulose hydrolysis system to possess multiple cellulases (such as the two CBH and up to six EG secreted by *T. reesei*) with different affinity towards different parts of a cellulose substrate.

The four cellulases we selected have been characterized in terms of their structure, mechanism, synergy, cellulose specificity and Langmuir adsorption onto cellulose (*3*, *4*, *11*, *12*, *18-20*, *25*, *37*, *38*, *44*, *45*, *50-55*). Hydrolysis rate saturation upon either cellulase or cellulose concentration has been previously reported (*22*, *42*, *56*, *57*). In this study, we used the saturation kinetics to probe the "productive" adsorption, from four cellulases onto two celluloses, that leads to β-1,4 glycosidic bond cleavage. Our data showed that for a given cellulase, amorphous PASC seemed to be at least four times more "productively" accessible than microcrystalline Avicel. *T. reesei* CBH-I had ten times more accessibility than its close structural analog *H. insolens* CBH-I towards the "productive" surface sites on Avicel, although their accessibilities towards PASC were comparable. The "productive" adsorption of *T. reesei* EG-I on PASC or Avicel was one-order of magnitude lower than *T. reesei* CBH-I, although the former had 80% higher overall adsorption than the latter. It appeared that only ~1% of initial EG-I adsorption could lead to subsequent hydrolysis. Under our conditions, *T. reesei* CBH-I and *H. insolens* CBH-I seemed to have overlapping adsorption sites, while *H. insolens* CBH-I and CBH-II, as well as *T. reesei* CBH-I and EG-I seemed to have different adsorption sites.

To improve enzymatic cellulose catalysis is of significant interest for not only the fundamental understanding of the structure-function relationship of cellulases but also various industrial applications, including the development of biomass-derived sustainable, environment-friendly industries (such as fuel ethanol production from lignocellulosics). Various protein engineering efforts have been made in attempt to enhance cellulase adsorption/catalysis by either modifying native CBM or creating hybrids with desirable core and CBM derived from different sources (*12*, *16*, *30*, *37*, *38*, *48*). Our preliminary study pointed out the need and usefulness of evaluating the "productive" adsorption onto cellulose in analyzing and evaluating cellulases. Conventional Langmuir-type isotherms alone are insufficient in probing the effective cellulase concentration at the cellulose surface. Future works with other celluloses (such as bacterial cellulose), cellulases (from the same or different family or functional group), and engineered variants (by site-directed mutagenesis or core-CBM fusion) should be conducted to further validate/apply the methodology.

Acknowledgement

We thank Alexander Blinkovsky and James Langston of Novozymes Biotech for critical reading of the manuscript.

References

1. Walker, L. P.; Wilson, D. B. *Bioresource Technol.* **1991,** *36*, 3–14.
2. Mosier, N. S.; Hall, P.; Ladisch, C. M.; Ladisch, M. R. *Adv. Biochem. Engineer. Biotechnol.* **1999,** *65,* 23-40.
3. Teeri, T. T.; Koivula, A.; Divne, C.; Linder, M.; Wohlfahrt, G.; Ruohonen, L.; Lehtiö, J.; Reinikainen, T.; Srisodsuk, M.; Kleman-Leyer, K.; Kirk, T. K.; Jones, T. A. In *Carbohydrases from Trichoderma reesei and Other Microorganisms;* Claeyssens, M.; Nerinckx, W.; Piens, K. Eds.; The Royal Society of Chemistry: Cambridge, UK, 1998; pp 3-12.
4. Teeri, T. T.; Divne, C.; Jones, T. A.; Kleywegt, G.; Koivula, A.; Linder, M.; Ståhlberg, J.; von Ossowski, I.; Wohlfahrt, G.; Zou, J.-Y. In *Recent Advances in Carbohydrate Bioengineering;* Gilbert, H. J.; Davies, G. J.;

Henrissat, B.; Svensson, B. Eds.; The Royal Society of Chemistry: Cambridge, UK, 1999; pp 302-308.

5. Mansfield, S. D.; Mooney, C.; Saddler, J. N. *Biotechnol. Prog.* **1999,** *15,* 804-816.
6. Gan, Q.; Allen, S. J.; Taylor, G. *Process Biochem.* **2003,** *38,* 1003-1018.
7. Gilkes, N. R.; Henrissat, B.; Kilburn, D. G.; Miller, R. C.; Warren, R. A. J. *Microb. Rev.* **1991,** *55,* 303-315.
8. Béguin, P. ; Aubert, J.-P. *FEMS Microbiol. Rev.* **1994,** *13,* 23-58.
9. Schülein, M. *Biochim. Biophys. Acta* **2000,** *1543,* 239-252.
10. Väljamäe, P.; Sild, V.; Pettersson, G.; Johansson, G. *Eur. J. Biochem.* **1998,** *253,* 469-475.
11. Boisset, C.; Pétrequin, C.; Chanzy, H. ; Henrissat, B. ; Schülein, M. *Biotechnol. Bioeng.* **2001,** *72,* 339-345.
12. Srisodsuk, M. ; Lehtiö, J. ; Linder, M. ; Margolles-Clark, E. ; Reinikainen, T. ; Teeri, T. T. *J. Biotechnol.* **1997,** *57,* 49-57.
13. Bayer, E. A.; Chanzy, H.; Lamed, R.; Shoham, Y. *Curr. Opin. Struct. Biol.* **1998,** *8,* 548-557.
14. Mandels, M.; Reese, E. T. *J. Industrial Microbiol. Biotechnol.* **1999,** *22,* 225-240.
15. Wilson, D. B.; Irwin, D. C. Adv. *Biochem. Eng. Biotechnol.* **1999,** *65,* 1-21.
16. Boraston, A. B.; McLean, B. W.; Kormos, J. M.; Alam, M.; Gilkes, N. R.; Haynes, C. A.; Tomme, P.; Kilburn, D. G.; Warren, R. A. J. In *Recent Advances in Carbohydrate Bioengineering*; Gilbert, H. J.; Davies, G. J.; Henrissat, B.; Svensson, B. Eds.; The Royal Society of Chemistry: Cambridge, UK, 1999; pp 202-211.
17. Bourne, Y.; Henrissat, B. *Curr. Opinion Struct. Biol.* **2001,** *11,* 593-600.
18. Zou, J.-Y.; Kleywegt, G. J.; Ståhlberg, J.; Driguez, H.; Nerinckx, W.; Claeyssens, M.; Koivula, A.; Teeri, T. T.; Jones, T. A. *Structure* **1999,** *7,* 1035-1045.
19. Varrot, A.; Schülein, M.; Davies, G. J. *Biochemistry* **1999,** *28,* 8884-8891.
20. Ståhlberg, J.; Johansson, G.; Pettersson, G. *Bio/Technol.* **1991,** *9,* 286-290.
21. Peitersen, N.; Ross, E.W. *Biotechnol. Bioeng.* **1979,** *21,* 997–1017.
22. Holtzapple, M. T.; Caram, H. S.; Humphrey, A. E. *Biotechnol. Bioeng.* **1984,** *26,* 775-780.
23. Fan, L. T.; Gharpuray, M. M.; Lee, Y. H. *Biotechnology Monographs—Cellulose Hydrolysis;* Springer: New York, NY, 1987; Vol. 3.
24. Sattler, W.; Esterbauer, H.; Glatter, H.; Steiner, W. *Biotechnol. Bioeng.* **1989,** *33,* 1221-1234.
25. Nidetzky, B.; Steiner, W.; Hayn, M.; Claeyssens, M. *Biochem. J.* **1994,** *298,* 705-710.
26. Nidetzky, B.; Zachariae, W.; Gercken, G.; Hayn, M.; Steiner, W. *Enzyme Microb. Technol.* **1994,** *16,* 43-52.

27. Kurakake, M.; Shirasawa, T.; Ooshima, H.; Converse, A. O.; Kato, J. Appl. *Biochem. Biotechnol.* **1995,** *50,* 231-241.
28. Nutt, A.; Sild, V.; Pettersson, G.; Johansson, G. *Eur. J. Biochem* **1998,** *258,* 200-206.
29. Bolam, D. N.; Ciruela, A.; McQueen-Mason, S.; Simpsons, P.; Williamson, M. P.; Rixon, J. E.; Boraston, A.; Hazlewood, G. P.; Gilbert, H. J. *Biochem. J.* **1998,** *331,* 775-781.
30. Carrard, G.; Koivula, A.; Soderlund, H.; Beguin, P. *Proc. Natl. Acad. Sci. USA* **2000,** *97,* 10342-10347.
31. Esteghlalian, A. R.; Srivastava, V.; Gilkes, N. R.; Kilburn, D. G.; Warren, R. A. J.; Saddler, J. N. *Appl. Biochem. Biotechnol.* **2001,** *91-93,* 575-592.
32. Linder, M.; Teeri, T. T. *Proc. Natl. Acad. Sci. USA* **1995,** *93,* 12251-12255.
33. Carrard, G.; Linder, M. *Eur. J. Biochem.* **1999,** *262,* 637-643.
34. Ooshima, H.; Burns, D. S.; Converse, A. O. *Biotechnol. Bioeng.* **1990,** *36,* 446-452.
35. Eriksson, T.; Karlsson, J.; Tjerneld, F. *Appl. Biochem. Biotechnol.* **2002,** *101,* 41-60.
36. Klyosov, A. *Biochemistry* **1990,** *29,* 10577-10585.
37. Tomme, P.; Driver, D. P.; Amandoron, E. A.; Miller, R. C.; Antony, R.; Warren J.; Kilburn D. G. *J. Bacteriol.* **1995,** *177,* 4356-4363.
38. Reinikainen, T.; Teleman, O.; Teeri, T. T. *Proteins: Struct., Function Genet.* **1995,** *22,* 392-403.
39. Schülein, M. J. *Biotechnol.* **1997,** *57,* 71-81.
40. Suurnäkki, A.; Tenkanen, M.; Siika-Aho, M.; Niku-Paavola, M.-L.; Viikari, L.; Buchert, J. *Cellulose* **2000,** *7,* 189-209.
41. Lever, M. *Anal. Biochem.* **1972,** *47,* 273-279.
42. Woodward, J.; Hayes, M. K.; Lee, N. E. *Biotechnol.* **1988,** *6,* 301–304.
43. Beldman, G.; Voragen, A. G. J.; Rombouts, F. M.; Searle-van Leeuwen, M. F. *Biotechnol. Bioeng.* **1987,** *30,* 251-257.
44. Tomme, P.; Heriban, V.; Claeyssens, M. *Biotechnol. Lett.* **1990,** *12,* 525-530.
45. Bothwell, M. K.; Daughhetee, S. D.; Chaua, G. Y.; Wilson, D. B.; Walker, L. P. *Biosource Technol.* **1997,** *60,* 169-178.
46. Linder, M.; Mattinen, M.-L.; Kontteli, M.; Lindeberg, G.; Stahlberg, J.; Drakenberg, T.; Reinikainen, T.; Pettersson, G.; Annila A. *Protein Sci.* **1995,** *4,* 1056-1064.
47. Hoffrén, A.; Teeri, T. T.; Teleman, O. *Protein Eng.* **1995,** *8,* 443-450.
48. Kim, H.; Goto, M.; Jeong, H.-J.; Jung, K. H.; Kwon, I.; Furukawa, K. *Appl. Biochem. Biotechnol.* **1998,** *75,* 193-204.
49. Kulshreshtha, A. K.; Dweltz, N. E. *J. Polym. Sci., Polym. Phys. Ed.* **1973,** *11,* 487-497.

50. Shoemaker, S.; Schweickart, V.; Ladner, M.; Gelfand, D.; Kwok, S.; Myambo, K.; Innis, M. *Bio/Technol.* **1983,** *1,* 691-699.
51. Chanzy, H.; Henrissat, B. *FEBS Lett.* **1985,** *184,* 285-288.
52. Converse, A. O.; Optekar, J. D. *Biotechnol. Bioeng.* **1993,** *42,* 145–148.
53. Baker, J. O.; Adney, W. S.; Thomas, S. R.; Nieves, R. A.; Chou, Y.-C.; Vinzant, T. B.; Tucker, M. P.; Laymon, R. A.; Himmel, M. E. In *Enzymatic Degradation of Insoluble Carbohydrates;* Saddler, J. N.; Penner, M. H. Eds.; American Chemical Society: Washington, DC, 1995; pp 113-141.
54. Imai, T. ; Boisset, C. ; Samejima, M. ; Igarashi, K. ; Sugiyama, J. *FEBS Lett.* **1998,** *432,* 113-116.
55. Medve, J.; Ståhlberg, J.; Tjerneld, F. *Appl. Biochem. Biotechnol.* **1997,** *66,* 39-56.
56. Bernardez, T. D.; Lyford, K. A.; Lynd, L. R. *Appl. Microbiol. Biotechnol.* **1994,** *41,* 620-625.
57. Ortega, N.; Busto, M. D.; Perez-Mateos, M. *Internatl. Biodeterioration Biodegradation* **2001,** *47,* 7-14.

Chapter 10

Cloning and Heterologous Expression of the Gene Encoding a Family 7 Glycosyl Hydrolase from *Penicillium funiculosum*

Yat-Chen Chou, William S. Adney, Stephen R. Decker, John O. Baker, Glenna Kunkel, David W. Templeton, and Michael E. Himmel

National Bioenergy Center, National Renewable Energy Laboratory, 1617 Cole Boulevard, Golden, CO 80401

The gene encoding for a family 7 glycosyl hydrolase, designated *cel7a*, from the filamentous fungus *Penicillium funiculosum* was cloned from genomic DNA and expressed in *Aspergillus awamori* under the control of the *A. niger* glucoamylase promoter using the *E. coli/Aspergillus* shuttle vector pFE2. Characteristics of the recombinant enzyme were compared to those of the native enzyme and to the Cel7A enzyme from *Trichoderma reesei.* Unexpectedly, the glycosyl hydrolase isolated from the filamentous fungus *Penicillium funiculosum* broth demonstrated performance superior to that of *Trichoderma reesei* Cel7A when tested as part of a two-cellulase mixture with *Acidothermus cellulolyticus* E1 endoglucanase in the saccharification of pretreated corn stover. The mature enzyme consists of a 504 residue glycoprotein with a molecular weight of 52,436 Daltons. In pursuit of the goal of increasing the heterologous production of this enzyme, the endogenous signal peptide was replaced genetically with the signal peptides from *T. reesei* Cel7A, *P. funiculosum* Cel7A and *A. niger* GA.

Introduction

Penicillium funiculosum, a common Deuteromycete isolated from soil and environmental sources, has been shown to produce a number of biologically active compounds of potential interest to the biotechnology industry[1-6]. The organism's ability to secrete a range of cellulases, hemicellulases, and other polysaccharide degrading enzymes is of interest to the emerging biomass conversion industry[7-10].

Corn stover is a promising feedstock for the development of future bio-refineries and for fuel ethanol production because it is both readily available and less expensive than either corn kernels or sugarcane[11]. Nevertheless, substantial problems must be overcome before these feedstocks can be utilized effectively and economically for the fermentative production of ethanol or other byproducts. Other viable biomass feedstocks may include wood processing waste, municipal solid waste, and agricultural residues, such as rice hulls, straw, and grass. A typical lignocellulosic feedstock is comprised of approximately 40-50% cellulose, 20-30% hemicellulose, 15-25% lignin, and 10% of other components. Saccharification of cellulose releases cellobiose and ultimately glucose, which may be converted into ethanol or other products by fermentation[12]. The hemicellulose fraction of biomass is comprised primarily of pentose sugars, including xylose and arabinose.

Fuel ethanol, when derived from lignocellulosic biomass, is relatively expensive compared to analogous fuels from other sources[11]. Cellulase enzymes are a significant, if not debilitating cost factor in this process. Commercial cellulases are produced today primarily by fungi[13]. Many fungi have demonstrated the ability to produce cellulose degrading enzymes and a number of family 7 glycosyl hydrolases (GH 7) have been purified and characterized from these sources[14]. The best-characterized fungal cellulase system today is undoubtedly that produced by *Trichoderma reesei.*[15] Enzymes belonging to GH family 7 are known to be the principal components in commercial cellulase formulations and account for most of the actual bond cleavage in the saccharification of cellulose[16]. Structurally, the GH family 7 cellobiohydrolases are members of the Class of beta proteins, the Superfamily of concanavalin A-like lectins/glucanases, and the Family of glycosyl hydrolase family 7 catalytic core proteins[17]. The GH 7 family of enzymes is known to differ from one another by various insertions, deletions, and alterations in the catalytic domain and linker peptides and by the extent and types of post-translational modifications that occur. These enzymes also contain non-catalytic domains, the carbohydrate-binding modules (CBMs), which target the enzyme to the appropriate substrate. The CBMs for GH family 7 enzymes are generally more highly conserved than other examples[18]. For specific examples of known GH family 7 glycosyl hydrolases see http://afmb.cnrs-mrs.fr/CAZY/acc.html.

Signal peptides are known to control the entry of proteins into the secretory pathway in eukaryotic organisms[19], and are thought to influence the initiation of protein folding, thus making them potential targets for engineering more effective expression of proteins in recombinant systems. Structurally, signal peptides are comprised of three distinct regions consisting of a positively charged n-region, followed by a hydrophobic h-region and a neutral but polar c-region[19]. The residues at positions -3 and -1 relative to the cleavage site are generally small neutral amino acids. Signal peptides are known to vary in length and composition. When the gene for a protein of interest is fused with DNA fragments encoding a signal peptide, a cleavable linker peptide, and a portion of a protein native to the fungal host, higher levels of secreted protein can often be obtained, although the functionality and stability of the recombinant protein may be affected.

In this study, we report comparisons of the activities and thermal stabilities of purified and recombinant *P. funiculosum* and *T. reesei* Cel7A enzymes. All assays were conducted in the presence of the catalytic domain of *A. cellulolyticus* EI endoglucanase (E1cd) as expressed in *Escherichia coli* and used the dialysis saccharification assay (DSA) format reported by Baker and coworkers[20]. We also report the cloning of the *cel7A* gene from *P. funiculosum* in *E. coli* and exploration of several alternative signal peptide configurations.

Material and Methods

Bacterial strains, culture conditions, and plasmids

Strains and plasmids used in this study are listed below. *E. coli* DH5α was cultured in LB (Luria Broth) at 37°C, 250 rpm whereas *Aspergillus awamori* ATCC22342 and *Penicillium funiculosum* ATCC62998 were grown in CM media (per liter: glucose 10 g, yeast extract 5 g, tryptone 5 g, and 50 mL Clutterbuck's salts solution, pH 7.5) at 29°C, 250 rpm. Clutterbuck's salts contain 120 g/L Na_2NO_3, 10.4 g/L KCl, 10.4 g/L $MgSO_4.7H_2O$ and 30.4 g/L KH_2PO_4. For the selection and maintenance of pFE2 and its derivatives, antibiotics were supplemented to the media at the following concentrations: Zeocin (Z or Zeo), 170 μg/mL for the initial selection of *A. awamori* transformants, 300 μg/mL for the growth of transformants for cellulase production; Ampicillin (Amp), 100 μg/mL for *E. coli*. For agar media, Bacto Agar (Difco) was added to CM at 2% and LB at 1.5%.

List of plasmids used in this study.

pFE2: Amp^{r}, Zeo^{r} (PanLabs)

pFcbhI(Tr): *Amp^{r}, Zeo^{r}* ; *T. reesei cel7a* expressed under native *T. reesei cel7a* signal sequence.

pFcbhI(Pf)3726: *Amp^{r}, Zeo^{r}; P. funiculosum cel7a* expressed under native *P. funiculosum cel7a* signal sequence.

pFcbhI(Pf)m142: *Amp^{r}, Zeo^{r}*; *P. funiculosum cel7a* expressed under *A. niger* GA signal sequence

pFcbhI(Pf)A5-2: Amp^{r}, Zeo^{r}; P. funiculosum cel7a expressed under T. reesei cel7a signal sequence.

List of strains and descriptions

***E. coli* DH5α:** *F^{-} F80d lacZD M15 D(lacZYA-argF)U169 hsdR17($r_k^- m_k^+$) deoR recA1 endA1 supE44l- thi-1 gyrA96 relA1* (Invitrogen)

***Penicillium funiculosum* ATCC62998**

***Aspergillus awamori* ATCC22342**

3726: pFcbhI(Pf)3726 transformed *A. awamori, Zeo^{r}*; *P. funiculosum cel7a* expressed under native *P. funiculosum cel7a* signal sequence and integrated in ATCC22342 genome

m142: pFcbhI(Pf)m142 transformed *A. awamori, Zeo^{r}*; *P. funiculosum cel7a* expressed under *A. niger* GA signal sequence and integrated in ATCC22342 genome.

A5-2: pFcbhI(Pf)A5-2 transformed *A. awamori, Zeo^{r}*; *P. funiculosum cel7a* expressed under *T. reesei cel7a* signal sequence and integrated in ATCC22342 genome

PCR amplification and cloning

All the PCR reactions were carried out using *Pfu* polymerase (Stratagene, La Jolla, CA) in a Gene Amp PCR System 9700 or 2400 machine (PE Applied

Biosystems). The Cel7A gene *(cel7a)* of *P. funiculosum* ATCC62998 was cloned in the vector pFE2 (Figure 1) and expressed in *A. awamori* under three different signal sequences including that of 1) native *P. funiculosum cel7a*, 2) *A. niger* glucoamylase (GA), and 3) *Trichoderma reesei cel7a*. The glucoamylase promoter (P_{ga}) and *TrpC* terminator (both present in pFE2) were used in all three constructs. Using overlap extension PCR technique[21], P_{ga}, signal sequence and *P. funiculosum cel7a* mature protein coding sequence were synthesized and precisely fused to each other without the use of restriction enzymes, thus the introduction of exogenous nucleotides was prevented. PCR primers are listed in Table 1.

Figure 1. Plasmid map and features of vector pFE2

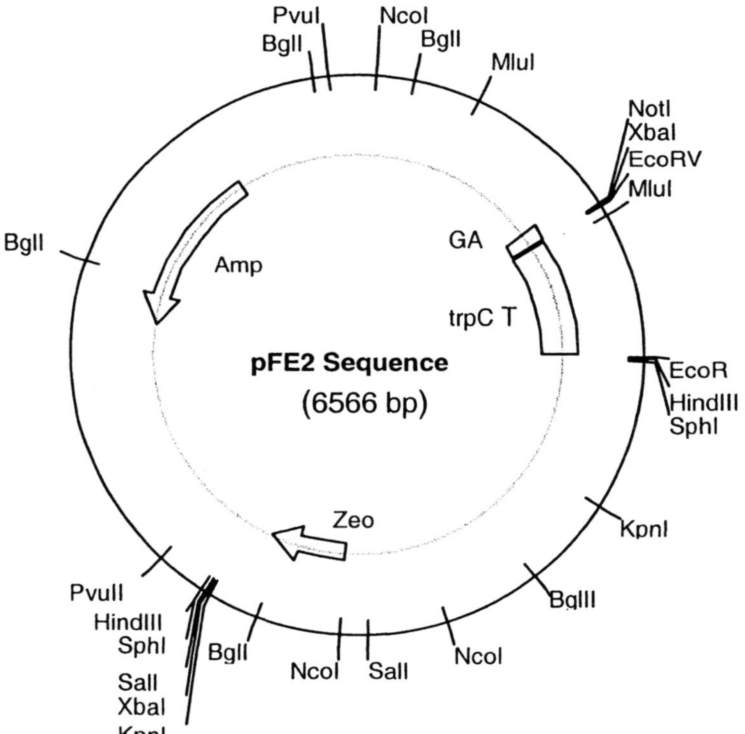

Three PCR reactions were carried out for each construct. In the first construct (Figure 2A and Table 1A), 824-bp P_{ga} region was amplified with

Primers 1 (forward) and 2 (reverse) using pFE2 as the template. Primer sequences were based on the sequence of pFE2 (Panlabs, private communication). Primer 2 comprised 34 nucleotides in which 19 nucleotides anneal to the 3' end of the P_{ga} region and 15 nucleotides (underlined) would anneal to the 5'end of the *P. funiculosum cel7a* starting at ATG codon. After PCR 1, P_{ga} fragment was synthesized with an addition of a short 15-bp DNA of *P. funiculosum cel7a*. The 1587-bp *P. funiculosum cel7a* was amplified with Primers 3 (forward) and 4 (reverse) using genomic DNA as template. Primer sequences were based on GenBank sequence of *P. funiculosum* xylanase/cellobiohydrolase gene (AJ312295). Primer 3 comprised 34 nucleotides in which 15 nucleotides (underlined) anneal to the 5'end of *P. funiculosum cel7a* starting at ATG codon and the 19 nucleotides anneal to the 3'end of the P_{ga}. After PCR 2, *P. funiculosum cel7a* was synthesized with an addition of a 19-bp DNA of 3' P_{ga}. The P_{ga} and *cel7a* PCR products were purified from agarose gels and used as templates in PCR3. Since Primers 2 and 3 are complementary, the last 34 bps of P_{ga} and the first 34 bps of the *cel7a* are identical. Therefore these overlapping ends denatured and annealed to each other and were extended by polymerase in PCR3. The result was a precise fusion between P_{ga} and *cel7a*. Using Primer 1 and 4, the full-length 2.4-kb DNA fragment containing P_{ga} + native *P. funiculosum cel7a* signal + mature protein coding sequences was further enriched. Note that a NotI site (boldfaced) was incorporated in Primer 4 for the cloning of PCR fragment in pFE2. The 2.4-kb PCR product was digested with BamHI (located in the P_{ga} region, see primer 1) and treated with Klenow Fragment of DNA Polymerase I followed by NotI digestion. This fragment was ligated with a 5.5-kb vector backbone of pFE2 (devoid of P_{ga} and GA signal sequences), which was digested with SacI and treated with T4 DNA polymerase followed by NotI digestion. Essentially, the P_{ga} and GA signal sequences in pFE2 were replaced with the 2.4-kb PCR fragment. The final construct was named pFcbhI(Pf)3726 (Figure 2A).

The second (pFcbhI(Pf)m142) and third (pFcbhI(Pf)A5-2) constructs were made according to the protocol used for the first construct. For pFcbhI(Pf)m142 (Figure 2B and Table 1B), a 916-bp DNA carrying P_{ga} and GA signal sequences was synthesized with Primers 1 and 2. A 1518-bp *cel7a* mature protein coding sequence was synthesized with Primers 3 and 4. Underlined sequences (Table 1B) were the starting of *cel7a* mature protein coding whereas the other part was the end of GA signal. As a result of PCR3, the precise fusion was between GA signal and *cel7a* mature protein coding sequence. GA signal sequence was based on the sequence of pFE2.

For pFcbhI(Pf)A5-2 (Figure 2C and Table 1C), a 874-bp DNA carrying P_{ga} and *T. reesei cel7a* signal sequences was synthesized with Primers 1 and 2. A 1518-bp *P. funiculosum cel7a* mature protein coding sequence was synthesized with Primers 3 and 4. Underlined sequences (Table 1C) were the starting of *P.*

Table 1. PCR primers for the construction of pFcbhI(Pf) plasmids

A. P_{ga} + native P. funiculosum *Cel7A signal + mature protein coding*

PCR	Primers	PCR product	Template
1	Primer 1 5'GTATACACGCTGGATCCGAACTCC Primer 2 5'GTTCAAGGCAGACATTGCTGAGGTGTAATGATGC	GA promoter (PGA)	pFE2
2	Primer 3 5'GCATCATTACACCTCAGCAATGTCTGCCTTGAAC Primer 4 5'ATAAGAAT**GCGGCCGC**CTACAAACATTGAGAGTAGTAAGGG	*P. funiculosum* Cel7A signal and mature protein coding sequence	*P. funiculosum* genomic DNA
3	Primer 1 and Primer 4	PGA + native *P. funiculosum* Cel7A signal + mature sequence	PCR 1 and 2

B. P_{ga} *+ GA signal +* P. funiculosum *Cel7A mature protein coding*

1	Primer 1 5'GTATACACGCTGGATCCGAACTCC Primer 2 5'ACCAATTTGCTGAGCGCGCTTGGAAATCAC	GA promoter and GA signal	pFE2
2	Primer 3 5'GTGATTTCCAAGCGCGCTCAGCAAATTGG Primer 4 5'ATAAGAAT**GCGGCCGC**CTACAAACATTGAGAGTAGTAAG GG	*P. funiculosum* Cel7A mature coding sequence	*P. funiculosum* genomic DNA
3	Primer 1 and Primer 4	PGA + GA signal + *P. funiculosum* Cel7A mature coding sequence	PCR 1 and 2

C. P_{ga} + T. reesei Cel7A signal + P. funiculosum Cel7A mature protein coding

1	Primer 1 5'GTATACACGCTGGATCCGAACTCC Primer 2 5'AATTTGCTGAGCAGCACGAGCTGTGG	GA promoter and reesei Cel7A signal	pFcbhI(Tr)
2	Primer 3 5'ACAGCTCGTGCTGCTCAGCAAATTGG Primer 4 5'ATAAGAAT**GCGGCCGC**CTACAAACATTGAGAGTAGTAAG GG	*P. funiculosum* Cel7A mature protein coding sequence	*P. funiculosum* genomic DNA
3	Primer 1 and Primer 4	PGA + *T. reesei* Cel7A signal + *P. funiculosum*Cel7A mature protein	PCR 1 and 2

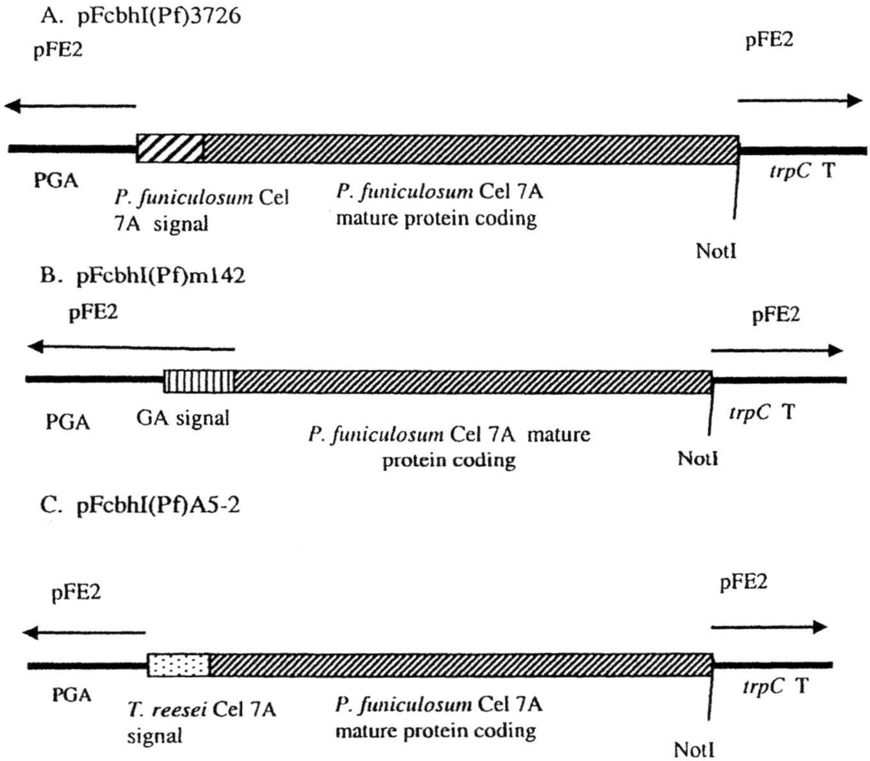

Figure 2. Cloning and expression of P. funiculosum *cel7a under three different signal sequences in pFE2.*

funiculosum cel7a mature protein coding whereas the other part was the end of *T. reesei cel7a* signal. The fusion point for this construct was between *T. reesei cel7a* signal and *P. funiculosum cel7a* mature protein coding sequences. *T. reesei cel7a* signal sequence was based on a previously constructed pFchbI(Tr)[21], in which P_{ga} and *T. reesei cel7a* were fused together precisely using overlap extension PCR.

Transformation

All the plasmids were initially constructed and cloned in the host *E. coli* DH5α. Transformation of *E. coli* was achieved by electroporation using a Bio-Rad Gene Pulser at 25 μFD, 200 ohms, and 16 KV/cm. SOC medium (BRL) was added to *E. coli* transformation mixture after electroporation and incubated at 37°C (250 rpm shaking) for one hour before plated on selection plates. After confirmed correct by either PCR screening or restriction digestions, the plasmids were purified from DH5α and used to transform *A. awamori*. A spheroplast transformation protocol[22] was used to transform *A. awamori*. Transformants were regenerated and selected on CM plates with 170 μg/mL Zeocin. Zeocin resistant colonies were streaked on CM plates containing 300 μg/mL Zeocin (CMZ300) to obtain the single isolates.

Activity assay for screening for Cel7A-positive transformants

To screen for the CelA-positive clones, an endpoint assay using p-nitrophenyl β-D-lactopyranoside as substrate was used to measure cellobiohydrolase activity in the supernatants of the grown transformants. Several single isolates from the transformation of each construct were inoculated in 40-mL CMZ300 in 250-mL baffled Erlenmeyer flasks and incubated at 29°C with shaking (250 rpm) for 3-4 days. Ten microliters of supernatant was assayed in a 50-μl reaction containing 40 mM sodium acetate buffer (pH 5) and 2.2 mM pNPL in a microtitre plate and incubated at 50°C for 3 hours. For the control, pNPL was substituted with buffer. The reactions were stopped by the addition of 150-μL 1 M sodium carbonate. The intensity of the yellow color was measured at 405 nm in a Spectra Max190 microtiter plate reader. The transformants with the highest pNPL activity were subjected to 1 L growth and were used for protein purification and characterization.

Cel7A enzyme production and activity change during the growth of recombinant *A. awamori*

To monitor the enzyme production and activity levels over the course of growth, supernatant from the culture broths of recombinant *A. awamori* were assayed using pNPL as a substrate. The reaction rates were measured by following the release of p-nitrophenol from timed fermentation samples. Frozen spores were inoculated in 100-mL CMZ300 in 500-mL baffled Erlenmeyer flasks and grown for 3-4 days and transferred into 1-L CMZ300 in baffled 2.8-L flasks. Culture broths were sampled routinely and supernatant generated by filtering through Miracloth followed by glass fiber filters. The assay was set up as followed: In a 1-mL total reaction mix containing 0.5 mM pNPL and 50 mM sodium acetate (pH 5), different volumes of supernatant were included. The reactions were incubated in a 50°C water bath. Fifty microliters of reaction was taken to a microtitre plate well at different time points and mixed with 150-uL 1 M sodium carbonate to stop the reaction. The yellow color was measured at 405 nm in a Spectra Max190 microtiter plate reader. The activity was determined against a standard curve generated by known concentrations of p-nitrophenol. Specific activity was calculated based on the total protein concentration in the supernatant. Total protein concentrations in supernatants were determined using the micro BCA method (Pierce Biochemicals). For purified native Cel7 solutions the protein concentrations were determined by absorbance at 280 nm using the extinction coefficient and molecular weight calculated for Cel7A by the ProtParam tool on the ExPASy website (http://www.expasy.ch/tools/protparam.html).

Protein purification

The purification of rCel7a from fungal broths was initiated by removal of the fungal hyphae from a 1 L shake flask growth by passing the broth first through miracloth and then through glass fiber filters. The broth was then concentrated to a final volume of approximately 20 mL using an Amicon stirred cell concentrator with a PM-10 membrane at 4°C. Buffer exchange was accomplished using a HiPrep 26/10 desalting column (Amersham Biosciences) equilibrated with 20 mM Bis-Tris pH 5.8 buffer. The column void volume fraction was collected and applied to a HiPrep 16/10 DEAE FF column (Amersham Biosciences) equilibrated with 50 mM Bis-Tris, pH 5.8 buffer with a flow rate of 10 mL/min. After the sample was loaded and the column washed extensively with equilibration buffer the bound fraction was eluted with a linear gradient of 0 to 0.5 M NaCl in the same equilibration buffer. The fractions containing activity on p-nitrophenol β-D lactopyranoside were pooled and concentrated to a final volume of 10 mL using Amicon stirred cell concentrators and PM-10 cutoff filters (10,000 kDa nominal molecular weight cutoff). The concentrated enzyme was then subjected to size exclusion chromatography using a Superose 12 Prep grade 35/600 column in 20 mM acetate, 100 mM NaCl pH 5 buffer.

The purification of the native Cel7 from *P. funiculosum* broth was started by filtration through glass fiber filters followed by concentration of the broth by using a 500 mL Amicon stirred cell concentrator with PM-10 cutoff membranes. The broth was first concentrated to a volume of approximately 50 mL followed by extensive diafiltration by successively diluting and concentrating the broth with 50 mM Bis-Tris pH 5.8 buffer to a point at which the conductivity of the solution was less than 2 mS/cm. The sample was then applied to a 6 mL Resource Q column (Amersham Biosciences) equilibrated with 50 mM Bis-Tris, pH 5.8 buffer with a flow rate of 6 mL/min. After the sample was loaded, the column washed extensively with equilibration buffer and the bound fraction was eluted with a linear gradient of 0 to 1.0 M NaCl in the same equilibration buffer. The fractions containing activity on p-nitrophenol β-D-lactopyranoside were pooled and concentrated to a final volume of 10 mL using Amicon stirred cell concentrators and PM-10 cutoff filters (10,000 kDa nominal molecular weight cutoff). The enzyme was further purified and the buffer exchanged by means of size exclusion chromatography using a HiLoad 26/60 Superdex 200 column (Amersham Biosciences) in 20 mM acetate, 100 mM NaCl, pH 5.0 buffer. The purity was confirmed as a single band using a NuPage 4-12 % Bis-Tris gradient gel and MOPS-SDS buffer (Invitrogen) using the manufactures recommended conditions.

Diafiltration Saccharification Assay (DSA) and Protein Stability Measurements

Diafiltration saccharification assays were carried out as described earlier[20]. Overall protein stability was measured by differential scanning microcalorimetry using a Microcal model VP-DSC calorimeter (Microcal, Inc., Northampton, MA), with data analysis by means of Origin for DSC software (Microcal).

Thermograms were collected for samples containing 50 μg/mL protein at pH 5.0 in 20 mM sodium acetate with 100 mM NaCl. Calorimeter scan rate was 60°C/h.

SDS-PAGE analysis of crude supernatant proteins

The supernatants from the culture broths were analyzed on PAGE. Electrophoresis was conducted using NuPAGE pre-cast gels (Invitrogen) according to the provided protocol.

Results and Discussion

Cellobiohydrolase enzymes from glycosyl hydrolase family 7 are considered key components in the suite of enzymes produced by fungi for cellulose hydrolysis. Historically, the CBH I (Cel7A) enzyme from *T. reesei* was among the first member of this family to be studied in detail23. Very early on, researchers began considering the mechanisms by which these enzymes decrystallized cellulose and presented a cellodextrin chain to the active site tunnel. The three distinct peptide domains of this enzyme have been considered in detail24-26, yet a mechanistic model describing the action of family 7 cellobiohydrolases on cellulose does not yet exist. Today, some 25 members of family 7 can be identified as cellobiohydrolases. We have chosen to study the cloning, expression, and characterization of a Cel7A enzyme isolated from P. funiculosum broth and cloned from genomic DNA. Among the Cel7A enzymes tested in our lab, the *P. funiculosum* Cel7A has the highest activity on pretreated biomass using a binary enzyme assay and diafiltration reactor. In this study, a new family 7 glycosyl hydrolase (cel7A) was successfully cloned and expressed in *A. awamori* using the native *P. funiculosum* signal sequence and the Aspergillus GA promoter.

Expression of *P. funiculosum cel7a* in pFE2 vector

Penicillium funiculosum cel7a mature protein coding sequence was expressed under three different signal sequences in pFE2 vector (Figure 1). These three signal sequences included the native *P. funiculosum cel7a*, *T. reesei cel7a* and *A. niger* glucoamylase signal sequences. P_{ga} and *trpCT* were used in all three constructs to promote and terminate the transcription. Overlap extension PCR was used for the construction of plasmids so that the only difference between these constructs was the signal sequences and no potential frameshift or extra amino acids were introduced due to the use of restriction enzymes. Three constructs (pFcbhI(Pf)3726, pFcbhI(Pf)m142 and pFcbhI(Pf)A5-2) were made and used to transform *A. awamori* expression host. An SDS-PAGE gel comparing these purified proteins is shown in Figure 3.

After transformation of *A. awamori* with each construct, at least 12 transformants were analyzed using pNPL assay. The assay was an endpoint assay and was used for the initial screening for the best Cel7A producers. In general, 3726 showed the highest activity among the three constructs. Although the majority of the transformants from the three constructs were active in pNPL assay, we observed various levels of activity within the same construct (data not shown). One possible cause might be the integration loci in the genome which could influence the expression level of the gene. It was indicated that the integration of pFE2-based plasmids is a random event in the genome of *A. awamori* [22]. Two to four best producers were further inoculated in 1-L CMZ300 for protein purification and further analysis.

Characterization by differential scanning calorimetry

Comparisons of *P. funiculosum* Cel7A to *T. reesei* Cel7A using differential scanning calorimetry showed a 1.8°C difference in thermal transition temperature (Figure 4). The higher thermal transition temperature for the *P. funiculosum* possibly indicates either differences in the peptide secondary structures or differences in the extent and nature of glycosylation of the two enzymes (or possibly a combination of both).. Additional analysis of the *N*-linked glycosylation motifs show similar *N*-linked glycosylation sites on the surface of the protein based on homology models generated by SWISS- MODEL (not shown).

Cel7A activity at different growth stages

We attempted to monitor the enzyme production and the stability of the enzyme for the three constructs over the course of growth. Strains 3726, m142 and A5-2 were cultured in 1-L CMZ300 for 10 days. The supernatants were analyzed using pNPL assay and specific activity was determined. A profile of specific activity is shown in Figure 5.

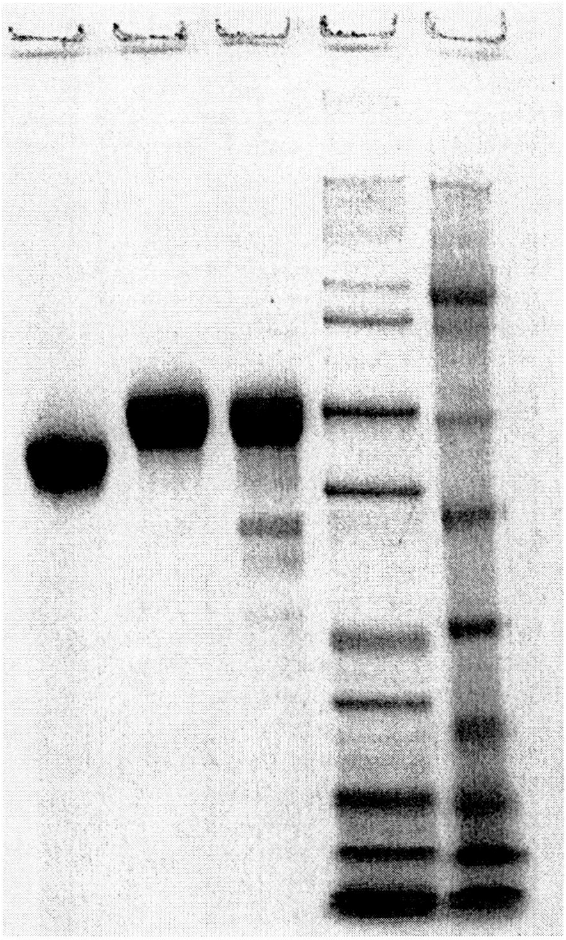

Figure 3. SDS-PAGE analysis of purified P. funiculosum *Cel7A enzymes. Lane 1- native* P. funiculosum *Cel7A purified from culture broth of ATCC62998, Lane 2-purified Cel7A enzyme from 3726 native signal, Lane 3 – purified Cel7A enzyme from A5-2* T. reesei *signal, Lane 4 – Mark 12 MW standard, Lane 5 – See Blue Plus2 MW standard*

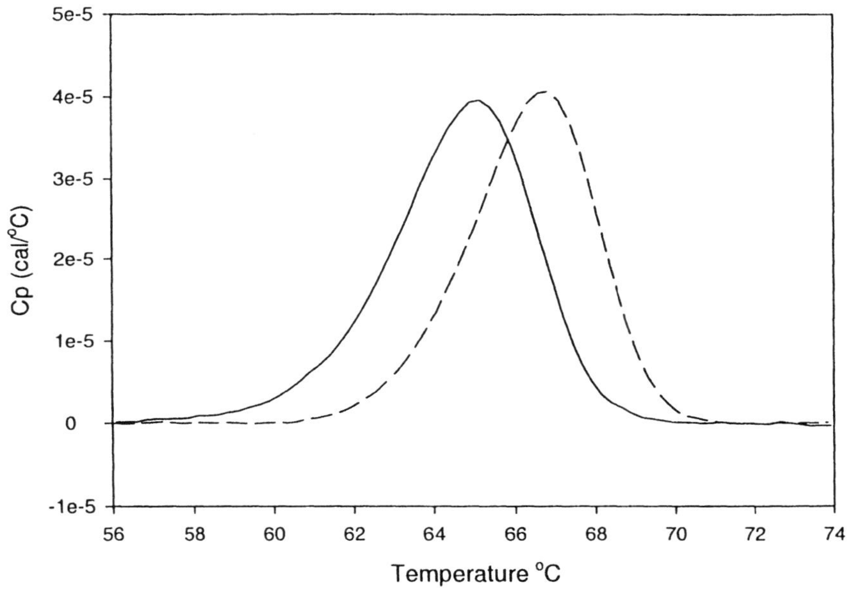

Figure 4. Co-plotted DSC thermograms of A. awamori *expressed* P. funiculosum *Cel7A (dashed line) to native* T. reesei *expressed Cel7A (solid line) by differential scanning calorimetry. Solid line shows the thermal denaturation of* T. reesei *Cel7A with a maximum at 65°C, the dashed line shows the thermal denaturation of* A. awamori *expressed* P. funiculosum *with a maximum at 66.8°C. Conditions: pH 5.0 in 20 mM acetate, 100 mM NaCl, 50 µg/mL protein, scanned at 60oC/h.*

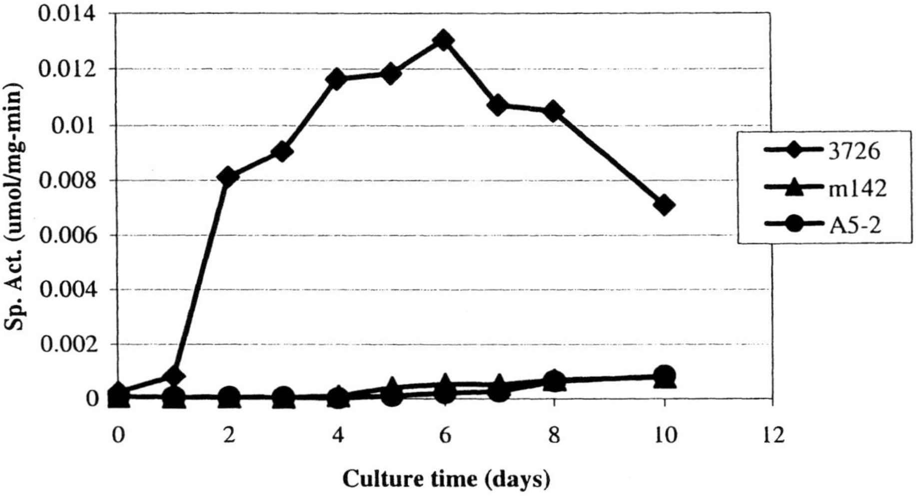

Figure 5. A. awamori *expressed* P. funiculosum *Cel7A activity on pNPL*

Strain 3726 (native *P. funiculosum* signal) produced the enzyme at the highest level during the 10-day growth period. After the 6th day, the specific activity started to drop. According to the SDS-PAGE analysis (Figure 6), the loss of activity was not caused by the protein degradation. The protein band pattern remained the same throughout the growth. The higher MW predominant band co-migrated with the purified Cel7A from strain 3726. The other bands including the lower MW predominant band were host background proteins (data not shown). No obvious degradation products were detected using this method. The low activity of strains m142 and A5-2 was confirmed by the lack of the Cel7A band. Similar amounts of the total proteins were loaded on the gel yet no bands were present in the m142 and A5-2 lanes to compensate the loss of Cel7A band indicating very small proteins may be present in the supernatants and not visible on the gel using Coomassie stain method. Whether those are degradation products of Cel7A requires further investigation. In this study, we intended to focus on the expression level of Cel7A by different signal sequences. It was shown clearly that native *P. funiculosum* expressed the *cel7a* at the highest level.

DNA sequence comparison between *P. funiculosum* ATCC62998 and GenBank sequence (AJ312295)

A GenBank sequence of *P. funiculosum* xylanase/cellobiohydrolase (AJ312295) was based on for the design of primer sequences for the amplification of *cel7a* using the genomic DNA of *P. funiculosum* ATCC62998 as template. We found significant differences in the nucleotide sequences between the GenBank sequence and the PCR fragments sequences we amplified from ATCC62998. The differences were also confirmed by sequencing the PCR products of *cel7a* from 3926 genomic DNA. The deduced amino acid is shown in Table 2. Thirty-seven amino acids were changed due to the difference in the DNA sequences. None of them changed the length of the protein. According to the GenBank interpretation of the sequence, there are 24 amino acids in the signal peptide. The cleavage site is between glycine and alanine leaving the N-terminus of mature protein with an alanine followed by a glutamine. This interpretation is different from the observation of the fact that most of the Cel7A enzymes' cleavage site is between a small neutral amino acid (such as alanine) and a glutamine and that mature proteins start with a glutamine. PCR primers for pFcbhI(Pf)m142 and A5-2 were designed according to the GenBank interpretation of the sequence. Whether the cleavage occurred before or after the alanine and the effect on the enzyme activity need to be further investigated.

GenBank cleavage site

↓

MSALNSFNMYKSALILGSLLATAG A QQIGTY---------

↑

general Cel7A cleavage site

There are three potential *N*-glycosylation sites in ATCC62998 whereas there are four in the GenBank sequence AJ312295. It was unclear whether the GenBank sequence was a xylanase or a cellobiohydrolase gene and there was no strain name designated for the *P. funiculosum* either. Therefore the sequence discrepancy might be the result of different strains or different genes.

Table 2. Amino acid sequence of *P. funiculosum* Cel7A from ATCC62998. Mature protein sequence shown in capital letters with the signal sequence designated by small script.

```
msalnsfnmyksalilgsllatagaQQIGTYTAETHPSLSWSTCKSGGSCTTNSGAITLDANWRWVH
GVNTSTNCYTGNTWNSAICDTDASCAQDCALDGADYSGTYGITTSGNSLRLNFVTGSNVGSRTYLMA
DNTHYQIFDLLNQEFTFTVDVSHLPCGLNGALYFVTMDADGGVSKYPNNKAGAQYGVGYCDSQCPRD
LKFIAGQANVEGWTPSANNANTGIGNHGACCAELDIWEANSISEALTPHPCDTPGLSVCTTDACGGT
YSSDRYAGTCDPDGCDFNPYRLGVTDFYGSGKTVDTTKPFTVVTQFVTNDGTSTGSLSEIRRYYVQN
GVVIPQPSSKISGISGNVINSDYCAAEISTFGGTASFSKHGGLTNMAAGMEAGMVLVMSLWDDYAVN
MLWLDSTYPTNATGTPGAARGTCATTSGDPKTVESQSGSSYVTFSDIRVGPFNSTFSGGSSTGGSTT
TTASRTTTTSASSTSTSSTSTGTGVAGHWGQCGGQGWTGPTTCVSGTTCTVVNPYYSQCL
```

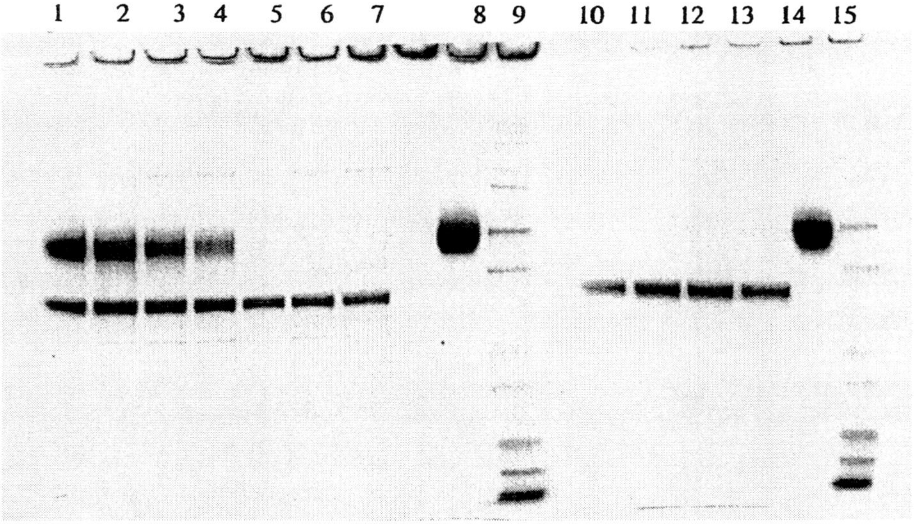

Figure 6. SDS-PAGE analysis for the supernatant of <u>A. awamori</u> *expressed* <u>P. funiculosum</u> *Cel7A. Lane 1- 3726 day 3, Lane 2- 3726 day 6, Lane3- 3726 day 8, Lane 4- 3726 day 10, Lane 5- m142 day 3, Lane 6- m142 day 6, Lane 7- m142 day 8, Lane 8- purified 3726 rCel7A, Lane9- Mark 12 MW Standard, Lane 10- A5-2 day 3, Lane 11- A5-2 day 6, Lane 12- A5-2 day 8, Lane 13- A5-2 day 10, Lane 14- purified 3726 rCel7A, Lane15- Mark 12 MW Standard.*

Activity of *P. funiculosum* Cel7A on pre-treated corn stover

In this study, a new family 7 glycosyl hydrolase (cel7A) was successfully cloned and expressed in *A. awamori* using the native *P. funiculosum* signal sequence and the *Aspergillus* GA promoter. The Diafiltration Saccharification Assay (DSA) was used to compare native and recombinant enzymes produced during this study. The recombinant Cel7A from *P. funiculosum* produced in *A. awamori* showed activity similar to that of the native enzyme when tested on pretreated corn stover in a binary mixture with the EI endoglucanase from *A. cellulolyticus* (see Figure 7). *P. funiculosum* Cel7A was cloned using PCR and expressed under three different signal sequences in *A. awamori*. With one-half loading in DSA, *A. awamori* expressed rCel7A performed significantly better than full loading of *T. reesei* Cel7A on pretreated corn stover (see Figure 8).

Our results on pretreated biomass using a binary enzyme system indicated that this non-*Trichoderma* Cel7A enzyme was superior to the *Trichoderma* enzyme, at least under conditions such that the entire process of saccharification must be carried out by a single endoglucanase and a single cellobiohydrolase. It is possible that the *P. funiculosum* Cel7A may perform differently when directly substituted for *T. reesei* Cel7A in a commercial enzyme mix containing larger numbers of at least partially overlapping endoglucanase and cellobiohydrolase activities. Furthermore, , even though our testing was performed in a buffer-swept membrane reactor designed to reduce end product inhibition and permit more direct testing of cellulases under conditions where this variable has been removed or reduced, neither the instantaneous concentration of cellobiose nor that of glucose is reduced to zero in the DSA reaction volume as set up in these experiments.. It therefore remains possible that some portion of the differences noted between the *P. funiculosum* and *T. reesei* Cel7 enzymes may arise from differences in susceptibility to end-product inhibition. . We continue to test this important family 7 enzyme under use-relevant conditions and on industrially important feedstocks.

Acknowledgment

This work was funded by the U.S. Department of Energy Office of the Biomass Programs.

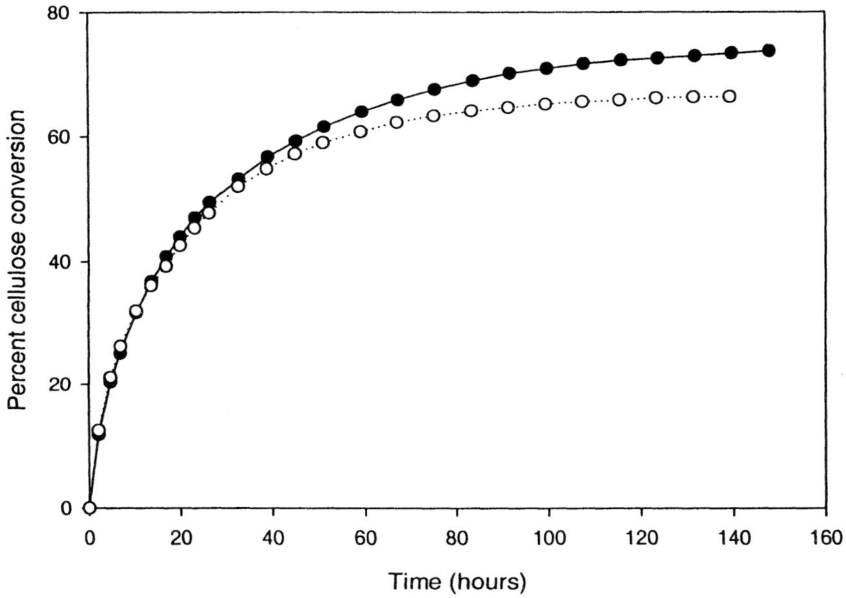

Figure 7. Hydrolysis of corn stover by binary mixtures of P. funiculosum *Cel7A and* A. cellulolyticus *E1. Open circles demonstrate the hydrolysis kinetics of the recombinant enzyme produced in* A. awamori *and closed circles demonstrate the kinetics of Cel7A enzyme purified from the native* P. funiculosum *host. Each cellobiohydrolase was loaded at 27.8 mg/g cellulose in the presence of* A. cellulolyticus *rE1cd loaded at 1.13 mg/g cellulose (95:5 molar ratio of cellobiohydrolase to endoglucanase). DSA at pH 5.0 (20 mM acetate/sodium acetate), 38°C, with pretreated corn stover loaded at 4.4% (w/v) total solids for 2.5% (w/v) cellulose.*

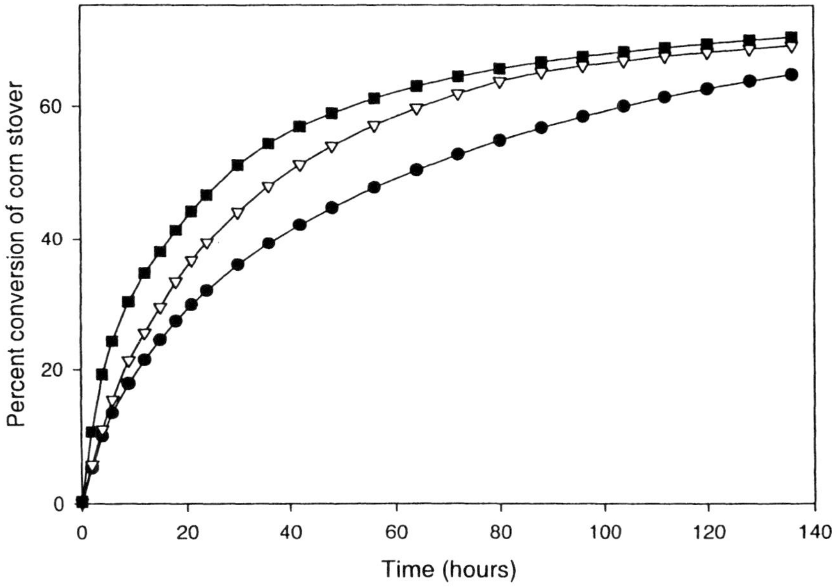

Figure 8. Comparison of the activities of P. funiculosum *Cel7A (filled square and open triangle) and* T. reesei *Cel7A (filled circle) on pre-treated corn stover, when used in combination with* A. cellulolyticus *rE1cd as the endoglucanase. The* P. funiculosum *Cel7A enzyme was loaded at two different levels, 27.8 mg/g cellulose (square) and at ½ loading, or 13.9 mg/g cellulose (triangle); the* T. reesei *Cel7A enzyme was loaded at 27.8 mg/g cellulose. The loading of* A. cellulolyticus *rE1cd was held constant at 1.13 mg/g cellulose for all curves shown.*

References

(1) Furniss, C. S. M.; Belshaw, N. J.; Alcocer, M. J. C.; Williamson, G.; Elliott, G. O.; Gebruers, K.; Haigh, N. P.; Fish, N. M.; Kroon, P. A. *Biochimica et Biophysica Acta* **2002**, *1598*, 24-29.

(2) Abdel-Naby, M. A.; Ismail, A. M. S.; Abdel-Fattah, A. M.; Abdel-Fattah, A. F. *Process biochemistry* **1999**, *34*, 391-398.

(3) Rao, M.; Mishra, C. *Applied microbiology and biotechnology* **1989**, *30*, 130-134.

(4) Linko, Y. Y.; Li, G. X.; Linko, P. *Annals of the New York Academy of Sciences* **1987**, *501*, 381-385.

(5) Kantham, L.; Vartak, H. G.; Jagannathan, V. *Biotechnology and bioengineering* **1985**, *27*, 781-785.

(6) Ando, K.; Suzuki, S.; Saeki, T.; Tamura, G.; Arima, K. *J ANTIBIOT* **1969**, *22*, 189-194.

(7) Kroon, P. A.; Williamson, G.; Fish, N. M.; Archer, D. B.; Belshaw, N. J. *Eur J Biochem* **2000**, *267*, 6740-6752.

(8) Van Wyk, J. P. H. *Bioresource technology* **1998**, *63*, 275-277.

(9) Wyk, J. P. H. v.; Leogale, P. B. *Biotechnology letters* **2001**, *23*, 1849-1852.

(10) Elshafei, A. M.; Vega, J. L.; Klasson, K. T.; Clausen, E. C.; Gaddy, J. L. *Bioresource technology* **1991**, *35*, 73-80.

(11) Sheehan, J.; Himmel, M. *Biotechnol Prog* **1999**, *15*, 817-827.

(12) Sheehan, J.; Himmel, M. *Biotechnol Progress* **1999**, *15*, 817-827.

(13) Himmel, M. E.; Ruth, M. F.; Wyman, C. E. *Current Opinion In Biotechnology* **1999**, *10*, 358-364.

(14) Henrissat, B.; Teeri, T. T.; Warren, R. A. J. *FEBS Lett.* **1998**, *425*, 352-354.

(15) Teeri, T. T.; Koivula, A.; Reinikainen, T.; Ruohonen, L.; Srisodsuk, M.; Divne, C.; Rouvinen, J.; Szardenings, M.; Jones, T. A. *Abstr. Pap. Am. Chem. Soc.* **1994**, *207*, 21-AGFD.

(16) Teeri, T. T. *Trends Biotechnol.* **1997**, *15*, 160-167.

(17) Coutinho, P. M.; Deleury, E.; Davies, G. J.; Henrissat, B. *J. Mol. Biol.* **2003**, *328*, 307-317.

(18) Henrissat, B.; Coutinho, P. M. In *Hyperthermophilic Enzymes, Pt A*, 2001; Vol. 330, pp 183-201.

(19) Nielsen, H.; Engelbrecht, J.; Brunak, S.; von Heijne, G. *Protein Eng.* **1997**, *10*, 1-6.

(20) Baker, J. O.; Vinzant, T. B.; Ehrman, C. I.; Adney, W. S.; Himmel, M. E. *Appl. Biochem. Biotechnol* **1997**, *63-65*, 585-595.

(21) Ho, S. N.; Hunt, H. D.; Horton, R. M.; Pullen, J. K.; Pease, L. R. *Gene* **1989**, *77*, 51-59.

(22) Adney, W. S.; Chou, Y.-C.; Decker, S. R.; Ding, S.-Y.; Baker, J. O.; Kunkel, G.; Vinzant, T. B.; Himmel, M. E., Eds. *Heterologous Expression of Trichoderma reesei 1,4-beta-D-glucan cellobiohydrolase (Cel 7A)*; American Chemical Society: Washington, DC, 2003; Vol. 855.

(23) Reinikainen, T. R.; Ruohonen, L.; Koivula, A.; Srisodsuk, M.; Jones, A.; Knowles, J. K. C.; Claeyssens, M.; Teeri, T. T. *Abstr. Pap. Am. Chem. Soc.* **1991**, *202*, 206-BIOT.

(24) Srisodsuk, M.; Reinikainen, T.; Penttila, M.; Teeri, T. T. *J. Biol. Chem.* **1993**, *268*, 20756-20761.

(25) Linder, M.; Teeri, T. T. *J. Biotechnol.* **1997**, *57*, 15-28.

(26) Divne, C.; Stahlberg, J.; Teeri, T. T.; Jones, T. A. *J. Mol. Biol.* **1998**, *275*, 309-325.

Chapter 11

Effect of Rational Mutagenesis of Selected Cohesin Residues on the High-Affinity Cohesin–Dockerin Interaction

Tal Handelsman[1], Adva Mechaly[2], Yoav Barak[2], David Nakar[2], Raphael Lamed[3], Yuval Shoham[1], and Edward A. Bayer[2,*]

[1]Department of Food Engineering and Biotechnology, Technion - Israel Institute of Technology, Haifa, Israel
[2]Department of Biological Chemistry, The Weizmann Institute of Science, Rehovot, Israel
[3]Department of Molecular Microbiology and Biotechnology, Tel-Aviv University, Ramat Aviv, Israel

The high-affinity cohesin-dockerin interaction that dictates cellulosome assembly was probed by site-directed mutagenesis of suspected recognition residues on the cohesin domain. The involvement of two loops that flank the 8,3,6,5 β sheet of a cohesin domain of the cellulosomal scaffoldin from *Clostridium thermocellum* was examined by their partial replacement with analogous portions of a cohesin domain from *Clostridium cellulolyticum.* Similarly, several amino acids located on this β sheet were replaced with matching residues on the counter species cohesin. The dockerin-binding specificity of the cohesin was not altered by those mutations. However, the binding affinity of certain mutants significantly decreased, thus corroborating the notion that the dockerin-binding site stretches along this particular face of the cohesin molecule and that some of the mutated surface residues play a significant role in the binding process.

Introduction

Many cellulolytic microorganisms produce extracellular multi-enzyme complexes that efficiently degrade cellulose (*1-3*). The bacterial cellulosomes consist of a suprastructural organizing component, termed scaffoldin, and various enzymatic components attached to it. The scaffoldin – a large multimodular, multifunctional protein subunit consists of a cellulose-binding module (CBM) and a series of highly homologous domains, termed cohesins. The hydrolytic subunits, including various cellulases and hemicellulases, bind to the scaffoldin via specialized inherent domains, termed dockerins, which interact with the cohesins on the scaffoldin. The scaffoldins of some bacterial species, e.g. *Clostridium thermocellum*, also contain a divergent type of dockerin that serves to attach the complex to the bacterial cell surface via complementary divergent cohesins of cell-surface anchoring scaffoldins.

The scaffoldin of the *C. thermocellum* cellulosome has been fully sequenced, as have other scaffoldins from a growing number of bacterial species. Dockerin domains of many cellulosomal enzymes have also been identified and sequenced. However, several important issues concerning cellulosome assemblage are still unclear. Previous experiments have shown that the cohesins of both *C. thermocellum* and *C. cellulolyticum* recognize nearly all the dockerins on their own enzymatic subunits (*4, 5*), which would seem to be consistent with the intrinsic similarity of their sequences. It thus seems that, within a given species, the incorporation of the cellulosomal enzymes into a complex is a non-selective process, and the catalytic subunits would appear to interact randomly along the scaffoldin. However, despite the high sequence homology, interaction among cohesins and dockerins between the two *Clostridia* was found to be species specific, i.e., the cohesin domain from one bacterium interacts with the dockerins from its own cellulosome but fails to recognize dockerins from the other species (*5*). Moreover, the C-terminal dockerin of the scaffoldin fails to recognize any of its own complement of "type-I" cohesins; rather, it binds specifically to a divergent type of cohesin (termed "type-II" cohesin) harbored by another type of scaffoldin located on the cell surface (*6*).

The crystal structures of three cohesins, two from *C. thermocellum* and one from *C. cellulolyticum*, have been reported (*7-9*). The cohesin domains form a nine-stranded β sandwich with a jelly-roll topology. The β sandwich results from the association of a four-stranded antiparallel β sheet, and a five-stranded mixed β sheet, stabilized by a hydrophobic core. The two β sheets are composed of strands 8,3,6,5 and strands 9,1,2,7,4 respectively. In addition, the three-dimensional structure of one dockerin (from the family-48 *C. thermocellum* cellulosomal enzyme, CelS) was solved by NMR spectroscopy (*10*). The dockerin structure consists of two Ca^{2+}-binding loop-helix motifs connected by a linker.

The objective of one direction of ongoing research in our group has been to define the molecular basis behind the highly specific, tenacious cohesin-dockerin interaction (*11*). During the course of our work, we have employed site-directed mutagenesis in attempts to assess the dockerin residues that may contribute to the observed interspecies specificity of cohesin binding.

Using a bioinformatics-based approach, we first compared sequences of various enzyme-borne dockerin domains from *C. thermocellum* and *C. cellulolyticum*. In this approach, we sought conserved residues among the intraspecies sequences versus divergent residues among the interspecies sequences. Using this strategy, we identified a series of positions in the dockerin sequences whose residues could serve as potential recognition codes in the selective interaction with the cohesins (*5, 11*). The primary recognition residues included two tandem pairs at positions 10 and 11 of the duplicated dockerin sequence that were conserved in one species but different in the other species. Thus, instead of conventional alanine scanning, the objective of this approach was to convert the specificity from one species to another by switching the divergent amino acids at the suspected positions. When comparing two very similar proteins, we consider this approach superior to alanine scanning, as it reduces the probability that the mutation would result in non-specific destabilization of the complex due to unfolding of the mutated protein. Indeed, mutagenesis of positions 10 and 11 of the duplicated dockerin sequences suggested their involvement in the cohesin-dockerin interaction (*12, 13*), but the results implied that additional residues would also play a role in the selective interaction with the cohesins. In this context, subsequent mutagenic studies (*14*) have indicated that additional dockerin amino acid residues (positions 17, 18 and 22 of the duplicated region) may serve as secondary recognition residues that would also be critical to the exclusivity of the interaction.

In a complementary manner, we have also endeavored to identify cohesin residues involved in the cohesin-dockerin interaction, with an aim to eventually control the specificity of interaction between different species by rational mutagenesis. In this case, however, the identification of cohesin recognition residues has proved more challenging, due to the larger size of the cohesin versus the dockerin (~140 versus ~70 amino acid residues) and the numerous (>30) divergent residues identified. Subsequent discovery (*15, 16*) of a single functionally similar cohesin but of divergent sequence (*olpA*) provided a means with which to reduce the number of suspected recognition residues, and a 6-residue cluster of putative cohesin recognition codes were thus implicated (*11*).

In the present work, we have systematically mutated the latter residues and have selected additional residues on the cohesin surface as targets for site-directed mutagenesis. Using an approach similar to that described above for the mutagenesis of the dockerin domain, we have tried to convert the specificity of the cohesin of one species to match that of the other (i.e., from C. *thermocellum* to C. *cellulolyticum*), by replacing the selected residues to the parallel residues of the cohesin from the counter species. According to the results of the present communication, we were, however, unable to convert the specificity of the interaction. Nevertheless, mutagenesis of certain combinations of cohesin surface residues resulted in a dramatic decrease in the binding capacity of the mutated cohesin to the dockerin domain. These results provide further insight into the cohesin interface responsible for dockerin recognition.

Experimental Procedures

Protein Constructs and Cloning

The protein construct containing the cohesin from *C. thermocellum,* consists of cohesin-2 and a cellulose-binding domain from CipA. The construct containing the cohesin from *C. cellulolyticum,* comprises a cellulose-binding domain, a hydrophilic domain, and cohesin-1 from CipC. Details of the cloning of cohesin constructs (termed Coh2CBD_t and miniCipC_c; where t and c denote domains derived from *C. thermocellum* and *C. cellulolyticum,* respectively) were described elsewhere (*4, 17*).

The dockerin constructs comprise the dockerin domain of CelS from *C. thermocellum* (*18*) or the dockerin domain of CelA from *C. cellulolyticum* (*19*), fused downstream of the noncellulosomal family-10 xylanase T-6 from *Geobacillus stearothermophilus* (*20, 21*). These constructs, termed XynDocS_t and XynDocA_c, respectively, were cloned using a specially designed cassette produced for this purpose. The cassette consisted of the gene for *G. stearothermophilus* xylanase T-6 with a His-tag and a *BspHI* site at the 5'-terminus and a *KpnI* site at the 3'-terminus. This construct was ligated at the *KpnI* site with the PCR product of a *C. thermocellum* CelS (Cel48A) dockerin (containing a 5'-terminal *KpnI* site and a 3'-terminal *BamHI* site) and inserted into the pET9d vector at the *NcoI* and *BamHI* sites. This plasmid allows replacement of the CelS dockerin with any other desired dockerin by digesting with *KpnI* and *BamHI,* and the resultant expressed product constitutes a His-tagged xylanase T-6 fusion-protein bearing a dockerin at the C-terminus.

Site-Directed Mutagenesis

Generally, mutated cohesins were produced in a sequential manner, in which one mutant served as a template for the subsequent one, as detailed in Table I.

For the preparation of mutants 18-20, site-directed mutagenesis was performed using overlap-extension PCR (*22, 23*). The primer for N122GD123T was used with the T7 promoter, and the primer for A89L was used with the T7 terminator. The resulting two overlapping fragments were then used to generate the entire mutated cohesin fragment, which was subsequently ligated into pET9d by restriction via *BamHI* and *NcoI.* The resultant mutant was then used as a template for further mutations in the same general manner, using S63N plus the T7 terminator and E81DG84 plus the T7 promoter. Mutant proteins 23-50 were prepared using the QuikChange site-directed mutagenesis kit (Stratagene, La Jolla, CA). Mutagenic primers are given in Table I.

The presence of the mutations was first verified (when possible) by restriction analysis utilizing the unique enzyme restriction sites designed within the primers, followed by sequencing of the coding region.

Expression and Purification of Proteins

All proteins were expressed in *Escherichia coli* BL21(DE3) grown overnight in Terrific Broth medium (*24*). For the production of miniCipC_c cloned in pET22b, the medium was supplemented with 0.1 mg/ml ampicillin, and protein expression was induced with 0.4 mM isopropyl β-D-thiogalactoside. For all other proteins cloned in pET9d, medium was supplemented with 25 μg/ml kanamycin, and growth was carried out without induction. Following growth, cells were harvested, resuspended in TrisNC buffer (50 mM Tris, 100 mM NaCl, 2 mM $CaCl_2$, 0.02% sodium azide, pH 7.5), disrupted by two passages through a French® press (Spectronic Instruments, Inc., Rochester, NY, USA), and centrifuged for the production of clear crude protein extracts, that were further purified as described below.

Xylanase-containing constructs (XynDocS_t and XynDocA_c) were purified by gel filtration using a Superdex 200 26/60 column, AKTA explorer (Pharmacia), running at 2.5 ml/min with TrisNC buffer. CBD-containing constructs (Coh2CBD_t and Coh1_c) were purified by affinity chromatography on cellulose. Microcrystalline cellulose (Avicel Type PH-101 FMC) was added to the crude protein extract, originating from a 1-liter cell culture. The ratio of cellulose to cells was 0.7 g per 1 unit OD_{600}. The resultant suspension was stirred for 1 h. After centrifugation the pellet was washed twice with TrisNC buffer, containing 0.1 M NaCl and twice with TrisNC buffer, containing 1 M NaCl. The CBD-containing proteins were eluted from the cellulosic matrix with 11 ml of 1% (v/v) triethylamine. The eluent fractions were neutralized with TrisNC buffer. Purity of all proteins was estimated by SDS–PAGE, and protein concentration was estimated by Bradford (*25*).

Table I. Primers Used for Site-Directed Mutagenesis of Cohesin-2 from *C. thermocellum*

Mutated Cohesin[a]	*Template*	*Primer*[b]	*Restriction Site Change*[c]
mut 18	wild-type	T7 promoter and CCTTCTGTTCTACCA**G**G**GT**A**CC**ATCTGCAAATCC	*KpnI*
		T7 terminator and GGAACAGGAGCGTAT**CTG**AT**C**ACTAAAGACGGAG	*BclI*
mut 20	mut 18	T7 promoter and AGATACGCTCCTGTGCTGTC**G**TCCGCAAACAGG	– *BpuAI*
		T7 terminator and CCCGAATCCTAC**G**AAGA**A**CTT**C**GATACTGC	*XmnI*
mut 23	wild-type	CCTACCAAGAGCTTT**AG**TACTGCAATATATCC	*ScaI*
mut 25	mut 23	CCAAAGGAAT**C**GC**G**AACTGC**A**ACTTTGTGTTC	*NruI*
mut 27	wild-type	GAGCTTTGATACTGCAATACC**C**GA**TC**GAAAGATA GTATTCCTGTTTGCGG	*PvuI*
mut 28	mut 27	GCTTTGATACTGCAATA**T**CC**A**AT**G**GAA**C**GATAGT ATTCCTG	– *PvuI*
mut 30	mut 25	GACAGAAAGATAATA**TC**ATTCCTGTTTGCGG	none
mut 32	mut 30	GAGCTTTAGTACTGCAATACC**C**GA**TC**GAAAGATA TCATTCCTGTTTGCGG	*PvuI*
mut 33	mut 32	GCTTTAGTACTGCAATA**T**CC**A**AT**G**GAA**C**GATATC ATTCCTG	– *PvuI*
mut 39	mut 30	GCTCCGGGCTATATTACTTT**TAAA**GAAGTAGGTG GATTTGC	*DraI*
mut 42	mut 30	GAATCGCGA**C**CTGCAACTTTGTGTTC	*BspMI*
mut 46	mut 42	TTCCTGTTT**CTC**GAAGACAGCGGAACAG	none
mut 49	mut 39	CGCGAACTGCAACTTT**TAC**TTCAGATATGATCCG	none
mut 50	mut 49	GGAATCGCGA**C**CTGCAACTTTTACTTC	*BspMI*

[a] Mutated amino acids relevant for the results are detailed in Tables II and III.

[b] Bold characters indicate mutated bases. Restriction sites are underlined.

[c] (–) indicates the deletion of an existing restriction site at the position of mutation.

Noncompetitive Enzyme-Linked Interaction Assay (ELIA)

Microtiter plates (MaxiSorp-immunoplates, NUNC A/S, Roskilde, Denmark) were coated overnight at 23°C with the cohesin test samples (200 µl/well, 270 nM of miniCipC_c, wild-type or mutated Coh2CBD_t). The plates were blocked for 2.5 h with blocking solution (300 µl/well 3% (w/v) bovine serum albumin in TrisNC buffer) and washed three times with TrisNC buffer (300 µl/well). The cohesin-dockerin interaction was initiated upon addition of dockerin samples (200 µl/well, 94 nM of XynDocA_c or XynDocS_t), and the plates were incubated for 2.5 h. After five washes, the bound dockerins were detected by means of the fused-xylanase activity: substrate solution (240 µl/well 2.9 mM *p*-nitrophenyl β-D-cellobioside) was added followed by incubation at 60°C. Optical density was determined at 420 nm on a VERSAmax microplate reader (Molecular Devices Corp., Sunnyvale CA).

Competitive Enzyme-Linked Interaction Assay (cELIA)

Microtiter plates were coated overnight with wild-type *C. thermocellum* cohesin samples (200 µl/well, 270 nM of Coh2CBD_t). Plates were blocked for 2.5 h with the above-described blocking solution, and washed three times with TrisNC buffer. The cohesin-dockerin interaction was carried out by the addition of 100 µl of the desired competitor cohesin sample (i.e., wild-type or mutant Coh2CBD_t at various concentrations, up to a maximum of 1.3 µM), immediately followed by the addition of dockerin solution (100 µl of XynDocS_t to final concentration of 47 nM). Dilutions of the competitor cohesins were carried out in TrisNC buffer containing BSA, to maintain a constant protein concentration. After incubation for 2.5 h, the wells were washed five times, and the amount of dockerin bound to the coating cohesin was detected by means of the fused-xylanase activity, as described above.

Results were expressed as percentage of binding, derived from the mean optical density values of five repetitions for each competitor concentration (percentage of binding = 100 x optical density of the test competitor concentration / optical density without competitor).

Results and Discussion

Rational Mutagenesis

In previous work, we employed a combined bioinformatics-based approach with site-directed mutagenesis, in order to identify and corroborate the involvement of dockerin residues in the recognition of the cohesin domain. This approach revealed a group of 8 positions on the dockerin domain suspected to be critical to the observed species-specific selectivity of the cohesin-dockerin interaction. In like fashion, we attempted to employ a similar approach for identification of recognition residues on the surface of the molecular counterpart – the cohesin.

For this purpose, multiple sequence alignment of cohesins was combined with the superposition of the three-dimensional structures of cohesins from *C. thermocellum* (*7*) and *C. cellulolyticum* (*9*). Using this strategy, we identified a series of putative residues, which may be important for dockerin recognition. Generally, we were guided by the following assumptions: (i) The cohesin domain has a compact structure, devoid of any obvious binding pocket or cleft. Hence, its binding to the dockerin domain most likely results from interactions between exposed surface residues. (ii) The overall dockerin-binding site of the cohesin domain would in general be considerably conserved, while amino acid residues specifically responsible for dockerin recognition would be conserved *within* a given species but would show a high degree of dissimilarity *between* the two divergent species. (iii) The inner hydrophobic/aromatic core of the two species of cohesins is very similar. Thus, the fold and stability of the mutated molecules would likely be retained.

In order to challenge our prediction, the suspected residues were subjected to site-directed mutagenesis with the intent of altering the binding specificity. The general idea was to mutate the cohesin molecule of one species (i.e., in this case, cohesin-2 of the *C. thermocellum* scaffoldin) and redirect its specificity to recognize and/or prefer the enzyme-borne dockerin of the competing species (i.e., *C. cellulolyticum*). Using this approach, mutated cohesins, which include different combinations of positions N32, D34, V36, D65, V76, A80, D114 and the loop of YPDRKI at positions 69-74 were prepared. These positions are all located in the highly conserved 8,3,6,5 β sheet. Several mutations were also designed to test the possibility that a recognition strip lies at the "crown" of the cohesin domain (*11*).

The Cohesin-Dockerin Interaction

The mutated cohesins were overexpressed, purified, and their binding specificities towards the dockerins from *C. thermocellum* and *C. cellulolyticum* were tested, using native cohesins from the two bacterial species as controls. Qualitative binding analysis (ELIA) showed that all the tested mutant cohesins, listed in Table II, still retained their innate binding specificity. That is, the mutated cohesins (all prepared from a *C. thermocellum* cohesin template) continued to recognize the *C. thermocellum* test dockerin to varying degrees, but they all failed to bind to the test dockerin from *C. cellulolyticum*. This indicates that the residues we selected did not comprise the entire complement of residues essential for recognition. Due to the nature of the ELIA assay employed, these results can only be viewed as qualitative and did not reflect a quantitative assessment of cohesin-dockerin interaction

Table II. Sites and Mutations of the Cohesin-2 from *C. thermocellum*

Mutated Cohesins	*Position[a] and Mutation*
mut 20	A89L+N122G+D123T+S63N+E81D+ΔG84
mut 28	YPDRKI(69-74)SNGT
mut 30	D65S+D34N+V76S
mut 33	D65S+ D34N+V76S+YPDRKI(69-74)SNGT
mut 46	D65S+D34N+V76S+N32T+A80L
mut 49	D65S+D34N+V76S+D114K+V36Y
mut 50	D65S+D34N+V76S+D114K+V36Y+N32T

[a] Numbering refers to the positions of the indicated residues in the cohesin crystal structure (7).

Consequently, in order to determine whether any of the combination of mutations did in fact include binding-site residues, the binding affinities of the mutated cohesins were also evaluated in a quantitative manner. The results are presented in Figure 1. In competitive enzyme-linked interaction assay, cELIA, the native cohesin was used as a standard to coat microtiter plates. The immobilized cohesin was then allowed to interact with an enzyme-linked dockerin solution together with a competitor cohesin (native or mutated) in the solution phase. The measured enzymatic activity, expressed as the percentage of activity detected in the absence of the soluble competitor, reflects the amount of dockerin bound to the immobilized cohesin standard. The IC_{50}, i.e., the

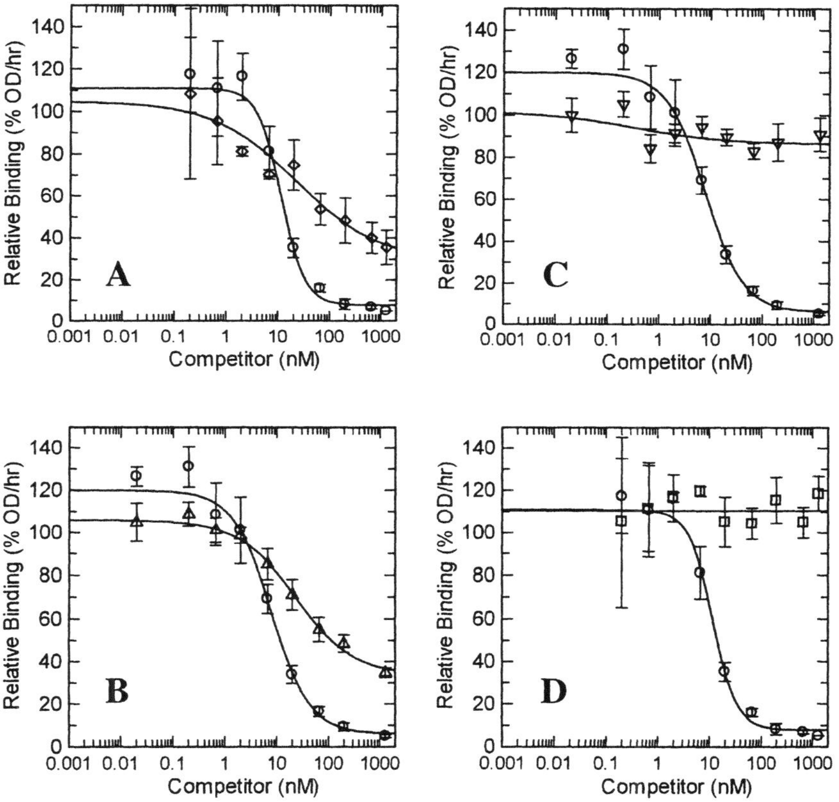

Figure 1. Competitive-ELIA of native and mutated cohesins from C. thermocellum interacting with a native dockerin of the same species. A solution containing the native cohesin was used to coat microtiter plates, and the immobilized cohesin was allowed to interact with an enzyme-linked dockerin in the presence of native (○) or mutated competitor cohesin [A, mut 20 (◇); B, mut 28 (△); C, mut 30 (▽); D, mut 33 (□)]. The observed enzymatic activity reflects the amount of dockerin bound to the immobilized cohesin.

Table III. IC_{50} Values of Mutated Cohesin-2 from *C. thermocellum*

Cohesin	*Position[a] and Mutations*	*IC_{50} (nM)*
wild type	none	$1x10^1$
mut 20	A89L+N122G+D123T+S63N+E81D+ΔG84	$2x10^2$
mut 28	YPDRKI(69-74)SNGT	$5x10^2$
mut 30	D65S+D34N+V76S	>> $2x10^3$
mut 33	D65S+ D34N+V76S+YPDRKI(69-74)SNGT	>> $2x10^3$

[a] Numbering refers to the positions of the indicated residues in the cohesin crystal structure (*7*).

concentration of competitor that resulted in 50% inhibition of the binding of the positive control, was determined for each mutated protein (Table III).

The high IC_{50} values obtained for the mutated cohesins reflect the decrease in their affinities towards the dockerin, compared to that of the native cohesin. This suggests that one or more of the mutated residues are located at or near the binding site. It is clearly shown that the maximum tested concentration of mutant 33 (Figure 1-D) failed to compete at all with the wild-type cohesin. Moreover, the apparent affinity of mutant 30 (Figure 1-C), in which only three residues were replaced, is at least two orders of magnitude lower than that of the native molecule, implying that D34, D65 and/or V76 may play a significant role in the dockerin binding, either individually or in any combination thereof.

As mentioned above, in mutant 20 the mutated residues (S63, E81, G84, A89, N122, D123) are located at the crown of the cohesin mainly facing the 8,3,6,5 β sheet. These combined residues were suspected earlier, mainly on the basis of bioinformatics analysis, to be critical to dockerin binding. The results of the cELIA revealed a decrease in the observed affinity of one order of magnitude, thus indicating a definite but minor involvement in dockerin binding. Mutant 28 was designed so as to replace a loop at the other side of this β sheet, i.e., the loop that connects β strands 5 and 6 – namely, replacement of residues Y69, P70, D71, R72, K73, I74 of *C. thermocellum* cohesin-2 with the corresponding loop (2-residues shorter) of cohesin-1 from *C. cellulolyticum*. Replacement of this loop resulted in an even greater impairment of binding.

The combined mutated residues and loops are positioned at or near the 8,3,6,5 β sheet of the cohesin domain. The observed reduction in affinity of the mutants would thus support the notion that the dockerin-binding site stretches along this particular face of the cohesin molecule. This assessment is in line with the strong negative potential and the conserved region associated with the 8,3,6,5 face (*7*), as well as previous mutagenesis experiments (*26*).

Concluding Remarks

In the work presented here, site-directed mutagenesis was applied in order to gain insight into the cohesin residues involved in the interaction with the dockerin. Whereas the commonly used alanine scanning is characterized by an innate uncertainty as to whether mutation-generated inhibition of activity would reflect a selective involvement (or lack thereof) of the mutated residue or a more general structural impairment of the molecule, a different approach is promoted in this communication. The alternative approach involves the swapping of divergent residues that occur at homologous positions in highly similar proteins that exhibit dissimilar specificities. The concept behind this approach is that successful and exclusive conversion of the specificity from one species to the other would intrinsically preclude any ambiguity regarding improper folding or non-specific interaction. This approach has been successfully applied in our laboratory for interspecies conversion of specificity of the dockerin domain.

We have hitherto been unable to convert the cohesin specificity, indicating that we have yet to identify the relevant recognition residues. However, as shown in this chapter, the mutation of relatively few residues (i.e., comparatively conservative mutations of only three residues in mut 30) was sufficient to effect complete or near-complete inhibition of binding. These results imply that one or more of the three mutated residues may be involved in the recognition and/or binding process. Future studies in our laboratory will thus concentrate on the individual contribution of the implicated residues. In addition, alternative strategies, e.g., gene swapping and/or directed evolution, will also be adopted to provide meaningful insight into the recognition and binding properties of the cohesin-dockerin interaction.

Acknowledgements

This research was supported by the Israel Science Foundation (Grant Nos. 394/03, 771/01 and 446/01) and by the United States-Israel Binational Science Foundation (BSF), Jerusalem, Israel. Additional support was provided by the Otto Meyerhof Center for Biotechnology, established by the Minerva Foundation, (Munich, Germany).

References

1. Bayer, E. A.; Chanzy, H.; Lamed, R.; Shoham, Y. Curr. Opin. Struct. Biol. **1998**, 8, 548-557.
2. Shoham, Y.; Lamed, R.; Bayer, E. A. Trends Microbiol. **1999**, 7, 275-281.

3. Schwarz, W. H. Appl. Microbiol. Biotechnol. **2001**, 56, 634-649.
4. Yaron, S.; Morag, E.; Bayer, E. A.; Lamed, R.; Shoham, Y. FEBS Lett. **1995**, 360, 121-124.
5. Pagès, S.; Bélaïch, A.; Bélaïch, J.-P.; Morag, E.; Lamed, R.; Shoham, Y.; Bayer, E. A. Proteins **1997**, 29, 517-527.
6. Leibovitz, E.; Béguin, P. J. Bacteriol. **1996**, 178, 3077-.3084
7. Shimon, L. J. W.; Bayer, E. A.; Morag, E.; Lamed, R.; Yaron, S.; Shoham, Y.; Frolow, F. Structure **1997**, 5, 381-390.
8. Tavares, G. A.; Béguin, P.; Alzari, P. M. J. Mol. Biol. **1997**, 273, 701-713.
9. Spinelli, S.; Fierobe, H.-P.; Bélaïch, A.; Bélaïch, J.-P.; Henrissat, B.; Cambillau, C. J. Mol. Biol. **2000**, 304, 189-200.
10. Lytle, B. L.; Volkman, B. F.; Westler, W. M.; Heckman, M. P.; Wu, J. H. D. J. Mol. Biol. **2001**, 307, 745-753.
11. Bayer, E. A.; Morag, E.; Lamed, R.; Yaron, S.; Shoham, Y. In *Carbohydrases from Trichoderma reesei and other microorganisms*; M. Claeyssens, W. Nerinckx, and K. Piens, Ed; The Royal Society of Chemistry: London, 1998; 39-67.
12. Mechaly, A.; Yaron, S.; Lamed, R.; Fierobe, H.-P.; Bélaïch, A.; Bélaïch, J.-P.; Shoham ,Y.; Bayer, E. A. Proteins **2000**, 39, 170-177.
13. Schaeffer, F.; Matuschek, M.; Guglielmi, G.; Miras, I.; Alzari, P. M.; Béguin, P. Biochemistry **2002**, 41, 2106-2114.
14. Mechaly, A.; Fierobe, H.-P.; Bélaïch, A.; Bélaïch, J.-P.; Lamed, R.; Shoham, Y.; Bayer, E. A. J. Biol. Chem. **2001**, 276, 9883-9888.
15. Fujino, T.; Béguin, P.; Aubert, J.-P. J. Bacteriol. **1993**, 175, 1891-1899.
16. Salamitou, S.; Lemaire, M.; Fujino, T.; Ohayon, H.; Gounon, P.; Béguin, P.; Aubert, J.-P. J. Bacteriol. **1994**, 176, 2828-2.834
17. Pagès, S.; Bélaïch, A.; Tardif, C.; Reverbel-Leroy, C.; Gaudin, C.; Bélaïch, J.-P. J. Bacteriol. **1996**, 178, 2279-2286.
18. Wang, W. K.; Kruus, K.; Wu, J. H. D. J. Bacteriol. **1993**, 175, 1293-1302.
19. Fierobe, H.-P.; Gaudin, C.; Bélaïch, A.; Loutfi, M.; Faure, F.; Bagnara, C.; Baty, D.; Bélaïch, J.-P. J. Bacteriol. **1991**, 173, 7956-7962.
20. Mechaly, A.; Teplitsky, A.; Belakhov, V.; Baasov, T.; Shoham, G.; Shoham, Y. J. Biotechnol. **2000**, 78, 83-6.
21. Lapidot, A.; Mechaly, A.; Shoham, Y. J. Biotechnol. **1996**, 51, 259-264.
22. Ho, S. N.; Hunt, H. D.; Horton, R. M.; Pullen, J. K.; Pease, L. R. Gene **1989**, 77, 51-56.
23. Meza, R.; Nufiez-Valdez, M.-E.; Sanchez, J.; Bravo, A. FEMS Microbiol. Lett. **1996**, 145, 333-339.
24. Sambrook, J.; Russell, D. W. *Molecular cloning: a laboratory manual*; Cold Spring Harbor Laboratories: Cold Spring Harbor, New York, 2001 ;
25. Bradford, M. Anal. Biochem. **1976**, 72, 248-254.
26. Miras, I.; Schaeffer, F.; Béguin, P.; Alzari, P. M. Biochemistry **2002**, 41, 2115-2119.

Chapter 12

Structure–Reactivity Studies of *Trichoderma reesei* Cellobiohydrolase Cel7A

Kathleen Piens[1,3], Jerry Ståhlberg[2], Wim Nerinckx[1], Tuula T. Teeri[3], and Marc Claeyssens[1]

[1]Department of Biochemistry, Physiology and Microbiology, Ghent University, K. L. Ledeganckstraat 35, B–9000 Ghent, Belgium
[2]Department of Molecular Biology, Swedish University of Agricultural Sciences, Box 590, SE–75124 Uppsala, Sweden
[3]Department of Biotechnology, Royal Institute of Technology, Albanova University Center, SE–106 91 Stockholm, Sweden

The major cellulase secreted by the filamentous fungus *Trichoderma reesei* is cellobiohydrolase Cel7A. It hydrolyzes the β-1,4-linkage of a cellulose chain by means of a double-displacement mechanism. A series of substituted aryl β-lactosides was used for reactivity studies. The Hammett plot shows that the formation of the glycosyl-enzyme intermediate is the rate-limiting step, also with highly activated substrates. A catalytic triad of carboxylate residues Glu212-Asp214-Glu217 is directly involved in the retaining mechanism. As previously revealed by protein X-ray crystallography, their specific function is presently confirmed by a detailed kinetic analysis of the mutants E212Q, D214N and E217Q.

Introduction

Filamentous fungi secrete enzymes capable of fast and efficient degradation of crystalline cellulose. The soft-rot fungus *Trichoderma reesei* (*Tr*) produces a well-studied and industrially important cellulase system. It consists of two cellobiohydrolases that remove cellobiose units from one or the other end of a cellulose chain, and a range of endoglucanases involved in the cleavage of internal bonds in cellulose chains (1). The cellobiohydrolase Cel7A, the most abundant component, plays a key role in the hydrolysis of the more inaccessible, crystalline regions of cellulose. This is attributed to the special active site design of Cel7A, a long tunnel created by four surface loops covering as many as ten glucose binding sites (i.e. sites -7 to +3) (2, 3). Structural studies and biochemical data with *Tr* cellobiohydrolases and several mutants support the model in which a single glucan chain enters the tunnel from one end and threads through the entire tunnel for bond cleavage; the products are then released from the opposite end of the tunnel (2-4). One of the loops, the exo-loop forming the roof of the tunnel at the catalytic center, was shown to be important for the processive action of Cel7A (5), i.e. the performance of multiple rounds of catalysis before dissociation of the enzyme-substrate complex.

Tr Cel7A is a member of the glycoside hydrolase (GH) family 7, which comprises both cellobiohydrolases and endoglucanases (6). While this family has an excellent correlation with the overall protein fold, the endoglucanases lack the active site covering loops and thus have a more open active site cleft (7-10). A close structural relationship was discovered between GH family 7 and GH family 16, which contains e.g. 1,3-1,4-β-glucanases, 1,3-β-glucanases, κ-carrageenases, 1,3-α-1,4-β-galactanases and xyloglucan endotransglycosylases (11). Both families – grouped in the GH clan-B (12)– share the β-sandwich topology and a limited number of residues in the active site, including a catalytic triad of carboxylate residues. These are directly involved in the retaining mechanism of the clan-B enzymes.

The reaction is a double-displacement mechanism, as shown in Figure 1 for the hydrolysis of aryl β-lactosides by *Tr* Cel7A. One carboxylate residue (Glu217) acts as the general acid in the first step (*glycosylation*) and protonates the glycosidic oxygen, while in the second step (*deglycosylation*) this residue acts as the general base, which deprotonates the incoming water molecule. A second carboxylate residue (Glu212) is the enzymatic nucleophile, which attacks the anomeric center (C1) to give a glycosyl-enzyme intermediate in the glycosylation step. Both steps proceed via transition states with considerable oxocarbenium ion character (13).

In the first Cel7A wt structure with *o*-iodobenzyl-1-thio-β-glucoside bound in the product sites +1/+2 (2), three carboxylate residues were proposed to play a catalytically important role: Glu212 as the nucleophile and Glu217 as the proton

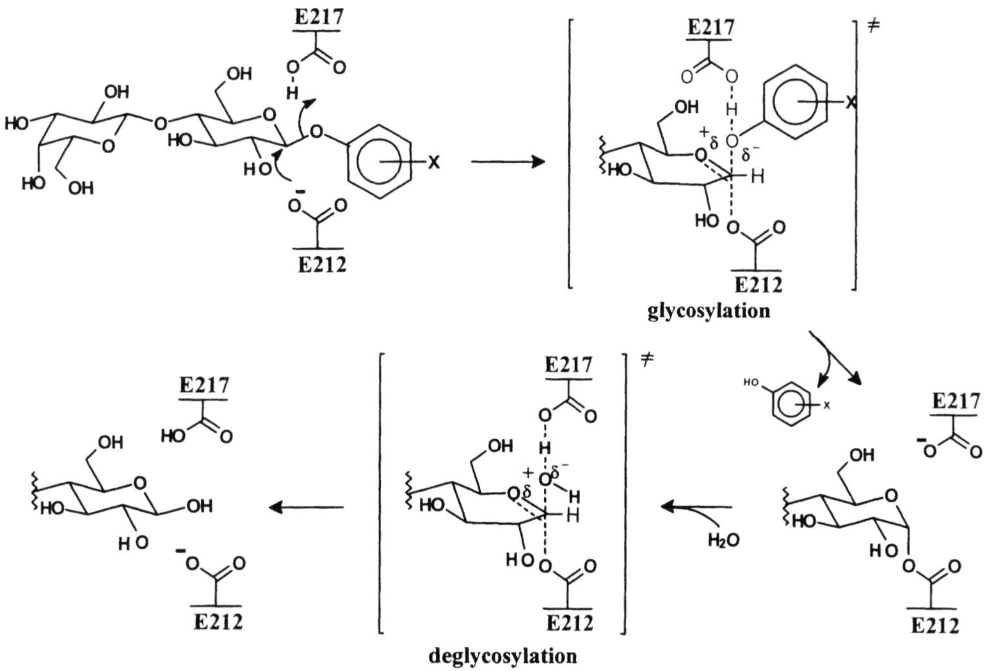

Figure 1. Retaining glycosidase mechanism showing the transition states of the glycosylation and deglycosylation steps (here illustrated for the hydrolysis of aryl β- lactosides by Tr *Cel7A).*

donor; the third residue, Asp214, which interacts closely with Glu212, was suggested to be a nucleophile assisting residue, influencing the position and/or the protonation state of the nucleophile. Results from activity measurements with catalytically impaired mutants of Cel7A lent support to this hypothesis (14). Three isosteric amide mutations were introduced to produce the mutants E212Q, D214N and E217Q. On insoluble crystalline cellulose no significant activity was detected for the E212Q and E217Q mutants, whereas the D214N mutant displayed some residual activity. On the chromophoric substrate CNPLac, the k_{cat} values were reduced to 1/2000 (E212Q), 1/85 (D214N) and 1/370 (E217Q) of the wild type activity, whereas the K_M values were unaffected. Crystal structures of the mutants proved an unchanged active site (3, 14). Crystallization of the variants E212Q and E217Q with different cellooligosaccharides revealed glucose moieties covering the subsites -7 to +4 (3). This allowed modelling of a continuous cellulose chain in the active site of Cel7A, with the sugar in subsite -1 residing in a non ground state conformation (Figure 2, PDB accession code 8CEL). This model is again in good agreement with the proposed catalytic functions for the carboxylate residues in the -1 site.

Thus, from several kinetic and crystallographic studies with *Tr* Cel7A, but also with other family 7 and family 16 glycosyl hydrolases (7, 8, 10, 11, 15-17), it is clear that Glu212 and Glu217 function as the catalytic nucleophile and the catalytic acid/base, respectively.

The aim of this study was to further investigate the catalytic mechanism of *Tr* Cel7A and the specific role of the catalytic carboxylate triad through detailed structure-reactivity experiments.

Experimental

Substrates

2-Chloro-4-nitrophenyl β-glycosides derived from lactose (CNPLac) and cellobiose ($CNPG_2$), and 4-methylumbelliferyl β-lactoside (MULac) were prepared as described (18, 19). The β-lactosides having a 4-bromophenyl (PBrPLac), 4-nitrophenyl (PNPLac) or 2,4-dinitrophenyl (2,4DNPLac) aglycon were synthesized according to published procedures for the corresponding β-D-glucopyranosides and β-cellobiosides (20, 21). The synthesis of (R,S)-3,4-epoxybutyl β-cellobioside (G_2-O-C_4) is described in (22).

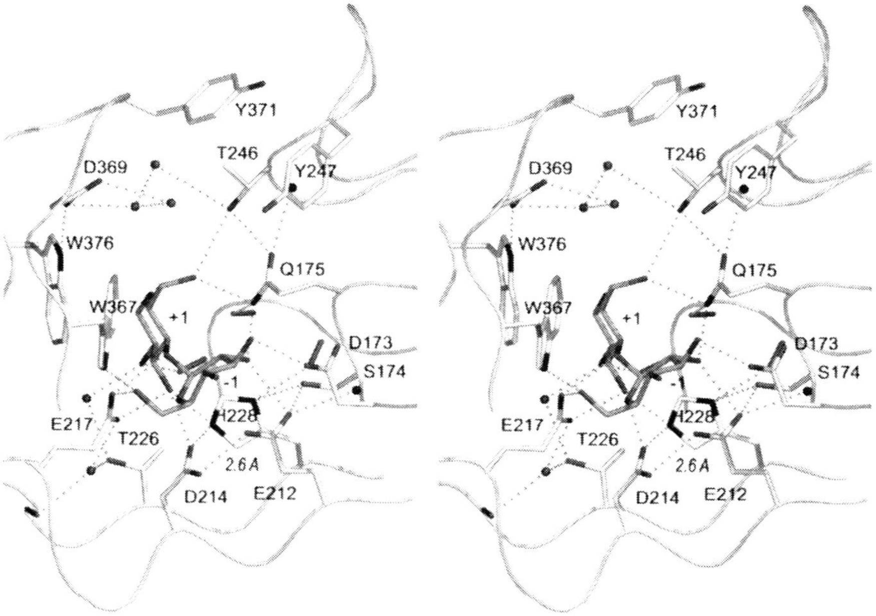

Figure 2. Divergent stereo view of subtrate binding and hydrogen bonding at the catalytic center of Tr *Cel7A. The glucose residues at sites -1 and +1 are from a model of a bound cellulose chain with 9 Glc residues (3; PDB code 8CEL) based on overlapping oligosaccharide complex structures with Cel7A mutants, and with the glucose residue at site -1 adapted from a complex of* Fusarium oxysporum *EG1 (Cel7B) with a nonhydrolysable substrate analog (9; PDB code 1OVW). The indicated distance is the short low-barrier hydrogen between D214 and the catalytic nucleophile E212 (2.50 to 2.67 Å in the structures 1CEL, 7CEL, 1DY4, 1EGN, 1Q2B, 1Q2E). With the mutants E212Q and D214N, respectively, the corresponding distance was longer (2.9-3.5 Å ; structures 2CEL, 3CEL, 4CEL, 5CEL, 6CEL).*

Enzymes

The mutagenesis experiments, the cultivation of the *T. reesei* wild-type (QM9414) and mutant strains (E212Q, D214N, E217Q), as well as the purification of intact Cel7A are reported in (14). Core proteins (i.e. catalytic domains) were prepared by partial proteolytic cleavage with papain (23, 24) and purified using affinity chromatography (25, 26). In order to remove minor Cel7B contamination the Cel7A samples were further purified on a diethylaminoethyl-trisacryl column (LKB) at pH 5.0 and 4.2 using an ammonium acetate gradient (10 to 300 mM). All samples were concentrated and dialysed in 10 mM sodium acetate, sterile filtered and stored at -20 °C. The concentration of Cel7A core (wild-type and mutants) was determined spectrophotometrically at 280 nm using an extinction coefficient of 73,000 $M^{-1}cm^{-1}$ (23).

Activity measurements

The catalytic domain of Cel7A (wild-type and mutants E212Q, D214N and E217Q) was used in all the activity measurements on small, soluble chromogenic substrates.

As the mutants were expressed in *T. reesei* (strain QM9414), all Cel7A samples were carefully checked for contamination with the homologous Cel7B enzyme which has a much higher specific activity than Cel7A for soluble substrates. The specific activity of all Cel7A samples on CNPLac and $CNPG_2$ was measured at room temperature by mixing 180 µl of 6 mM substrate in 50 mM phosphate buffer (pH 5.7) and 20 µl enzyme solution. The release of CNP was monitored continuously during 10 minutes for wt Cel7A and during two hours for the mutants at 405 nm in a microplate reader (Benchmark Easy Reader, Bio-Rad). A ratio of 1/12 for the specific activity on $CNPG_2$ *versus* CNPLac is indicative for pure Cel7A (27).

To check for potential contamination with wt Cel7A in the mutant enzyme preparations, the irreversible inhibition of these enzymes with (R,S)-3,4-epoxybutyl β-cellobioside (G_2-O-C_4) was followed in function of time (28). A solution of 90 µM Cel7A wt and 180 µM E212Q, D214N and E217Q in 10 mM acetate buffer (pH 5.0) was added to 0.5 mM G_2-O-C_4 in the same buffer. The reaction was incubated at room temperature during nine hours. At different time intervals, aliquots were withdrawn to measure the residual activity on 4 mM CNPLac in 50 mM phospate buffer, pH 5.7 following the procedure described above.

The kinetic parameters K_M and k_{cat} for the different chromogenic substrates were determined in triplicate at 37 °C in 50 mM phosphate buffer, pH 5.7. Substrate concentrations between 0.05 and 2.0 mM were used in all the experiments, and standard curves for the released phenols were obtained under the same conditions as in the enzyme kinetic experiments. With the substrates CNPLac and 2,4 DNPLac the experiments were performed in 1 ml cuvettes and the release of the aglycon was followed continuously at 405 nm (Vitatron DCP colorimeter). Discontinuous assays were used with the other chromogenic substrates. With PNPLac aliquots were taken into a microtiterplate at regular time intervals and diluted in 10 % Na_2CO_3. The released PNP was measured at 405 nm in a microplate reader (Model 450, Biorad). The hydrolysis of PBrPLac was followed by measuring the released PBrP at different time points. A colorimetric (490 nm) assay with 4-amino-antipyrine in the presence of an alkaline oxidant was used (29). A discontinuous fluorometric measurement (emission wavelenght > 435 nm, extinction 366 nm) was used to measure the MUF released from MULac (19).

Values for K_M and k_{cat} were derived from the linearized Hanes equation.

Results

Activities of the purified Cel7A wt and mutant preparations

Kinetic studies of Cel7A, especially the low activity catalytic mutants, require a thorough control of the purity of the protein samples. The mutated proteins studied in this work were expressed in *T. reesei* entailing possible minor contamination with cellulases (e.g. Cel7B), which could have activity on the small chromogenic substrates. The catalytic activity of the three mutants E212Q, D214N and E217Q was initially determined on CNPLac and this revealed a drastic effect on the k_{cat} values, while the K_M values remained essentially unchanged (average K_M on CNPLac = 0.5 mM at pH 5.7 and 37 °C) (14). Since the K_M value of the endoglucanase Cel7B on this substrate is considerably higher, namely 2.4 mM at pH 5.7 and 37 °C, it is unlikely that this cellulase would be responsible for the residual activity observed with the Cel7A mutants. Furthermore, the Cel7A mutants have a lower specific activity on the chromogenic β-cellobioside than on the β-lactoside (data not shown), indicating again that the mutant samples are essentially free of Cel7B contamination (30).

Given the similarity of the K_M values measured for the mutants to those of the wild-type enzyme, it was necessary to further investigate the possibility that residual activity of the active-site mutants arises from traces of wild-type enzyme, either by contamination during the purification process, by spontaneous deamidation or translational misreading (31, 32). Active site titration with the mechanism-based inactivator 2,4-dinitrophenyl-2-deoxy-2-fluoro β-cellobioside did not allow to determine the initial concentration of active wt Cel7A, as no initial burst could be observed (14). Therefore, irreversible inactivation of Cel7A wt with the inhibitor (R,S)-3,4-epoxybutyl β-cellobioside (G_2-O-C_4) was used to eliminate the possibility of wild-type contamination (28). Cel7A wt in the presence of 0.5 mM G_2-O-C_4 is exponentially inactivated in function of time with a pseudo-first-order rate constant k_{app} of 0.0030 min^{-1}. Inactivation of the catalytic mutants under the same conditions allowed to derive the following k_{app} values: 0.00022, 0.00066 and 0.00060 min^{-1} for E212Q (and control sample Cel7A wt without G_2-O-C_4), D214N and E217Q respectively (data not shown). The five to fourteen times slower inhibition of the mutants indicates that their residual activty is not, or only to a very small extent, due to wild-type contamination.

Structure-reactivity studies with chromogenic substrates

The aryl β-lactosides 2,4DNPLac, CNPLac, PNPLac, MULac and PBrPLac were used as substrates for the cellobiohydrolase-catalysed reaction to evaluate the effect of the aglycon leaving group on the reaction rates. This study was performed with the lactoside substrates and not with the corresponding cellobioside substrates, as the former exhibit much higher K_M and k_{cat} values and are thus better substrates for a reliable kinetic analysis of the Cel7A mutants (30). The Michaelis-Menten parameters were determined with wild-type Cel7A as well as with the three active-site mutants E212Q, D214N and E217Q at 37 °C and pH 5.7 (Table I). This pH, above the optimum of 4.2 for Cel7A wt (33), allows the continuous measurement of CNP (pK_a = 5.45) at 405 nm.

Mutation of Glu212, Asp214 and Glu217 to their respective amide functions has a significant effect on the catalytic constant k_{cat} for the hydrolysis of all the aryl β-lactosides. The smallest effect is observed with the D214N mutant for which a 20- to 80-fold reduction in k_{cat} could be measured. The E212Q mutant retained 1/2000, 1/850 and 1/750 of the wild-type activity against CNPLac, PNPLac and MULac respectively. A strong dependence on the aglycon acidity is observed with the E217Q mutant, with hardly a 10-fold reduction of k_{cat} for the hydrolysis of 2,4DNPLac but a 6400-fold reduction for the hydrolysis of MULac. The very low residual reactivity of the mutant enzymes on PBrPLac (aglycon pK_a = 9.34) precluded a reliable determination of the kinetic constants. The specific activity was measured with a 2 mM solution of PBrLac: 0.67 min^{-1} for Cel7A wt, 0.0082 min^{-1} for D214N, 0.00029 min^{-1} for E217Q and 0.00065 min^{-1} for E212Q.

Table I. Michelis-Menten parameters for the hydrolysis of aryl β-lactosides by *Tr* Cel7A (wt, E212Q, D214N and E217Q)

Phenol Substrate	pK_a[a]	*Cel7A*	K_M *(mM)*	k_{cat} *(min^{-1})*	k_{cat}/K_M *(mM^{-1}min^{-1})*
2,4DNP	3.96	wt	0.55 ± 0.03	104 ± 11	190
		E212Q	nd[b]	nd	nd
		D214N	0.67 ± 0.05	2.9 ± 0.6	4.3
		E217Q	0.61 ± 0.07	12.2 ± 0.2	20
CNP	5.45	wt	0.46 ± 0.02	12.8 ± 0.2	28
		E212Q	0.68 ± 0.10	0.0063 + 0.0005	0.0096
		D214N	0.57 ± 0.07	0.15 ± 0.05	0.26
		E217Q	0.78 ± 0.10	0.035 ± 0.005	0.045
PNP	7.18	wt	0.22 ± 0.01	3.5 ± 0.1	16
		E212Q	0.27 ± 0.05	0.0041 ± 0.0005	0.015
		D214N	0.37 ± 0.02	0.042 ± 0.002	0.11
		E217Q	0.21 ± 0.08	0.00066 ± 0.00005	0.0031
MUF	7.50	wt	0.52 ± 0.02	47 ± 5	90
		E212Q	0.53 ± 0.07	0.063 ± 0.005	0.12
		D214N	0.79 ± 0.05	2.3 ± 0.5	2.9
		E217Q	0.65 ± 0.10	0.0073 ± 0.0006	0.011
PBrP	9.34	wt	0.52 ± 0.05	1.7 ± 0.5	3.3

All measurements were performed at 37 °C in 50 mM phosphate buffer, pH 5.7

[a] Phenol pK_a values were taken from (21)

[b] nd = not determined

The most drastic effect was observed for E217Q (a factor 1/2300 compared to Cel7A wt). A reduction of the specific activity for the mutants D214N and E212Q with a factor 1/80 and 1/1000 respectively is comparable to the data observed with the other aryl β-lactosides. For all three mutants, the Michaelis constant K_M on the lactoside substrates remains essentially the same as that of the wild-type enzyme.

Values of k_{cat} and k_{cat}/K_M for Cel7A wt, D214N and E217Q are plotted as a function of the aglycon pK_a in the form of Hammett-plots. For k_{cat}/K_M a linear dependence is observed over the entire pK_a area, except for the hydrolysis of MULac (Figure 3A-C). Excluding this data point a slope (Brønsted coefficient β) of -0.3, -0.5 and -1.2 can be deduced for Cel7A wt, D214N and E217Q respectively. The pK_a dependence of k_{cat} is more difficult to interpret, although the Hammett-plot does show a trend: a linear relationship for substrates with a good leaving group (pK_a phenol < 7.2). The β-values are -0.5, -0.6 and -1.3 for Cel7A wt, D214N and E217Q respectively (Figure 3D-E). Again the value for

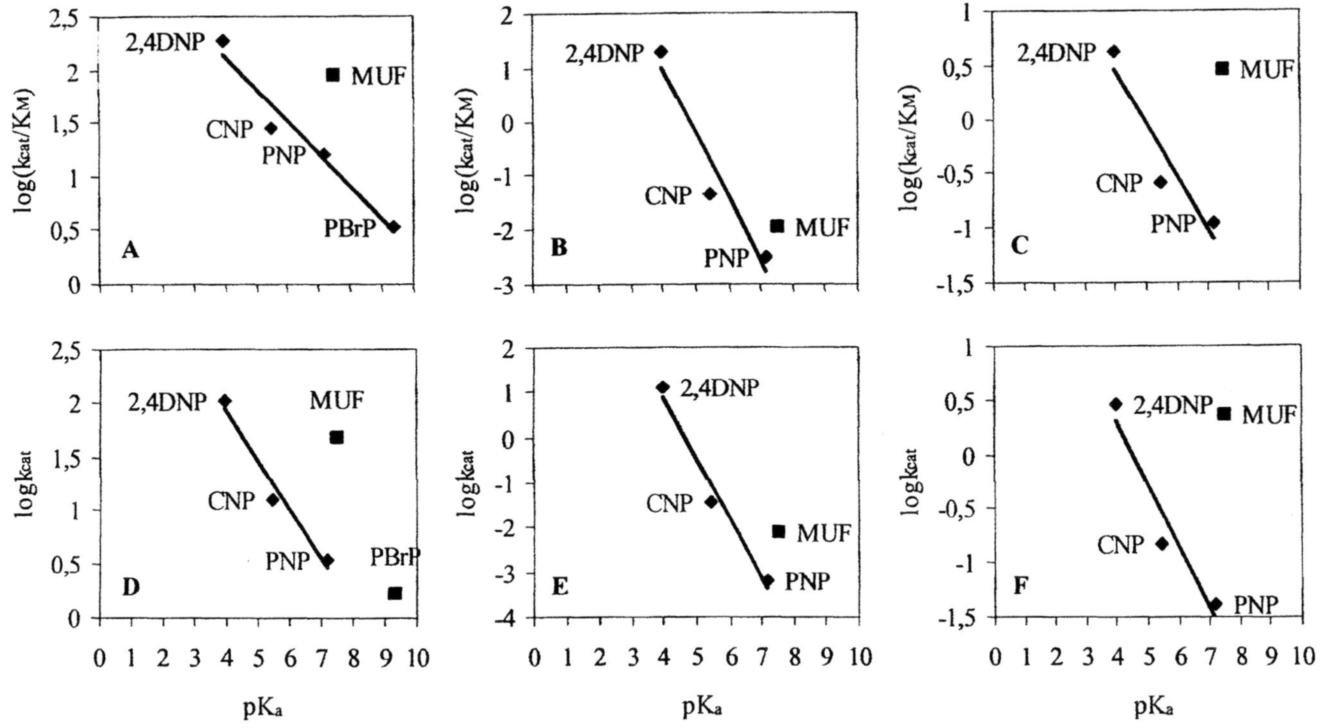

Figure 3. Hammett-plot log k_{cat}/K_M and log k_{cat} versus pK_a for the hydrolysis of aryl β-lactosides by Tr *Cel7A wt (A and D), and the mutants E217Q (B and E) and D214N (C and F).*

MULac clearly deviates from the linear trend observed with the other aryl β-lactosides. In the case of Cel7A wt, the logk_{cat} *versus* pK_a plot seems to level off at the higher pK_a values (PBrP).

Discussion

A series of substituted aryl β-lactosides were used as substrates of *Tr* Cel7A wt to evaluate the effect of the aglycon leaving group on reaction rates. The study was extended to the catalytic mutants E212Q, E217Q and D214N to check if their kinetic behaviour could be correlated with the proposed roles for Glu212, Glu217 and Asp214 in a double-displacement mechanism: nucleophile, proton donor and nucleophile assisting residue, respectively.

Enzyme mechanism: structure-reactivity studies with *T. reesei* Cel7A wt

Substrate reactivity

The small chromogenic substrates used in this study are only cleaved at the heterosidic bond with release of the aryl aglycon. In a double-displacement mechanism, deglycosylation will be the same for all lactoside substrates, while glycosylation will be influenced by the substituents on the phenol group. The effect of these substituents on the rate constants can be evaluated by means of a Hammett analysis (See e.g. (20, 21, 34, 35) for application in glycosidase research). The reaction constant β (the slope in a log k_{cat} or a log k_{cat}/K_M versus pK_a plot) reflects the sensitivity of the reaction for the electronic effects of the substituents, the latter expressed as the pK_a of the free phenol.

The second order rate constant k_{cat}/K_M contains rate constants up to the *glycosylation* step and thus a leaving group dependency is expected. For Cel7A wt the Hammett plot for k_{cat}/K_M (Figure 3A) is linear over the entire range of pK_a values. However, the slope for log k_{cat}/K_M, the Brønsted constant, is considerably smaller (β = -0.3) than that seen for retaining β-glycosidases (β between -0.7 and -1.0) (20, 34-36). The Brønsted constant is dependent on two factors: a large degree of C-O bond cleavage at the transition state gives a larger negative β-value, whereas efficient proton donation lowers the β-value. Independent results from hydrolysis experiments with 1-fluoro-lactosyl fluorides (hardly detectable hydrolysis) and lactosyl fluorides point towards a transition state with substantial oxocarbenium ion character, and thus extensive C-O bond cleavage (Weber, M., Becker D., Vasella, A., and Sinnott, M. L., unpublished results, 1998). Therefore the rather small negative β-value could possibly be ascribed to efficient proton donation by the acid/base residue Glu217 in *Tr* Cel7A. X-ray studies have revealed that Glu217 is solvated and has a polar microenvironment (3, 33). These factors possibly contribute to the proton donation capacity of this residue.

The Hammett plot log k_{cat} versus pK_a (Figure 3D) is curved and cannot be interpreted in detail due to the limited number of data. Still, the general trend is a decrease in k_{cat} with increasing pK_a (β = -0.5 for 2,4DNPLac, CNPLac and PNPLac) and indicates that the glycosylation step is rate determining even for the most activated substrates. For deglycosylation to become rate limiting, k_{cat} would level off in the low pK_a range, as observed in the biphasic plots for many retaining glycosidases (20, 21, 34). It is remarkable that a biphasic log k_{cat} versus pK_a plot with an upward curvature at low pK_a values was observed with the *Bacillus licheniformis* (*Bl*) 1,3-1,4-β-glucanase, a family 16 glycosyl hydrolase, belonging to the same clan-B as *Tr* Cel7A (17, 35). A change in transition-state structure depending on the aglycon was suggested to explain this behavior (35). In this respect it is also interesting to point out the divergent result with MULac, which indicates that the type of aglycon can give rise to an independent structure-activity relationship. The interaction of MUF (a bicyclic system) in the +1 site might be very different from this with the mono-and di-substituted phenols.

Glycosylation is the rate-limiting step even with activated substrates

For both *Tr* Cel7A and *Bl* 1,3-1,4-β-glucanase, glycosylation is rate determining for the hydrolysis of aryl β-glycosides, also for the reactive 2,4DNP-substrate. Remarkably, other Hammett studies for retaining β-glycosidases (e.g. family 1 β-glucosidases and *Cellulomonas fimi* (*Cf*) Xyn10A) report a change in rate-limiting step from glycosylation to deglycosylation with highly activated substrates (20, 21, 34). This different behavior might be related to the high degree of oxocarbonium ion character in the transition state of *Tr* Cel7A. Indeed, when using a substrate with a good leaving group (e.g. 2,4 DNP), which will inherently have a higher oxocarbenium ion character at the transition state than one with a bad leaving group (e.g. PNP), the effect on the glycosylation step will be less pronounced in an enzyme which is able to stabilize the oxocarbenium ion-like transition state more efficiently. Specific protein-ligand interactions in the different binding sites and the polar microenvironment will be important factors in this stabilization.

The position of the proton donor with respect to the O5-C1 bond can also be a determining factor. In 1999, Heightman and Vasella introduced the *syn-anti* mechanistic concept: the orientation of the lone pair of the glycosidic heteroatom in the plane of the sugar ring requires the interaction with the catalytic acid in this plane and defines a protonation trajectory: either *syn* or *anti* to the endocyclic O5-C1 bond. Crystal structures of β-glycosyl hydrolases with ligands in their active site allow to classify the enzymes as *syn*- or *anti*-protonators (37). The position of the proton donor in a *syn*-protonating glycosyl hydrolase is close to the O5-C1 bond. The developing anion character during the protonation step will allow stabilization of the developing oxocarbenium ion character at the O5-C1 bond on its way to and during the transition state. In *anti*-protonators the acid/base catalyst is closer to the C1-C2 bond and will thus have a less pronounced effect on oxocarbenium ion stabilization.

Both *Tr* Cel7A and *Bl* 1,3-1,4-β-glucanase (Clan-B) are *syn*-protonators, whereas the family 1 β-glucosidases and *Cf* Xyn10A (Clan-A) are *anti*-protonators (37). The fact that glycosylation remains the rate-limiting step with the clan-B enzymes might be related to their *syn*-protonation trajectory, but whether this is a general phenomenon among β-glycosyl hydrolases cannot be stated due to the limited number of Hammett studies reported in literature.

Reactivity of the Cel7A mutants E212Q, D214N and E217Q

A kinetic analysis of the three mutants E212Q, D214N and E217Q with a series of chromogenic β-lactosides allows to interpret more accurately the earlier obtained kinetic constants k_{cat} and K_M for CNPLac (14). As discussed above for wild-type Cel7A, glycosylation is rate-determining even for highly activated substrates. This also seems to be the case with the mutants since the K_M values do not change significantly compared to those of the wild-type enzyme. Indeed, a drop in K_M would be expected if deglycosylation becomes rate limiting (38).

Glu217 as the acid/base catalyst

For an acid/base mutant, a strong dependence on aglycon leaving group ability can be expected. The k_{cat} value of the E217Q mutant drops with only a factor 9 for the activated substrate 2,4DNPLac (pK_a 2,4DNP 3.96), whereas a large decrease in activity (factor 6400) is observed for MULac (pK_a MUF 7.50). The change of the Brønsted-constant β from -0.3 for Cel7A wt to -1.2 for E217Q (in the plot log k_{cat}/K_M versus pK_a, Figure 3A, B) indicates a larger degree of negative charge accumulation on the phenolate oxygen in the transition state, exactly as expected in the absence of the proton donor. The same effect was observed for the acid base mutant E127A from *Cf* Xyn10A (39).

A direct comparison of the kinetic data for the E217Q mutant with those reported for other acid/base mutants of retaining β-glycosidases is difficult, due to different mutations (e.g. alanine instead of glutamine) and due to the fact that the rate-limiting step might be different for a specified substrate. Modification of the acid/base residue will have an effect on both steps of the retaining mechanism, but not necessarily to the same extent (39, 40). For both *Tr* Ce7A and *Bl* 1,3-1,4-β-glucanase, glycosylation is the rate limiting step when a 2,4DNP substrate is used, thus the effect of the catalytic acid/base mutation can be compared for the same step in the reaction mechanism. With the *Tr* E217Q mutant, the activity is reduced by a factor 9, whereas the *Bl* E138A mutant shows a 2400-fold reduction in activity. As the amide function in the E217Q mutant can still be involved in electrophilic assistance towards the glycosidic oxygen, these data might reflect the importance of this interaction for the observed activity with the glutamine mutant, especially on substrates with a good leaving group (e.g. 2,4 DNP and CNP).

Glu212 as the catalytic nucleophile

For retaining β-glycosidases it is generally stated that modification of the nucleophilic carboxylate to an amide or an alanine residue is accompanied by a dramatic loss in activity (at least a factor 10^6 compared to the wild-type enzyme) (38). Indeed, the nucleophile mutants E358Q from *Af* β-glucosidase and E233A from *Cf* Xyn10A are essentially inactive towards chromogenic substrates (32, 39).

The *Tr* Cel7A mutant E212Q has a pronounced effect on the hydrolysis of chromogenic substrates, but only with an average factor of 10^3 (Table I). As discussed above, contamination by wild-type or endoglucanase activity is unlikely to be responsible for the observed activity. Even though it is nucleophilic rescue experiments with an alanine mutant or covalent labeling of the catalytic nucleophile with a mechanism-based inactivator, such as an activated fluoro glycoside, that should be used to obtain an undisputed kinetic proof for the role of Glu212, all crystallographic observations of TrCel7A are consistent with the role of Glu212 as the nucleophile in a double-displacement mechanism. Furthermore, the limited amount of kinetic data with the E212Q mutant do seem to agree with its function as catalytic nucleophile since the leaving group ability has little effect on the observed activity.

As compared with related endoglucanases (30), *Tr* Cel7A has a very low k_{cat} value on small, soluble substrates while it is far superior in the solubilization of crystalline cellulose. It seems that there has been little evolutionary pressure to optimize the active site of the cellobiohydrolase for fast catalysis, which is apparently not the rate-limiting step on crystalline cellulose degradation. Therefore, even a "non-optimal" catalytic nucleophile can be efficient enough to achieve the catalytic rate required on insoluble substrates. If the nucleophile is not that crucial to begin with, it can be mutated without a complete loss of activity, as hereby observed with the E212Q mutant of *Tr* Cel7A. A similar apparent lack of catalytic efficiency has been observed with the inverting cellobiohydrolase, *Tr* Cel6A, which apparently exercises little base catalysis (41).

A possible, although hypothetical molecular explanation for the residual activity of the E212Q mutant of *Tr* Cel7A could be formulated as follows: *syn*-protonating β-glycosyl hydrolases may be inherently able to stabilize the oxocarbenium-ion-character more efficiently than *anti*-protonating enzymes (see above). Therefore, both the deprotonated acid/base residue and the nearby glutamine residue may contribute to the oxocarbenium-ion stabilization in the *syn*-protomating *Tr* Cel7A E212Q, thereby giving rise to the observed residual activity. A dramatic drop in the k_{cat} value by a factor 10^5 has been reported for the nucleophile mutant, E134A, of another syn-protonating enzyme in the clan-B, *Bl* 1,3-1,4-β-glucanase (17). Even though this is a substantial difference to the decrease of a factor of 10^3 observed for the *Tr* Cel7A E212Q, it is quite possible that an alanine residue cannot be involved in the stabilization of the oxocarbenium ion similar to the glutamine residue. Furthermore it needs to be stressed again that also the polar character of the active site will be important for stabilization of the oxocarbenium ion-type transition state, and this factor can be very different, even for enzymes within the same family or clan.

Asp214 as the nucleophile assisting residue

Mutation of Asp214 to the isosteric asparagine has a significant, but not a dramatic effect on the hydrolysis of the aryl β-lactosides (a factor 1/85 compared to Cel7A wt for CNPLac). The Hammett-plot for this mutant is comparable to the one obtained with Cel7A wt with nearly the same β-value (Figure 3C, F). Thus, with the D214N mutant glycosylation remains the rate-limiting step with all tested aryl β-lactosides, and the glycosylation transition state must be nearly identical to the one in the wild-type Cel7A with regard to C-O bond cleavage and/or proton donation to the glycosidic oxygen.

In the active site of Cel7A wt, a short hydrogen bond is formed between the carboxyl oxygen atoms of Asp214 and Glu212 (distance between 2.5 and 2.7 Å in different Cel7A structures) (2, 3, 33). The oxygen atom Glu212 $O^{\varepsilon 2}$, which performs the nucleophilic attack on the anomeric carbon, is involved in this interaction and thus shares a proton with Asp214 (Figure 2 and 4A). As described earlier, Asp214 will most probably control the protonation state and/or position of the nucleophile Glu212 and is also well positioned to influence the pK_a of the acid/base residue Glu217 (2, 3, 14, 33). But, the kinetic data with the D214N mutant also indicate a role for Asp214 as determining factor for the

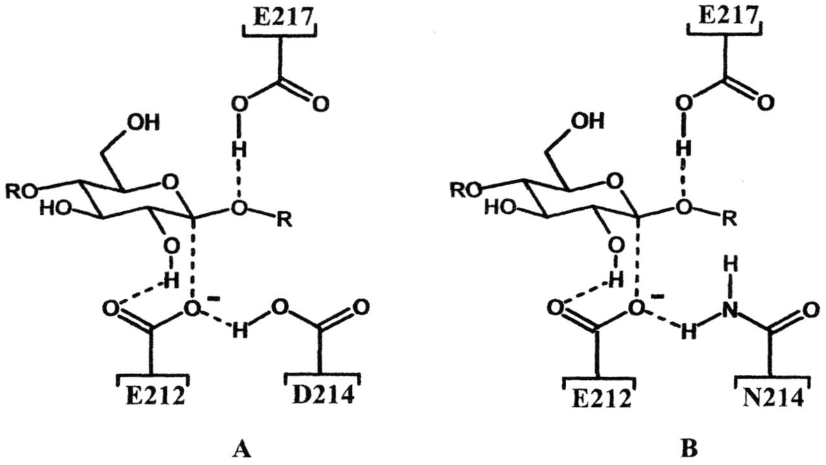

Figure 4. Schematic representation of the interaction of the nuclophile residue Glu212 with the nucleophile assisting residue Asp214 in Tr Cel7A wt (A) and with Asn214 in the mutant D214N (B).

nucleophilicity of Glu212. In the D214N crystal structure, the hydrogen bond between Asp214 and Glu212 is maintained, but longer (3.0 Å) (14), and the proton cannot be shared and belongs to Asn214 (Figure 4B). The negative charge will thus be localized more strongly on the carboxylic side chain of Glu212 and intuitively one could expect this residue to act as a better nucleophile in the first step of the double-displacement mechanism. Still, the mutant has lower activity, while the K_M value is unchanged. For a possible explanation, the interaction between Glu212 $O^{\varepsilon 1}$ and the C2-OH from the glucosyl unit in the -1 site has to be taken into account. This interaction is observed in the structure of the glycosyl-enzyme intermediate of some retaining β-glycosidases (42, 43) and is likely to be important in *Tr* Cel7A too, based on the modeled structure (Figure 2). Thus, Glu212 is positioned close to Asp214/Asn214 and the C2-OH group and might experience a competition between its capacity as a nucleophile (attack on the anomeric center) and its capacity to form a hydrogen bond with C2-OH. The Glu212-Asp214 diad in Cel7A wt will reduce the ability to form a hydrogen bond with C2-OH and will preferentially attack the anomeric center, in accordance with the HSAB (Hard and Soft Acid Bases) principle (44): the interacting residues Glu212/Asp214 can be regarded as a soft base and thus as a good nucleophile, because of the delocalization of the negative charge and will prefer the attack on a stabilized carbocation (a soft Lewis acid) over the interaction with a proton (a hard Brønsted acid). In the D214N mutant the negative charge is concentrated on Glu212, making the carboxylate more basic (a harder base), and thus reducing its nucleophilicity and making the interaction with the proton of C2-OH (hard proton) more important.

The nucleophile assisting residue is strictly conserved throughout the clan-B glycosyl hydrolases. In all crystal structures of other family 7 endoglucanases (7-9) and a family 16 1,3-1,4-β-glucanase (45) and κ-carrageenase (11), the close interaction between the nucleophile and the assisting carboxylate residue is observed and will most probably fulfill the same function in the catalytic mechanism.

Conclusions

Our kinetic data with *Tr* Cel7A on soluble chromophoric substrates confirm the earlier proposed roles for the three carboxylate residues in the -1 subsite: Glu212 is the nucleophile and Glu217 is the acid/base in a double displacement mechanism, while Asp214 is well positioned to assist the nucleophile. The reactivity studies unambiguously show that – even with the catalytic mutants – the formation of the glycosyl-enzyme is rate-determining. The large negative β-value with the acid/base mutant indicates that the *deglycosylation* step must be very fast compared to the *glycosylation* step, even when there is no general base.

However, the key to understanding the catalytic behaviour of *Tr* Cel7A and other cellobiohydrolases is to realize that these enzymes have not been optimized for efficient catalysis of bond cleavage but have instead evolved for breaking down cellulose crystals (3, 5, 46). The following factors are important in this process: (1) a mechanism to extract the glucan chain from the cellulose surface, and to feed it into the special tunnel-shaped active site; (2) dynamic binding of the cellulose chain over althogether 10 subsites enclosed in the tunnel, and (3) strong binding in the product sites +1 and +2. The inhibition resulting from the strong binding of the product is apparently tolerated as it probably facilitates the necessary processive action of the enzyme on the crystalline substrate surface (3, 5, 30). As it is the interactions with the crystalline surface that are rate-limiting, there has been little evolutionary pressure towards rate enhancement in the *Tr* Cel7A mechanism. This must be taken into account when interpreting the data obtained with catalytic site mutants of crystalline cellulases.

Acknowledgements

This work has been supported by the European Commission (BIO2-CT94-3030, BIO4-CT96-0580 and a Marie Curie fellowship HPMF-CT-2002-02009), the Fund for Scientific Research - Flanders (G.0249.97N) and the Ghent University Research Counsil (BOF 01107201). We are grateful to Ronan Van Rijsbergen and Madeleine Cambier for skillful technical assistance. Michael S. Sinnott is thanked for helpful comments on the manuscript.

References

1. Teeri, T. T. *Trends Biotechnol.* **1997,** *15*, 160-167.
2. Divne, C.; Ståhlberg, J.; Reinikainen, T.; Ruohonen, L.; Pettersson, G.; Knowles, J. K. C.; Teeri, T. T.; Jones, A. *Science* **1994,** *265*, 524-528.
3. Divne, C.; Ståhlberg, J.; Teeri, T. T.; Jones, T. A. *J. Mol. Biol.* **1998,** *275*, 309-325.
4. Zou, J.-Y.; Kleywegt, G. J.; Ståhlberg, J.; Driguez, H.; Nerinckx, W.; Claeyssens, M.; Koivula, A.; Teeri, T. T.; Jones, T. A. *Structure* **1999,** *7*, 1035-1045.
5. von Ossowski, I.; Ståhlberg, J.; Koivula, A.; Piens, K.; Becker, D.; Boer, H.; Harle, R.; Harris, M.; Divne, C.; Mahdi, S.; Zhao, Y.; Driguez, H.;

Claeyssens, M.; Sinnott, M. L.; Teeri, T. T. *J. Mol. Biol.* **2003,** *333*, 817-829.
6. Henrissat, B.; Bairoch, A. *Biochem. J.* **1996,** *316*, 695-696.
7. Kleywegt, G. J.; Zou, J.-Y.; Divne, C.; Davies, G. J.; Sinning, I.; Ståhlberg, J.; Reinikainen, T.; Srisodsuk, M.; Teeri, T. T.; Jones, T. A. *J. Mol. Biol.* **1997,** *272*, 383-397.
8. Sulzenbacher, G.; Driguez, H.; Henrissat, B.; Schülein, M.; Davies, G. J. *Biochemistry* **1996,** *35*, 15280-15287.
9. Mackenzie, L. F.; Sulzenbacher, G.; Divne, C.; Jones, T. A.; Wöldike, H. F.; Schülein, M.; Withers, S. G.; Davies, G. J. *Biochem. J.* **1998,** *335*, 409-416.
10. Sulzenbacher, G.; Schülein, M.; Davies, G. J. *Biochemistry* **1997,** *36*, 5902-5911.
11. Michel, G.; Chantalat, L.; Duee, E.; Barbeyron, T.; Henrissat, B.; Kloareg, B.; Dideberg, O. *Structure* **2001,** *9*, 513-525.
12. Davies, G.; Henrissat, B. *Structure* **1995,** *3*, 853-859.
13. Sinnott, M. L. *Chem. Rev.* **1990,** *90*, 1171-1202.
14. Ståhlberg, J.; Divne, C.; Koivula, A.; Piens, K.; Claeyssens, M.; Teeri, T. T.; Jones, T. A. *J. Mol. Biol.* **1996,** *264*, 337-349.
15. Davies, G. J.; Ducros, V.; Lewis, R. J.; Borchert, T. V.; Schülein, M. *J. Biotechnol.* **1997,** *57*, 91-100.
16. Juncosa, M.; Pons, J.; Dot, T.; Querol, E.; Planas, A. *J. Biol. Chem.* **1994,** *269*, 14530-14535.
17. Viladot, J.-L.; de Ramon, E.; Durany, O.; Planas, A. *Biochemistry* **1998,** *37*, 11332-11342.
18. Claeyssens, M. In *Fems symposium N° 43: biochemistry and genetics of cellulose degradation*; Aubert, J.-P.; Béguin, P.; Millet, J., Eds.; Academic Press Limited: London, 1988; pp. 393-397.
19. van Tilbeurgh, H.; Loontiens, F. G.; De Bruyne, C. K.; Claeyssens, M. *Methods Enzymol.* **1988,** *160*, 45-59.
20. Kempton, J. B.; Withers, S. G. *Biochemistry* **1992,** *31*, 9961-9969.
21. Tull, D.; Withers, S. G. *Biochemistry* **1994,** *33*, 6363-6370.
22. Høj, P. B.; Rodriguez, E. B.; Stick, R. V.; Stone, B. A. *J. Biol. Chem.* **1989,** *264*, 4939-4947.
23. van Tilbeurgh, H.; Tomme, P.; Claeyssens, M.; Bhikhabhai, R.; Pettersson, G. *FEBS Lett.* **1986,** *204*, 223-227.
24. Tomme, P.; van Tilbeurgh, H.; Pettersson, G.; Van Damme, J.; Vandekerckhove, J.; Knowles, J. K. C.; Teeri, T. T.; Claeyssens, M. *Eur. J. Biochem.* **1988,** *170*, 575-581.
25. van Tilbeurgh, H.; Bhikhabhai, R.; Pettersson, G.; Claeyssens, M. *FEBS Lett.* **1984,** *169*, 215-218.
26. Tomme, P.; McRae, S.; Wood, T. M.; Claeyssens, M. *Methods Enzymol.* **1988,** *160*, 187-193.

27. Claeyssens, M.; Aerts, G. *Bioresource Technol.* **1992,** *39,* 143-146.
28. Klarskov, K.; Piens, K.; Ståhlberg, J.; Høj, P. B.; Van Beeumen, J.; Claeyssens, M. *Carbohydr. Res.* **1997,** *304,* 143-154.
29. Pilz, W.; Johann, I. *Z. Anal. Chem.* **1965,** *212,* 410-419.
30. Claeyssens, M.; van Tilbeurgh, H.; Tomme, P.; Wood, T. M.; McRae, S. I. *Biochem. J.* **1989,** *261,* 819-825.
31. Schimmel, P. *Acc. Chem. Res.* **1989,** *22,* 232-233.
32. Withers, S. G.; Rupitz, K.; Trimbur, D.; Warren, R. A. J. *Biochemistry* **1992,** *31,* 9979-9985.
33. Becker, D.; Braet, C.; Brumer III, H.; Claeyssens, M.; Divne, C.; Fagerström, B. R.; Harris, M.; Jones, T. A.; Kleywegt, G. J.; Koivula, A.; Mahdi, S.; Piens, K.; Sinnott, M. L.; Ståhlberg, J.; Teeri, T. T.; Underwood, M.; Wohlfahrt, G. *Biochem. J.* **2001,** *356,* 19-30.
34. Vallmitjana, M.; Ferrer-Navarro, M.; Planell, R.; Abel, M.; Ausín, C.; Querol, E.; Planas, A.; Pérez-Pons, J.-A. *Biochemistry* **2001,** *40,* 5975-5982.
35. Planas, A.; Abel, M.; Millet, O.; Palasí, J.; C., P.; Viladot, J.-L. *Carbohydr. Res.* **1998,** *310,* 53-64.
36. Lawson, S. L.; Wakarchuk, W. W.; Withers, S. G. *Biochemistry* **1996,** *35,* 10110-10118.
37. Heightman, T. D.; Vasella, A. *Angew. Chem., Int. Ed.* **1999,** *38,* 750-770.
38. Ly, H. D.; Withers, S. G. *Annu. Rev. Biochem.* **1999,** *68,* 487-522.
39. MacLeod, A. M.; Tull, D.; Rupitz, K.; Warren, R. A. J.; Withers, S. G. *Biochemistry* **1996,** *35,* 13165-13172.
40. Wang, Q.; Trimbur, D.; Graham, R.; Warren, R. A. J.; Withers, S. G. *Biochemistry* **1995,** *34,* 14554-14562.
41. Koivula, A.; Ruohonen, L.; Wohlfahrt, G.; Reinikainen, T.; Teeri, T. T.; Piens, K.; Claeyssens, M.; Weber, B.; Vasella, A.; Becker, D.; Sinnott, M. L.; Zou, J.-Y.; Kleywegt, G. J.; Szardenings, M.; Ståhlberg, J.; Jones, T. A. *J. Am. Chem. Soc.* **2001,** *124,* 10015-10024.
42. White, A.; Tull, D.; Johns, K.; Withers, S. G.; Rose, D. R. *Nat. Struct. Biol.* **1996,** *3,* 149-154.
43. Notenboom, V.; Birsan, C.; Nitz, M.; Rose, D. R.; Warren, R. A. J.; Withers, S. G. *Nat. Struct. Biol.* **1998,** *5,* 812-818.
44. Pearson, R. G. *Coord. Chem. Rev.* **1990,** *100,* 403-425.
45. Hahn, M.; Keitel, T.; Heinemann, U. *Eur. J. Biochem.* **1995,** *232,* 849-858.
46. Teeri, T. T.; Koivula, A.; Linder, M.; Wohlfahrt, G.; Divne, C.; Jones, T. A. *Biochem. Soc. Trans.* **1998,** *26,* 173-178.

Chapter 13

Protein Engineering of Xylanases

Dawn E. Stephens[1], Kugen Permaul[1], Bernard A. Prior[2], and Suren Singh[1,*]

[1]**Department of Biotechnology, Durban Institute of Technology, P. O. Box 1334, Durban 4000, South Africa**
[2]**Department of Microbiology, University of Stellenbosch, Private Bag X1, Matieland 7602, South Africa**

Industrial applications of xylanases require enzymes that are thermostable with a wide pH and temperature range. Since the availability of such biocatalysts is limited, the use of protein engineering to study and redesign xylanases on the genetic level, has gained momentum. X-ray crystallography, site-directed and random mutagenesis studies have revealed a considerable amount of information about the basic structure and function of xylanases and have provided the basis for the genetic manipulation of the gene for desired enzymatic traits.

As man makes significant advancement in science, he has become more aware of the importance of the conservation of energy and natural resources, after repeatedly ravaging the environment for commercial gain. The depletion of natural resources, the advent of the threat of the greenhouse effect and other such catastrophes, led to the idea of utilizing lignocelluloses and other agricultural wastes for bioconversion. Lignocellulose is an abundant and renewable resource, and is therefore being exploited for the generation of numerous products. In addition, microbial degradation of this complex polymer has the capability of expanding the potential for this biomass.
The end products of such hydrolysis may be later transformed into single cell protein, liquid fuels, and other solvents or chemicals through the utilization of particular fermentative microbes. In doing so, this can greatly contribute to the elimination of agricultural wastes *(1, 2)*.

Increasing environmental concerns and governmental mandates have encouraged the development of greener chemistry to perform catalytic tasks in a variety of industrial processes. Consequently, enzymes have been introduced into consumer products such as laundry detergents and are finding increasing application in industrial processes in the food, textile and pulp and paper industries *(3)*. Initially, suitable enzymes were identified by screening strains from various culture collections. However, relatively poor enzyme stabilities, decreased catalytic activities under the conditions that characterize industrial processes, insufficient enzyme production and the availability or cost of producing the enzyme itself have proven to be formidable impediments to their large-scale application *(4)*.

Hemicelluloses, second only to cellulose, represent an enormous renewable agricultural resource, and have immense potential to be wielded as alternative sources of energy. Xylans are heteropolysaccharides that represent the most abundant type of hemicellulose synthesized in the biosphere. In recent years, there has been a growing interest in xylanases due to their potential application in the pulp, paper and fibre-processing industries. Biologically, xylanases are synthesized by microorganisms and secreted to degrade the surrounding xylan as food supply. Xylan degradation has been reported in bacteria, fungi and yeasts *(5)*. Xylanase-producing microorganisms include the thermophilic Deuteromycete, *Thermomyces lanuginosus*, which is reported to rank amongst the best cellulase-free xylanase producers in nature *(6)*.

Chlorine-free bleaching is usually carried out at acidic pH whilst a hot, caustic treatment of the wood precedes chlorine bleaching. Thus, stability and activity at high temperatures and extreme pH values are coveted attributes for a potential biotechnologically useful xylanase. For the commercial realization and economic viability of xylanase production, it is necessary to identify organisms that can hyper-produce the enzymes. Therefore, improvement in the hydrolyzing efficiency of xylanases has long been anticipated and trials, aided by modern recombinant DNA technologies, have been undertaken at an ever-increasing pace over the last few years *(2)*. Clearly, the best route for the improvement of xylanases to withstand the rigours of pulping processes (alkaline pH, high temperatures) would be to genetically tailor the xylanase gene to overcome such bottlenecks. Most investigations have focused on the external parameters (environmental pH, temperature, etc.) that maximize xylanase production, but improving the enzyme on a genetic level seems to be a much more feasible approach for the over-production of a multi-faceted enzyme with unique capabilities.

Protein Engineering – A Broad Perspective

A primary goal of protein engineering is to generate proteins with new or improved properties as a means to complement the limitations of known natural enzymes. The three main targets for optimization of enzymes for use as industrial biocatalysts are the volumetric productivity or activity of the enzyme, the stability of the enzyme under process conditions, and the availability or cost of producing the enzyme itself *(7)*. The key to protein engineering is in understanding how enzymes adapted to their unique niches. If these adaptive mechanisms are understood, then this insight could be used to engineer enzymes for non-natural, biotechnological conditions. Unfortunately, this understanding has proven elusive because these systems are both innumerable and intricate. Sometimes, it is impossible to determine if the variability between enzymes from polar extremes are the results of adaptation or the results of neutral drift. Another serious difficulty facing comparative studies is the identification of which enzyme properties have evolved under selective pressure. In addition, organisms are subject to complex combinations of selective pressures. Furthermore, not all differences in enzyme properties reflect adaptation, and biological relevance plays an important role in the selection of thermostable enzymes *(8, 9, 10)*.

Although evolution is a superb means to design proteins, it is a slow process. A process that naturally occurs over millions of years is impracticable for laboratory work and, even site-directed mutagenesis in rational protein design has its impediments. Despite many advances in protein engineering, altering the specificity of an enzyme proves to be a major stumbling block. When amino acids are altered to engineer a desired change, it is nearly impossible to predict all the small structural changes that occur to resident neighbouring amino acids. Even alteration of amino acids far from the active site has been shown in numerous systems to have large effects on enzyme capability. Orbital steering, which postulates that very small changes in the orientation of active site residues and substrates play a large role in substrate binding and catalysis, can explain some of these difficulties *(11)*. Also, proteins are surprisingly resilient toward mutation. Protein tolerance is defined as the ability of a protein to undergo mutation without disrupting its fitness or structure. Within a protein there is a distribution of tolerances. Some sites that are essential for function may not accept any mutations, while other positions can accept any amino acid substitutions with limited effect. Functional tolerance is a significant factor for the success of protein engineering. A protein that is functionally tolerant allows many mutations without disrupting its structure, making it more likely that there is a connected path in sequence space of single mutations that leads to regions of higher fitness. Tolerance also affects the quality of the mutant library. If the protein is functionally intolerant, the mutant library will consist of mostly inactive proteins *(12)*.

Enzyme engineering is undergoing the most profound and exciting transformation in its history and promises unprecedented expansion in the scope and applications of modified or improved enzymes with desired physical and catalytic properties. Two complementary strategies are currently available: rational design and directed evolution. Although both approaches have been met with great success, each has its limitations *(13)*.

Rational Design

In rational design precise mutations in amino acid sequence are determined based on detailed knowledge of protein structure, function and mechanism, and are then introduced using site-directed mutagenesis. This technology holds much promise for optimizing the desired properties of enzymes for commercial applications. It also greatly enhances the basic understanding of enzyme binding and catalytic mechanisms, thus increasing the success of future enzyme engineering efforts and lays the foundation for functional prediction of new protein sequences in databases *(14)*. The power of rational design has been demonstrated by the complete inversion of coenzyme specificities for isopropylmalate dehydrogenases *(15)*, where individual amino acid substitution or secondary structure engineering generated enzymes with desirable properties.

However, despite some of the sensational successes of rational protein engineering, numerous endeavours to genetically redesign enzymes have met with limited success. These failures might result, to some extent, from a deficient knowledge of the fundamental mechanisms required to enhance the solicited enzyme properties. Many original 'rational' engineering experiments were abortive since a significant number of attempts were based on primary amino acid sequence homologies as the only paradigm for amino acid replacements. In many instances, these substitutions were made without regard to the structural features of the protein. Such 'homology-based' engineering frequently leads to substituting rigidly conserved amino acids that do not affect the desired enzyme properties and renders the enzyme inactive because of changes in protein structure. This process overlooks key amino acid residues, particularly in the comparison of highly divergent enzymes *(15)*. The conventional approach of rational design requires confirmation of the mutation by sequencing and subsequent purification of the mutant enzymes following each round of mutagenesis so that kinetic and functional properties can be determined. Such an approach is tedious and expensive, and might be impractical for multiple cycles of mutagenesis. An efficient strategy for identifying beneficial mutants using kinetics has been put forward to greatly facilitate the rational design of enzymes that require many cycles of mutagenesis to improve their properties *(14)*.

Notwithstanding intensive investigations into the basic aspects governing protein folding and function, there are colossal disparities in the comprehension of two pivotal processes: the relationship between sequence and structure and the relationship between structure and function. As a result, the rational design of novel proteins by the classical 'reductionist' approach can be an extremely discouraging exercise (16). The directed evolution of enzymes, however, is described as a new and elegant approach to generate and identify new enzyme variants.

Directed evolution

Many clues as to how to engineer better enzymes came from drawing parallel comparisons with nature and studying how nature has created authentic enzymes. By studying the evolution of bona fide proteins, it has been learned that they are highly adaptable, incessantly changing molecules. They can sometimes acclimatize to different environments and can even adopt unique functions, at least over evolutionary time scales. It is now known that the natural processes of mutation, recombination and selection resulted in the creation of enzymes with varied functions through the evolution of a common ancestral protein of the same general structure *(16)*.

Directed evolution does not require information about how enzyme structure relates to function. Experiments that facilitate the evolution of enzymes in the laboratory under controlled conditions and well-defined selection pressures can help clarify the bedlam introduced by natural evolution. Directed evolution allows for the propagation of functional changes of enzymes with only small changes in sequence. This approach also allows for the creation of different evolutionary possibilities and the monitoring of the adaptation process. Moreover, it can be determined whether the mutations are similar to those that are found in natural homologous enzymes or whether there are multiple evolutionary pathways that culminate in the same desired functional result. Laboratory evolution is relatively immune from biological constraints and free to access sequence space for all possible enzymes that mutation, recombination and selection can generate. The only biological requirement is that the mutant enzymes must be functionally expressed in a suitable host organism *(17)*.

Directed enzyme evolution generally begins with the creation of a library of mutated genes. This is accomplished using a variety of methods including the synthesis of degenerate oligonucleotides *(18)*. Alternative methods are based on the use of chemical mutagens like sodium bisulphate, nitrous acid, formic acid, hydrazine or hydroxylamine. However, these methods do not yield highly mutagenized fragments and cannot generate all the possible base substitutions. Exposure of DNA to ultraviolet light and propagation of a particular gene in mutator strains can be used to produce a particular genetic segment containing multiple mutations allows one to mutagenize a gene of interest. Alternatively, 'poisoned' primers can be used to introduce mutations into a small region of a gene *(19)*. Random point mutagenesis by error-prone PCR (ep-PCR) involves a modified PCR protocol that uses variations in magnesium chloride ($MgCl_2$) and manganese chloride ($MnCl_2$) concentrations to achieve a mutation level of 2-5 base substitutions per gene, corresponding to an average exchange of one amino acid per mutated protein. Because of the inherent mutation bias of ep-PCR and the restrictions imposed by the genetic code, this method is not suitable for introducing all 20 amino acid residues at each position of the protein, i.e., the mutations are not truly random. In fact, an average of 5.7 amino acid substitutions is accessible for any given amino acid residue using this method *(10)*. The beginning of the 'modern era' of directed evolution can be essentially defined by the invention of DNA shuffling. This technology accesses an important facet of natural evolution that was lacking in previous formats: the ability to recombine mutations from individual genes akin to natural sexual recombination *(20, 21)*. Gene products that show improvement with respect to the desired property or set of properties are identified by selection or screening and the gene(s) encoding those enzymes are subjected to further cycles of mutation and screening in order to accumulate beneficial mutations. This evolution can involve few or many generations, depending on how far one wishes to progress and the effects of the mutations observed in each generation of mutant enzymes *(22)*.

Xylanases

Commercial potential

To really appreciate the rationale behind current studies of microbial xylanases, it is necessary to consider a broad view that takes into account not just their intrinsic interest, but also their undoubted commercial potential. The structural polysaccharides cellulose and hemicellulose together account for greater than 50% of plant biomass and are consequently the most abundant terrestrial organic molecules. The value of plant biomass as a renewable resource is thus immediately apparent *(23)*.

Xylan-degrading enzymes, especially xylanases, have considerable potential in several biotechnological applications. In some processes, the use of purified enzymes is required. However, in other applications, the presence of additional enzyme activities is desired. Commercial applications suggested for xylanases involve the conversion of xylan, which is present in wastes from the agricultural and food industry, into xylose *(2)*. Similarly, xylanases could be used for the clarification of juices, for the extraction of coffee, plant oils and starch and for the production of fuel and chemical feedstocks *(3)*.

Another application of xylanases is the use of this enzyme in poultry diets. Depression in weight gain and feed conversion efficiency in rye-fed broiler chicks has been associated with intestinal viscosity *(24)*. The efficiency of xylanases in improving the quality of bread has also been demonstrated. The introduction of *A. niger* var. *awamori* xylanase into bread dough resulted in an increase in specific bread volume. This is further enhanced when amylase in combination with xylanase is used *(25)*.

Over the last two decades, much research has been conducted on xylanases due to their potential application in the pulp, paper and fibre-processing industries. In the jute fibre industry, the xylanase pre-treatment of low grade fibres before milling may be an alternative to conventional chemical softening because the former enables the selective removal of xylan, which is assumed to be an economic and environmentally safe process. Jute fibre is a natural biodegradable product and could replace the usage of plastics and synthetic fibres *(26)*.

During kraft pulping, pulp xylan is first solubilized and later, some of it is redeposited back onto the pulp fibres. Xylanases act on these reprecipitated xylans by partially hydrolyzing them to facilitate the extraction of lignin during pulp bleaching *(27)*. The benefits of using xylanases have been mostly the economic and environmental advantages, which include:

- savings in bleaching chemicals;
- increased throughput;
- improved pulp properties such as brightness and strength;
- marketing advantage;
- easy adaptation to different bleaching sequences with minimal investment costs; and
- improved effluent with reduced AOX content *(28)*.

Thermostable Hemicellulases

Mounting social, political and environmental pressures on industry to provide alternatives to chemical-based methods, have given added impetus to the search for novel enzymes with unique and industrially significant traits. Thus, numerous organisms have been isolated from extreme environments. From the frosty Antarctic ice fields to the fiery volcanic pools, these organisms possess unique survival kits that allow them to survive under these extreme conditions. In most cases, adaptation to such extreme environments has not required completely new molecular machinery: in fact many ‘extremophilic’ enzymes are similar to their mesophilic counterparts found in more hospitable environments. Sequence comparisons indicate that these enzymes are derived from a common ancestral enzyme and have accumulated mutations that allowed them to adapt over millions of years *(29)*.

Even though there are so many advantages to using enzymes as substitutes for chemical catalysts, practical applications of enzyme catalysis are few and far between. This is largely due to their relatively poor stabilities and catalytic activities under the conditions that characterize industrial processes: high temperatures, extremes of pH or non-aqueous solvents. Enzymes evolved for the survival benefit of an organism may not exhibit features essential for *in vitro* application *(30)*.

Given the natural abundance of hemicellulose, it is not surprising that many microorganisms have enzyme systems for its hydrolysis. Moreover, given the variety and complexity of hemicelluloses, several biocatalytic steps are required to hydrolyze specific polysaccharides completely into simpler sugars that can be readily used as carbon and/or energy sources by particular microorganisms. Thermophilic microorganisms are sources of thermostable saccharolytic and hemicellulolytic enzymes that could either replace those currently used at less than optimal processing temperatures or be used in new biocatalytic applications *(31)*.

The rapid growth of the enzyme industry reflects the advantages of using enzymes as industrial catalysts, and the real and potential advantages of using enzymes from thermophiles have been well documented. The properties that allow thermostable enzymes to withstand high temperatures also confer resistance to denaturing agents *(32)*, solvents *(33)* and proteolytic enzymes *(34)*. In general, the higher the growth temperature of the organism, the more stable are its enzymes. Thus, the most stable enzymes are likely to be extracellular enzymes from the most extreme thermophiles. Extracellular enzymes are more stable than their intracellular counterparts, since they are not limited by the requirements for rapid turnover as a mechanism for controlling metabolite fluxes. Cells producing extracellular enzymes may also be subject to evolutionary pressure to minimize the loss of exported carbon; high specific activities and enhanced molecular stability would be positive consequences of such pressures *(35)*.

The concept of thermodynamic stability applies to the equilibrium between the native and unfolded state of a protein. Thermostability is a desired property in biotechnological applications for a number of reasons. Substrate solubility may be increased, the risk of microbial contamination may be minimized and the reaction rates may not only be increased in general, but may favour some side reactions over others *(35)*. Biotechnological processes may require extremes of pH, or the presence of chelators, proteases and detergents. Stability in aprotic environments would make protein catalysts interesting for a wide range of applications. Unfortunately, most proteins denature only a few degrees above the physiological temperature and this is frequently an irreversible process that rapidly draws folded protein out of the equilibrium and into the unfolded state. In general, the reason for irreversible inactivation at high temperatures or under other adverse environmental conditions is aggregation of the unfolded state *(36)*.

Biotechnological processes, such as pulping, are best carried out at elevated temperatures. The increase in temperature has a significant influence on the bioavailability and solubility of organic compounds, and is accompanied by a decrease in viscosity and an increase in the diffusion coefficient of organic compounds. Consequently, higher reaction rates due to smaller boundary layers are expected. The bioavailability of insoluble environmental pollutants can also be improved dramatically at elevated temperatures allowing efficient bioremediation *(37)*.

The use of xylanase in biobleaching of wood pulp produced in the sulphite (acid) and kraft (alkaline) processes has received much attention *(38)*. As the cooking of wood is conducted at temperatures above 145^{0}C, thermostable xylanases would be particularly useful in biobleaching directly after this process. Many of the xylanases produced by mesophilic fungi lack thermostability and this limits their industrial application. Xylanases produced by thermophilic fungi are usually more thermostable than those of mesophilic fungi *(8)*. The use of thermostable xylanases for enzymatic hydrolysis or pre-treatment of pulp at high temperatures might help in achieving technical and economic feasibilities. Moreover, as the thermophilic fungus is grown at high temperatures, there is less risk of microbial contamination and diminished cooling requirements for the fermentation process *(39)*.

The most thermostable xylanase reported to date is that from the extremely thermophilic anaerobic bacterium *Thermotoga*, which has a half-life of more than 20 min at 105^{0}C *(40)*. Despite offering a potentially rich source of a variety of very stable enzymes, the commercial development of xylanase production by archaebacteria has been impeded by difficulties in the cultivation of these organisms. Therefore, thermophilic fungi such as *T. lanuginosus* are an attractive alternative source of thermostable xylanases *(39, 41)*.

Genetically engineered Xylanases

Recombinant DNA techniques offer the means to not only enhance xylanase production but also to improve the stability and activity of the enzymes. Xylanase genes have been cloned from different microbial genera into various suitable hosts, of which *E. coli* features most commonly. The expression in *E. coli* is generally found to be lower than the parent organism, and confined to the cytoplasmic or periplasmic fractions. The absence of post-translational modifications such as glycosylation in *E. coli* and the intracellular accumulation of the recombinant xylanases have been suggested to be the key reasons for low levels of activity *(3)*.

In the last five years, a few attempts were made to improve xylanases at the genetic level. In an attempt to improve the thermostability and activity of xylanase for industrial biobleaching, the *N*-terminal region of xylanase B from *S. lividans* was replaced with the corresponding region from *T. fusca* xylanase A *(42)*. The latter xylanase retains 96% of its activity after 18 h at 75^{0}C and its catalytic domain has a 72% sequence homology to that of *S. lividans*. The stability of the *S. lividans* xylanase, however, decreases gradually above 37^{0}C. DNA shuffling of both xylanases however yielded two promising mutants that exhibited significant thermostabilities at 70^{0}C and had markedly higher activities than the parent enzymes. This study was a hallmark in the study of recombination between xylanases since it proved the theoretical concept that random shuffling between a mesophilic enzyme and its thermophilic counterpart is a promising approach for the improvement of the thermostability of a mesophilic enzyme.

To reveal structure-function relationships of family 10 glycanases, an *in vitro* molecular level shuffling experiment was conducted to accumulate useful amino acid residues from two homologous family 10 xylanases, *fxyn* of *Streptomyces olivaceoviridis* E-86 and *xylA* of *Thermomonospora alba* ULJB1, into a single chimeric enzyme. The parent genes were shuffled by crossovers at selected module borders using self-priming PCR. The shuffled constructs were cloned and two chimeras demonstrated activity against RBB-xylan and were over-expressed as His-tag fusion proteins. The chimeric proteins showed improved thermal and pH profiles compared to those of parent *fxyn*. This was apparently due to the influence of amino acids inherited from the thermophilic *xylA (2)*.

A procedure for gene shuffling between four thermophilic and two mesophilic family 11 bacterial xylanases called degenerate oligonucleotide gene shuffling (DOGS) was first described in 2001 *(18)*. This method used degenerate primers that allowed control of the relative levels of recombination between the shuffled genes by reducing the regeneration of unshuffled parental genes. This landmark procedure revolutionized traditional DNA shuffling and demonstrates the possibility of shuffling members of a gene family that are not particularly closely related, without compromising the frequency of chimeric molecules. These gene shuffling methodologies were developed with a desire to increase the temperature optimum of XynB from *Dictyoglomus thermophilum* Rt46B.1 and this study is still in progress.

Another highly successful example of modern recombinant DNA technology being used to improve xylanases was when the catalytic domain of the xylanase from the anaerobic fungus, *Neocallimastix patriciarum* was made more alkalophilic through directed evolution using error-prone PCR and site-directed mutagenesis. This was the first reported xylanase to be improved using directed evolution. Alkalophilic variants, grown on LB agar, produced large clear zones when overlaid with alkaline, xylan-containing agar. Whereas the wild-type xylanase exhibited no activity at pH 8.5, the relative and specific activities of six alkalophilic mutants were higher at pH 8.5 than pH 6. Eight amino acid substitutions were identified in the selected mutant xylanases and seven of these substitutions were assembled in a single enzyme (xyn-CDBFV) by site-directed mutagenesis. This composite xylanase not only had a relatively high specific activity, but was also more thermostable at 60^0C and alkaline tolerant at pH 10 than the wild-type xylanase. The composite mutant xylanase was a promising and suitable candidate for pulp biobleaching. This study showed a combination of rational and evolutionary protein design can greatly enhance the success of any protein engineering endeavour *(43)*.

Site-directed mutagenesis was also used to study the key amino acid residues responsible for the alkali-tolerance of the *B. pumilis* xylanase. Asparagine 71 was observed to be highly conserved in the alkaline family 11 xylanases. Substitution of this residue led to a decrease in the specific activity of the xylanase, especially in the alkaline pH range. It was noted that although other residues could also influence the pH optimum of the xylanase, they did not affect their activity in the alkaline pH range as much as that of asparagine 71. In addition, all mutant xylanases studied, changed their pH optima to a more acidic value. Thus, it was found that this residue was crucial for alkali-tolerance of the *B. pumilis* xylanase *(44)*.

The effect of increasing the number of surface arginines on the enzymatic activity and stability of *Trichoderma reesei* xylanase II was also studied *(45)*. The number of arginines was increased in two mutant series. The first set contained six arginines on different sides of the protein surface. These arginines had no significant effect on the thermostability. However, the optimal pH range became narrower. Another series of five arginines was engineered into the Ser/Thr surface of the enzyme. These mutations shifted the activity profile to the alkaline region by 0.5-1 pH units. In addition, the arginines on the Ser/Thr surface increased the enzyme activity at high temperature, although the enzyme stability in the absence of substrate decreased significantly at 50-55^0C. However, in the presence of the substrate, the thermostability increased 4-5 fold at 60-65^0C. Thus, the substrate was found to neutralize the destabilizing effect of Ser/Thr surface arginines and showed a stabilizing effect of the same mutations. The stabilizing effect of arginines at high temperatures was seen clearly only when five arginines were introduced into the Ser/Thr surface. This study paved the way for future site-directed mutagenesis studies of the family 11 xylanases.

In a recent study, *xynA* from *T. lanuginosus* DMS 5826 was mutated using error-prone PCR to improve its thermostability and catalytic activity *(46)*. The mutations were random and not restricted to a particular region of the protein and it was impossible to correlate amino acid changes with the different trends displayed by the mutants in terms of thermal stability and activity. A very significant change though was seen with mutant 1D2 where 4 amino acid substitutions occurred within the α-helix of the protein. This region was strongly conserved with the more thermostable mutant xylanases generated in this study, implying that it is entirely possible that this structural motif plays a more significant role in the stability of *xynA* than previously thought. The most profound mutation was observed with mutant 2B7-10, where a glycine residue disrupted the disulphide bridge. This disulphide bridge was partly credited with the high stability of the wild type *xynA (47)*. The mutant 2B7-10 xylanase was, however, more stable and active than *xynA*, proving that the disruption of the disulphide bridge did not have a deleterious effect on the enzyme, and somehow contributed to the improvement in its overall stability and catalytic activity. The substitution of 3 proline residues could also have contributed to the stability of the enzyme *(48)*.

Conclusion

Thus, the xylanolytic genes of many microorganisms have been sequenced and cloned in a bid to further understand the basic building blocks that contribute to the overall structure and function of these enzymes. Knowledge of the tertiary structure of an enzyme can facilitate the understanding of its function and its relationship with substrates and inhibitors. The advent of sophisticated and advanced molecular genetics heralded the beginning of a new era in enzyme technology. The availability of suitable methodology and the improved prospects for commercialization of the hemicellulases have prompted the application of recombinant DNA techniques to the study of xylan-degrading enzymes, with the result that a number of genes have been cloned and fully sequenced, to further improve them for eventual industrial use.

Industrial biocatalysis is on the verge of significant growth. Directed evolution has rapidly become the method of choice for developing enzyme- and microorganism–based biocatalysts. The recent improvement in new screening technologies will further accelerate biocatalyst development. Furthermore, the ever-expanding capabilities of rational design will lead to more powerful biocatalyst design strategies that combine the best of both approaches. Advances in other fields such as bioinformatics, functional genomics and functional proteomics will also extend the applications of directed evolution and rational design to more industrial biocatalysts.

It would thus seem that the speed and quality of evolutionary solutions to protein engineering problems is truly impressive because ever less knowledge is required about the system being optimized. This is a good indication that evolutionary enzyme engineering is rapidly moving biotechnology into a new phase. Thus, the solutions obtained by directed evolution offer a remarkable opportunity to unravel the mysteries of these amazing molecular machines. With evolution, it is clear that the future is now only limited by imagination.

References

1. Thomson, J.A. *FEMS Microbiol. Rev.* **1993,** *104*, 65-82.
2. Ahsan, M.M.; Kaneko, S.; Wang, Q.; Yura, K.; Go, M.; Hayash, K. *Enzyme Microb. Technol.* **2001,** *28*, 8-15.
3. Kulkarni N.; Shendye, A.; Rao, M. *FEMS Microbiol. Rev.* **1999,** *23*, 411-456.
4. Schellenberger, V. *ASM News.* **1998**, *64*, 634-638.
5. Dekker, R.F.H.; Richards, G.N. *Adv. Carb. Chem. Biochem.* **1976**, *32*, 277-352.
6. Singh, S.; Pillay, B.; Dilsook, V.; Prior, B.A. *J. Appl. Microbiol.* **2000**, *88*, 1-9.
7. Marrs, B.; Delagrave, S.; Murphy, D. *Curr. Opin. Microbiol.* **1999**, *2*, 241-245.
8. Jaenicke, R. *Ann. Rev. Biophys. Bioeng.* **1981**, *10*, 1-67.
9. Adams, M.; Kelly, R. *Trends Biotechnol.* **1998**, *16*, 329-332.
10. Miyazaki, K.; Arnold, F.H. *J. Mol. Evol.* **1999**, *49*, 716-720.
11. Doyle, S.A.; Fung, S.Y.F.; Koshland, D.E. *Biochemistry* **2000**, *39*, 14348-14355.
12. Voigt, C.A.; Kaufmann, S.; Wong, Z.G. *Adv. Protein Chem.* **2001**, *55*, 79-160.
13. Chen, R. *Trends Biotechnol.* **2001**, *19*, 13-14.
14. Chen, R. *Trends Biotechnol.* **1999**, *17*, 344-345.
15. Chen, R.; Greer, A.; Dean, A.M. *Proc. Natl. Acad. Sci. U.S.A.* **1996**, *93*, 12171-12176.
16. Arnold, F.H. *Chem. Eng. Sci.* **1996**, *51*, 5091-5102.
17. Arnold, F.H.; Wintrode, P.L.; Miyazaki, K.; Gershenson, A. *Trends Biochem. Sci.* **2001**, *26*, 100-106.
18. Gibbs, M.D.; Nevalainen, K.M.H.; P.L. Bergquist. Gene **2001**, *271*, 13-20.
19. Fromant, M.; Blanquet, S.; Plateau, P. *Anal. Biochem.* **1995**, *224*, 347-353.
20. Stemmer, W.P.C. *Proc. Natl. Acad. Sci. U.S.A.* **1994**, *91*, 10747-10751.
21. Stemmer, W.P.C. *Nature* **1994**, *370*, 389-391.
22. Kuchner, O.; Arnold, F.H. *Trends Biotechnol.* **1997**, *15*, 744-749.
23. Coughlan, M.P. *Biochem. Soc. Trans.* **1985**, *13*, 405-406.
24. Bedford, M.R.; Classen, H.L. In *Xylans and xylanases*; Visser, J.; Beldman, G.; Kusters van Someren, M.A.; Voragen, A.G.J., Eds.; Elsevier, Amsterdam, 1992; pp 361-370.
25. Maat, J.; Roza, M.; Verbakel, J.; Stam, H.; Santos da Silva, M.J.; Bosse, M.; Egmond, M.R.; Hagemans, M.L.D.; van Gorcom, R.F.M.; Hessing, J.G.M.; van Der Hondel, C.A.M.J.J.; van Rotterdam, C. In *Xylans and xylanases*; Visser, J.; Beldman, G.; Kusters van Someren, M.A.; Voragen, A.G.J., Eds.; Elsevier, Amsterdam, 1992; pp 349-360.

26. Hoq, M.M.; Hempel, C.; Deckwer, W. *J. Biotechnol.* **1994**, *37*, 49-58.
27. Kantelinen, A.; Hortling, B.; Sundquist, P.; Linko, M.; Viikari, L. *Holzforschung* **1993**, *47*, 318-324.
28. Christov, L.P.; Prior, B.A. *S. A. J. Sci.* **1998**, *94*, 195-200.
29. Jaenicke, R.; Bohm, G. *Curr. Opin. Struct. Biol.* **1998**, 8, 738-748.
30. Chen, K.; Arnold, F.H. *Proc. Natl. Acad. Sci. U.S.A.* **1993**, *90*, 5618-5622.
31. Duffaud, G.D.; McCutchen, C.M.; Leduc, P.; Parker, K.N.; Kelly, R.M. *Appl. Environ. Microbiol.* **1997**, *63*, 169-177.
32. Cowan, D.A.; Daniel, R.M. *Biochim. Biophys. Acta* **1982**, *705*, 293-305.
33. Zaks, A.; Klibanov, A.M. *Science* **1982**, *224*, 1249-1251.
34. Daniel, R.M.; Cowan, D.A.; Morgan, H.W.; Curran, M.P. *Biochem. J.* **1982**, *207*, 641-644.
35. Cowan, D.A. *Comp. Biochem. Physiol.* **1997**, *118*, 429-438.
36. Van den Berg, B.; Vriend, G.; Veltman, O.R.; Venema, G.; Eijsink, V.G. *Proc. Natl. Acad. Sci. U.S.A.* **1998**, *95*, 2056-2060.
37. Becker, P.; Abu-Reesh, I.; Markossian, S.; Antranikian, G.; Märkl, H. *Appl. Microbiol. Biotechnol.* **1997**, *48*, 184-190.
38. Christov, L.P.; Prior, B.A. *Biotechnol. Progress* **1997**, *13*, 695-698.
39. Singh, S.; Pillay, B.; Prior, B.A. *Enzyme Microb. Technol.* **2000**, *26*, 502-508.
40. Bragger, J.M.; Daniel, R.M.; Coolbear, T.; Morgan, H.W. *Appl. Microbiol. Biotechnol.* **1989**, *31*, 556-561.
41. Singh, S.; Madlala, A.M., Prior, B.A. *FEMS Microbiol. Rev.* **2003,** *27*, 3-16.
42. Shibuya, H.; Kaneko, S.; Hayashi, K. *Biochem. J.* **2000**, *349*, 651-656.
43. Chen, Y.L.; Tang, T.Y.; Cheng, K.J. *Can. J. Microbiol.* **2001**, *47*, 1088-1094.
44. Liu, X.; Qu, Y.; You, F.; Liu, Y. *J. Mol. Catalysis* **2002**, *18*, 307-313.
45. Turunen, O.; Vuorio, M.; Fenel, F.; Leisola, M. *Protein Eng.* **2002**, *15*, 141-145.
46. Stephens, D.E., M.S. thesis, Durban Institute of Technology, Durban, South Africa, 2003.
47. Gruber, K.; Klintschar, G.; Hayn, M.; Schlacher, A.; Steiner, W.; Kratky, C. *Biochemistry* **1998**, *37*, 13475-13485.
48. Kumar, S.; Bansai, M. *Biophys. J.* **1998**, *75*, 1935-1944.

Chapter 14

Microbial α-Glucuronidases

B. J. M. de Wet and B. A. Prior

Department of Microbiology, University of Stellenbosch, Private Bag X1, Matieland 7602, South Africa

Microbial α-glucuronidases play a crucial role in the enzymatic degradation of xylan in nature. They specifically hydrolyze the α-1,2-linkage between 4-*O*-methyl glucuronic acid and the xylose units of small oligosaccharides, liberated from polymeric xylan by the action of *endo*-β-1,4-xylanases. This review will focus on the assay for α-glucuronidase activity, the control of production of α-glucuronidase activity by microorganisms, their biochemical properties and substrate specificity, as well as the recently described crystal-structure of a microbial α-glucuronidase.

Introduction

Xylan is one of the main non-cellulosic polysaccharides found in plant cell walls and makes up between 10 and 35% of the dry weight. It is an abundant and renewable carbon source and its hydrolysis is important to many commercial processes. The hydrolysis of xylan is also central to the carbon cycle in nature, making the sugars locked up in the intact plant cell wall available for growth (*1*). Xylan consists of a backbone of 1,4-linked β-xylose units substituted with arabinofuranose, 4-*O*-methyl glucuronic acid and acetyl esters (*2*). The complete degradation of xylan requires the concerted action of the main chain cleaving *endo*-β-1,4-xylanase, β-xylosidase as well as the side-chain hydrolyzing enzymes, α-arabinofuranosidase, α-glucuronidase and acetylxylan esterase. Several α-glucuronidase enzymes have been purified and studied to date and the reaction mechanism has been elucidated. Sequence data has led to the clustering of microbial α-glucuronidases into a single family and the first crystal structure of a microbial α-glucuronidase has resulted in additional insight into the biochemical properties of these enzymes.

Enzyme assay

The assay for α-glucuronidase activity is based on the hydrolysis of the α-1,2 bond between 4-*O*-methylglucuronic acid and the xylose unit of small xylo-oligosaccharides (*3*). These substrates are prepared from acid or enzymatic hydrolyzates of xylan using anion exchange chromatography to separate the acidic and neutral oligosaccharides followed by gel filtration to separate the different aldouronic acids (*4*). Alternatively polymeric xylan or native lignocellulosic substrates are used in conjunction with high amounts of xylanase activity, generating the aldouronic acid substrates *in situ* (*2*). The colorimetric detection of the 4-*O*-methylglucuronic acid released is the most commonly used method for determining α-glucuronidase activity (*5*). This method employs a modification of the Somogyi-Nelson procedure for the determination of reducing sugars (*6*), using high salt and low copper concentrations, for the specific detection of uronic acids (*7*). α-Glucuronidase activity has also been measured by determining the amount of product released by high-performance liquid chromatography (*3, 8*), and by gas-liquid chromatography (*4*).

p-Nitrophenyl-glycosides have been used extensively as synthetic substrates in the assay of several glycosidase activities. Microbial α-glucuronidases are; however, unable to hydrolyze *p*-nitrophenyl-glucuronide and only α-glucuronidases from mollusks have been shown to be able to use it as a substrate (*4,9-11*). A novel assay has been proposed using 4-nitrophenyl 2-(4-O-methyl-α-D-glucopyranuronosyl)-β-D-xylopyranoside as a substrate. Its hydrolysis by α-glucuronidase is coupled with the hydrolysis of the *p*-nitrophenyl xylopyranoside product by excess exogenous β-xylosidase activity. The aldobiouronic acid part of the structure serves as the recognition element in the substrate and the presence of the *p*-nitrophenyl group does not appear to hinder substrate binding or hydrolysis of the α-1,2 glycosidic bonds. This assay offers an improvement over the colorimetric procedure, routinely used, as it is not sensitive to the presence of salts or other reducing sugars in reaction mixture. The assay has the additional advantage of being highly specific, as no *p*-nitrophenol is produced in the absence of α-glucuronidase activity (*12*), and it has already been used successfully in the characterization of *Cellvibrio japonicus* α-glucuronidase (*13*). The utility of the assay is at present limited by the availability of the substrate and the requirement for a pure β-xylosidase.

Production of α-glucuronidase activity by microorganisms

Several studies have been performed on different microorganisms in which the effect of carbon source on the production of α-glucuronidase activity has been investigated. These studies seldom use the same collection of carbon sources, making direct comparisons difficult. There are definite differences between microorganisms in terms of induction of α-glucuronidase activity, although a few general principles apply. The production of α-glucuronidase by most microorganisms requires growth in the presence of xylan-based or lignocellulosic carbon sources, while low levels of α-glucuronidase activity are constitutively produced in the absence thereof. This notion is supported by the fact that every α-glucuronidase purified so far was from an organism cultivated on polymeric xylan or lignocellulosic material. *C. japonicus* cultivated in Luria-Bertani broth produced α-glucuronidase activity in the presence of several xylan polymers, but no activity in the absence thereof (*14*). The same effect was observed for *Streptomyces olivochromogenes* and *Streptomyces flavogriseus* cultivated in the presence and absence of 1% oat spelt xylan or lignocellulosic material, although low activity was detectable without the inducers (*15*).

The oligomeric breakdown products of xylan, rather than the constituent monosaccharides, act as inducers for the production of α-glucuronidase activity (*14,16*). A sevenfold increase in α-glucuronidase activity in the culture filtrate of a *Thermoanaerobacterium* sp. was noted when polymeric xylan was used as carbon source instead of xylose (*16*). Similarly, no α-glucuronidase activity was found when *C. japonicus* was cultivated on glucuronic acid. Notably, induction of α-glucuronidase is repressed by the presence of glucose. α-Glucuronidase activity was induced when *C. japonicus* was grown on 1% oat spelt xylan as sole carbon source. However, production of the enzyme was completely repressed when the medium contained 1% glucose in addition to 1% oat spelt xylan (*14*). In the case of *Fibrobacter succinogenes*, growth on pure cellulose leads to poorer induction of α-glucuronidase activity compared to growth on lignocellulose (*17*). Contrary to these findings, no difference in activity was observed in the rumen anaerobic fungus *Piromonas communis* when grown on cellulose compared to when grown on xylan (*18*).

Northern analysis supports the enzyme activity data. The *T. reesei glr1* gene was the most highly induced by growth on xylobiose, an oligosaccharide breakdown product of xylan, and not induced at all by growth on xylose. This induction was repressed by the presence of glucose. The glucose repressor protein Cre1 is most likely not involved in glucose-induced repression of the transcription of the *glr1* gene, as the *glr1* transcript was not detectable in the *T. reesei cre*1-1-mutant strain Rut-C30 when cultivated on glucose. Induction of the *glr1* gene was also observed on glucuronoxylan from beech, arabinoxylan from oat spelts and unsubstituted xylan from birch, as well as cellulose and arabitol. Induction of the *glr1* gene was also observed in glucose-depleted medium.

A study of the effect of different carbon sources on the transcript levels of the *aguA* gene of *Aspergillus tubingensis* showed that transcription was induced by the presence of xylose, xylobiose and birchwood xylan but not glucuronic acid in the culture medium (*19*). The transcription of the *aguA* gene was attenuated in the presence of glucose in the medium in addition to xylose and completely inhibited when glucose was added with xylan. In *A. niger*, induction of *aguA* by xylose and xylan is mediated through XlnR, a transcriptional activator that controls the expression of several xylanolytic enzymes. No induction of the *aguA* gene was observed in a strain containing a functional mutant of *xlnR* (*20*). The transcription factor was found to recognize a divergent XlnR binding site in the promoter region of the *aguA* gene. Two functional CreA binding sites were additionally found in the areas upstream of the *aguA* coding sequence, indicating that glucose repression is mediated through CreA in *A. tubingensis*. Evidence was also found for an XlnR-independent regulatory system that induces the transcription of the *aguA* gene in response to glucuronic acid or galacturonic acid, the former being a product of xylan degradation (*21*).

The production of the complement of enzymes that degrade xylan to catabolizable monosaccharides requires a substantial energy investment on the part of the organism that produces them. The cellular location of the accessory enzymes produced by each organism is therefore likely a reflection of the strategy it follows to gain preferential access to the nutrients released from the degradation of xylan and its lifestyle. Most of the α-glucuronidases studied thus far appear to be secreted although investigation of the exact localization of the enzyme activity is not always performed or reported.

There are exceptions where the activity remains cell-associated or intracellular. Two notable examples are the intracellular α-glucuronidase from

Bacillus stearothermophilus (*22*) and the cell surface associated α-glucuronidase from *C. japonicus* (*14*). *B. stearothermophilus* protects substrate access by transporting aldotetraouronic acid, produced by the action of an extracellular xylanase (*xynA*), across the cell wall via a putative four protein membrane transporter (*orf1-4*) that also demethylates *O*4 of the aldouronic acid. It is subsequently cleaved to xylose and glucuronic acid by an intracellular α-glucuronidase (*aguA*), xylanase (*xynA2*) and β-xylosidase (*xynB*). All of the genes for glucuronic acid utilization were found on a single cluster in the genome of the organism. *C. japonicus* follows a different strategy by localizing the xylanolytic accessory enzymes on its cell surface, thereby hydrolyzing the small oligosaccharides, generated by the action of a secreted xylanase, to monosaccharides in close proximity to the membrane transporters that carry them across the cell wall. Other cell-associated α-glucuronidases have been purified from *Aspergillus niger* (*23*), *Thermoanaerobacterium* spp. (*9,16*) and *Clostridium stercorarium* (*9*).

Biochemical properties of purified α-glucuronidases

The existence of an enzyme capable of releasing glucuronic acid from glucuronoxylan was first observed in the culture supernatant of *T. reesei* (*24*) and several microbial α-glucuronidases have been purified and studied to date (Table 1). Based on primary sequence, all of the known α-glucuronidases are assigned to family 67 of the Carbohydrate Active enZYmes, and they are the sole members of this family (http://afmb.cnrs-mrs.fr/CAZY/). There are two distinct groups in this family based on sequence similarity and quaternary structure. The fungal α-glucuronidases are monomeric proteins with molecular weights between 90 and 150 kDa and acidic pH optima, ranging from 3.0 to 4.8. In contrast, bacterial α-glucuronidases are dimeric proteins with subunits with monomeric molecular weights of around 70 kDa. An exception is the recombinantly expressed *Thermotoga maritima* enzyme that shows a variable oligomeric structure in response to changing salt concentrations. At low salt concentrations the enzyme occurs as an oligomer with a molecular weight in excess of 630 kDa, while in the presence of high salt concentrations the enzyme shows both a hexameric (450 kDa) and a dimeric (140 kDa) conformation (*25*). The pH optima for the bacterial α-glucuronidases, although acidic, are generally higher than those for the fungal α-glucuronidases, ranging from 5.4 to 6.5.

The kinetic properties of several of the purified α-glucuronidases have been investigated. Several aldouronic acids have been used as substrates, which makes direct comparisons between enzymes from different species difficult.

Table 1. Properties of purified α-glucuronidases

Organism	M_r (kDa)	pH[c]	T[c] (°C)	V_{max} (U/mg)	K_m (mM)	Substrate specificity*
Eukaryotes						
Aspergillus niger (23)	150[a], 130[b]	4.8	60	1.4	0.37^{4}	tetrao, not pNP-GlcU
Aspergillus niger (23)	150[a], 130[b]	4.8	60	4.7	0.47^{4}	tetrao, not pNP-GlcU
Aspergillus tubingensis (19)	100[a], 107[b]	4.5-6.0	70	52	$0.14^{2/3}$	trio, trace xylan
Phanerochaete chrysosporium (8)	112[b]	3.5	nr	4.5	nr	trio>tetrao, not bio, trace xylan
Shizophyllum commune (26)	125[b]	4.5-5.5	nr	18	nr	trio, xylan
Thermoascus aurianticus (5)	118[a], 117[b]	4.5	65	4.0	0.14^{4}	bio to octao, xylan
Trichoderma reesei (27)	91[b]	4.5-6.0	nr	28	nr	trio>tetrao>pentao=bio, trace xylan
Helix pomatia (28)	180[a], 97[b]	3.0	50	2.8	17.6^{2}	pNP-GlcU, trio to pentao
Prokaryotes						
Bacillus stearothermophillus (11)	150[a], 78[b]	6.0	65	42	0.2^{4}	tetrao, not pNP-GlcU, not xylan
Bacillus stearothermophillus (29)	161[a], 78[b]	6.5	40	15.3	0.78^{3}	trio, not xylan
Cellvibrio japonicus (14)	150[a], 83[b]	6.3	nr	84.5	nr	pentao>tetrao>trio>bio, not xylan
Clostridium stercorarium (9)	124[a], 72[b]	6.0	nr	1.7	nr	tetrao, not pNP-GlcU, not xylan
Thermoanaerobacterium sp. *(16)*	130[a], 74[b]	5.4	60	8.4	0.76^{3}	bio>trio>tetrao, trace xylan
Thermoanaerobacterium saccharolyticum (9)	118[a], 71[b]	6.0	nr	10	nr	tetrao, not pNP-GlcU, not xylan
Thermotoga maritima (25)	140[a], 79[b]	6.3	85	31	0.95^{3}	trio

[a] determined by gel filtration, [b] determined by SDS-PAGE, [c] pH and temperature optima

nr – not reported

the superscripted number above the K_m-value indicates the degree of polymerisation of the aldouronic acid substrate used

* substrates tested and substrate preference where known, only prefix of aldouronic acids given

The specific activities for the purified enzymes range between 1.4 and 84.5 U/mg protein and the α-glucuronidases have K_m-values for aldouronic acids oligomers in the millimolar range. As a result of the complex structure of the xylan polymer and the variety of possible hydrolysis products, a number of issues arise in terms of substrate specificity.

Most of the microbial α-glucuronidases appear to have only very little or no activity against polymeric xylan compared to aldouronic acid oligomers. The only two exceptions are the α-glucuronidases from *Schizophyllum commune* (*26*) and *Thermoascus aurantiacus* (*5*). The latter enzyme was capable of releasing 4-*O*-methyl glucuronic acid from birchwood xylan at half the rate observed for aldotriouronic acid. Paradoxical results have been found by two studies that investigated the effect of the presence of substituents on the release of 4-*O*-methyl glucuronic acid from birch glucuronoxylan (*5,27*). *T. reesei* enzyme was capable of releasing about five times more 4-*O*-methyl glucuronic acid from acetylated than deacetylated birch xylan. Conversely, *S. commune* enzyme was capable of releasing five times more 4-*O*-methyl glucuronic acid from deacetylated than acetylated glucuronoxylan. The latter result would fit better with the current paradigm on xylan degradation, as it is believed that the presence of substituents limits accessibility of the enzyme to neighbouring glycosidic linkages.

The length of the aldouronic acid oligosaccharide may also have a marked effect on activity. The *C. japonicus* enzyme shows a preference for longer substrates and turnover number increases two-fold in the range from aldobiouronic to aldopentaouronic acid (*30*). The opposite trend is observed for the α-glucuronidases from *Thermoanaerobacterium* sp., *Phanerochaete chrysosporium* and *T. reesei* (*8,16,27*). This group of enzymes shows increasing activity as the chain length decreases to aldotriouronic acid. However, all three enzymes differ in their preference for aldobiouronic acid. *Thermoanaerobacterium* sp. α-glucuronidase shows its highest reaction rate on aldobiouronic acid, while the *T. reesei* enzyme has its lowest reaction rate on this substrate and *P. chrysosporium* enzyme cannot hydrolyze it at all. Chain length may also not influence activity to a significant degree, as is the case for the *T. aurantiacus* enzyme. This α-glucuronidase hydrolyzes aldobiouronic to aldooctaouronic acid at the same rate and maintains half of the rate of activity it has on aldouronic acids on the xylan polymer (*5*).

It appears to be a general rule that α-glucuronidases have a preference for 4-*O*-methyl glucuronic acid substituents on the terminal non-reducing end xylose unit of the oligosaccharide and almost all of the studies on substrate specificity have used substrates of this nature. Only one study so far shows direct evidence to this effect. The α-glucuronidase from *A. tubingensis* was found to lack activity against two aldouronic acid substrates containing non-terminal 4-*O*-methyl glucuronic acid substituents (*12*). Additional evidence for this fact comes from studies on the effect of combinations of α-glucuronidase and other xylanolytic enzymes on the amount of 4-*O*-methyl glucuronic acid released from xylan. In a study on the *T. reesei* α-glucuronidase, it was observed that only the combination of α-glucuronidase, xylanase and β-xylosidase resulted in release of significant amounts of 4-*O*-methyl glucuronic acid. Very little 4-*O*-methyl glucuronic acid was released by the combination of only α-glucuronidase and xylanase (*27*). Xylanases of family 11, such as those used in this study, release oligosaccharides with the substituents on non-terminal xylose units as they cannot access the xylosydic bond adjacent to substituted xylose units. Xylosidase activity is strictly required to remove the unsubstituted non-reducing end xylose units, resulting in the production of aldouronic acid oligomers carrying the 4-*O*-methyl glucuronic acid linked to the terminal non-reducing end xylose unit.

Reaction mechanism of α-glucuronidase

Glycoside hydrolases follow either a retaining mechanism, involving two steps and an enzyme-bound glycoside intermediate, or an inverting mechanism involving a single step direct displacement by water at the anomeric center. Both mechanisms involve an acid/base catalytic pair. The distance between the catalytic residues differs, being approximately 5Å in enzymes following a retaining mechanism, and between 9Å and 10Å in enzymes following an inverting mechanism, allowing for the presence of nucleophilic water for direct attack at the anomeric center (*31*). The *A. tubingensis* α-glucuronidase enzyme is the only α-glucuronidase that has been investigated specifically in terms of its reaction mechanism. The authors used proton-NMR to show that the α-anomer of the 4-*O*-methyl glucuronosyl residue in the aldotetraouronic acid substrate was converted to the β-anomer in the free 4-*O*-methyl glucuronic acid that was produced, indicating an inverting reaction mechanism (*12*). The highly conserved nature of the family 67 α-glucuronidases makes it likely that they all follow the same mechanism and this notion is supported by crystallographic data on the distance between the catalytic carboxylates in the *C. japonicus* α-glucuronidase, as well as the anomeric configuration of the reaction products co-crystallized with the enzyme (*30*).

Crystal structure of α-glucuronidase

The crystal structure of GlcA67A, the α-glucuronidase from *C. japonicus* has recently been solved and has added significant insight into the reaction mechanism and substrate specificity of this enzyme. The enzyme is a dimer, consisting of two identical subunits, each containing three domains. The enzyme can be described as being butterfly-shaped when it is oriented so that the dimer-surface runs vertically along the central axis, with the N-terminus at the top and the C-terminus at the bottom. The N-terminal domain forms the "top wing of the butterfly" and contains a two-layer β-sandwich, with each β-sheet made up of five strands. It also contains two small helices that interact with the central catalytic domain. The central domain itself is situated below the N-terminal domain and makes up the "bottom wing of the butterfly". It has a $(\beta/\alpha)_8$ barrel structure and contains the catalytic residues on the C-terminal side of the protein. The C-terminal domain is situated behind the catalytic domain, covering one of its faces, and is primarily made up of α-helices. It forms the largest part of the dimer surface and also has interactions with the α-helices of the N-terminal domain (*30*).

Co-crystallization of native GlcA67A with the reaction products glucuronic acid, xylobiose and a combination of glucuronic acid and xylotriose revealed the position and identity of the catalytic acid and several of the interactions that contribute to substrate specificity (Figure 1) (*30*). A catalytically inactive mutant of GlcA67A, E292A was used to characterize the enzyme-substrate complex and kinetic characterization of mutants of evolutionarily invariant amino acids in GlcA67A further confirmed their role in substrate recognition (*13*). The catalytic acid involved in the hydrolysis reaction was identified as Glu292 based on its position and distance relative to *O*2 of the terminal xylose unit. No acidic side-chain apart from Glu292 was found to be close enough to donate a proton to the glycosidic-bond oxygen. A greater than million-fold decrease in activity was observed after mutation of this residue to alanine or cysteine. The identity of the catalytic base that activates water for nucleophilic attack at the anomeric carbon is uncertain. Both Asp365 and Glu393 are ideally positioned at the β-face of the glucuronic acid to fulfill this role and mutation of either amino acid to alanine results in a greater than a million-fold reduction in activity. The position of the glucuronic acid product in the native enzyme is very similar to that of the 4-*O*-methyl glucuronic acid substrate in the E292A mutant. Both are in a $^4C_1$ chair conformation and apart from the change from the α-anomer in the substrate to the β-anomer in the product, the pyranoid ring is also rotated by 27° around the position of the C4 carbon. The position of the C1 atom in the product, 1.4 Å further away from the catalytic base/s and the α-position of the hydroxyl, create the ideal position for a water molecule to attack

C1. In this position the water molecule can be activated by either of the catalytic bases (*13,30*).

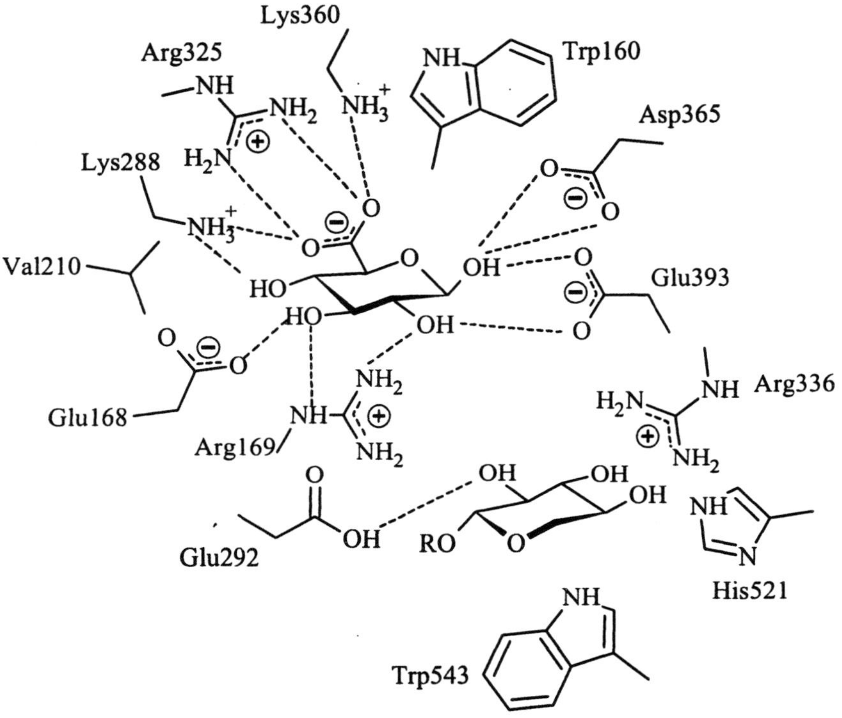

Figure 1. Schematic representation of the recognition elements involved in the binding of the reaction products to *C. japonicus* α-glucuronidase, GlcA67A (*14*).

The substrate binding site of the enzyme consists of a deep, partially hydrophobic pocket, with the specific recognition elements for glucuronic acid and the terminal non-reducing end xylose unit situated in the deepest recesses of the pocket. The terminal xylose unit is completely enclosed at the non-reducing end by Tyr329 and Arg336, the side-chain of the latter being able to form hydrogen bonds with *O*3 and *O*4 of this xylose unit. This feature explains the inability of the *C. japonicus* α-glucuronidase to hydrolyze 4-*O*-methyl glucuronic acid side-chains from internally positioned xylose units. The terminal xylose unit is also in an ideal position to form π-stacking interactions with Trp543, stabilizing its position in the active center. In the E292A mutant co-crystallized with aldotriouronic acid, the position of the non-reducing end xylose unit could not be determined, as it had no ordered structure. This finding might challenge the importance of the role of the xylose-Trp543 interaction in substrate recognition. Kinetic analysis of the W543A mutant of GlcA67A, however, revealed a 10^5-fold reduction in catalytic efficiency, confirming the essential nature of this interaction (*13,30*).

The 4-*O*-methyl glucuronic acid binding site is situated in the deepest part of the pocket and the carboxylate group is stabilized by three basic amino acids, namely Lys288, Arg325 and Lys 360. The importance of these interactions was confirmed by studying the kinetic properties of the three mutants K288A, R325A and K360A. Removal of any of these positively-charged side-chains led to at least 10^4-fold reduction in the catalytic efficiency. The most important of these interactions is that of the guanidinium group of Arg325 with the 4-*O*-methyl glucuronic acid carboxylate. It showed the largest reduction in catalytic efficiency when mutated, is geometrically and spatially in the most complementary position relative to the carboxylate group, and also has a higher pK_a in solution than the lysine amino groups. Further evidence for the importance of the 4-*O*-methyl glucuronic acid carboxylate in substrate recognition comes from the observation native GlcA67A has a twenty-fold higher K_i for glucose than for glucuronic acid (*13,30*).

Another structural recognition element is the methyl-group of 4-*O*-methyl glucuronic acid. In the crystal structure of the E292A mutant in association with its substrate, it was noted that the methyl-group is buried in a hydrophobic sheath made up by Trp160 and Val210. The role of the conserved Val210 is not clear as mutation of this amino acid to polar residues such as serine or asparagine did not have the same impact on catalytic efficiency as many of the other mutants studied. The inhibition constant of the native enzyme for glucuronic acid was, however, ~20 times lower than either V210N or V210S confirming the importance of a non-polar side-chain in close proximity to the methyl group. Trp160 is also ideally positioned to form π-stacking interactions with the surface of the pyranoid ring and the W160A mutant showed a 10^6-fold reduction in catalytic efficiency. The positioning of the glucuronic acid deeper in the active center in relation to the xylo-oligosaccharide implies the sequential release of first the xylo-oligosaccharide and thereafter the glucuronic acid and may have important implications in terms of the kinetic properties of the enzyme (*13,30*).

The role of α-glucuronidase in xylan degradation

Few in-depth studies have investigated the synergistic role of α-glucuronidase in the enzymatic degradation of xylan (*8,19,26,27,32*). All of the studies addressed the effect of xylanase and xylosidase activity on the release of 4-*O*-methyl glucuronic acid from xylan. As expected, the addition of either xylanase or xylosidase activity to the reaction significantly increases the amount of 4-*O*-methyl glucuronic acid released from the xylan polymer by α-glucuronidase. The addition of both xylanase and xylosidase to α-glucuronidase resulted in a far greater release of 4-*O*-methyl glucuronic acid from xylan than the addition of either only xylanase or only β-xylosidase to α-glucuronidase. The combined action of xylosidase and xylanase leads to the release of small xylo-oligosaccharides substituted with 4-*O*-methyl glucuronic acid on the non-reducing end xylose unit. This type of compound has been shown to be the preferred substrate for most α-glucuronidases and the above-mentioned findings are thus in line with this view. It stands to reason that other substituents on the xylan backbone, such as acetyl and arabinofuranose groups, could hinder the access of the α-glucuronidase to the α-1,2 glycosidic bond. The addition of side-chain cleaving enzymes increased the amount of 4-*O*-methyl glucuronic acid released by α-glucuronidase, whether α-glucuronidase was applied alone or in combination with β-xylosidase or *endo*-β-1,4-xylanase or both (*8,26*).

The effect of α-glucuronidase activity on the release of reducing sugars from xylan by the action of xylanase, xylosidase or a combination of the two has also been investigated (*19,27,32*). The degree to which α-glucuronidase enhances the release of reducing sugars from xylan is dependent on the nature of the xylanase that is used. The addition of *T. reesei* α-glucuronidase to *T. reesei endo*-β-1,4-xylanase I enhanced the release of reducing sugars from xylan to a much greater degree than the addition of *T. reesei* α-glucuronidase to *T. reesei endo*-β-1,4-xylanase II. This result is unexpected as both *endo*-β-1,4-xylanases belong to family 11 of the carbohydrate active enzymes and share a similar active site architecture (*27*). The effect of α-glucuronidase on the release of reducing sugars from xylan is also more pronounced when β-xylosidase acts alone than when *endo*-β-1,4-xylanase acts alone or in combination with β-xylosidase. This can be attributed to the difference in the nature of the bonds that β-xylosidase and *endo*-β-1,4-xylanase attack. β-xylosidase is an exo-acting enzyme and the presence of any substituents linked to the terminal non-reducing end xylose unit will prevent it from hydrolyzing the terminal xylose, whereas the endo nature of xylanase will allow it to hydrolyze β-1,4-xylosidic linkages that might exist further down the chain. The relatively low degree of substitution of xylan with 4-*O*-methyl glucuronic acid (1/10 xylose units in hardwood xylan and 1/5 xylose units in softwood xylan) likely allows for a high degree of depolymerization by a combination of *endo*-β-1,4-xylanase and β-xylosidase. This leaves few xylosidic linkages to be hydrolyzed after removal of the 4-*O*-methyl glucuronic acid side-chains (*19,27,32*).

The complete hydrolysis of xylan is dependent on the removal of all substituents from the backbone. α-Glucuronidases play a critical role in this process by making additional sites available for the further action of *endo*-β-1,4-xylanases and β-xylosidases. The accumulated knowledge of the enzymatic properties and functioning of α-glucuronidases, as well as the elucidation of the crystal structure of a bacterial α-glucuonidase have added another piece into the puzzle of lignocellulose biodegradation.

References

1. Beg, Q. K.; Kapoor, M.; Mahajan, L.; Hoondal, G. S. *Appl. Microbiol. Biotechnol.* **2001,** *56*, 326-338.
2. Puls, J. In *Xylans and Xylanases*; Visser, J.; Beldman, G.; Kusters-van Sommeren, M. A.; Voragen, A. G. J., Eds.; Elsevier: Amsterdam, 1992; pp. 213-224.
3. Puls, J.; Schmidt, O.; Granzow, C. *Enzyme Microb. Technol.* **1987,** *9*, 83-88.
4. Fontana, J. D.; Gebara, M.; Blumel, M.; Schneider, H.; MacKenzie, C. R.; Johnson, K. G. *Methods Enzymol.* **1988,** *160*, 560-571.
5. Khandke, K. M.; Vithayathil, P. J.; Murthy, S. K. *Arch. Biochem. Biophys.* **1989,** *274*, 511-517.
6. Nelson, N. *J Biol Chem* **1944,** *153*, 375-380.
7. Milner, Y.; Avigad, G. *Carbohydr. Res.* **1967,** *4*, 359-361.
8. Castanares, A.; Hay, A. J.; Gordon, A. H.; McCrae, S. I.; Wood, T. M. *J. Biotechnol.* **1995,** 43, 183-194.
9. Bronnenmeier, K.; Meissner, H.; Stocker, S.; Staudenbauer, W. L. *Microbiology* **1995,** *141*, 2033-2040.
10. Kawabata, Y.; Ono, K.; Gama, Y.; Yoshida, S.; Kobayashi, H.; Kusakabe, I. *Biosci. Biotechnol. Biochem.* **1995,** *59*, 1086-1090.
11. Zaide, G.; Shallom, D.; Shulami, S.; Zolotnitsky, G.; Golan, G.; Baasov, T.; Shoham, G.; Shoham, Y. *Eur. J. Biochem.* **2001,** *268*, 3006-3016.
12. Biely, P.; Hirsch, J.; la Grange, D. C.; van Zyl, W. H.; Prior, B. A. *Anal. Biochem.* **2000,** *286*, 289-294.

13. Nagy, T.; Nurizzo, D.; Davies, G. J.; Biely, P.; Lakey, J. H.; Bolam, D. N.; Gilbert, H. J. *J. Biol. Chem.* **2003,** *278*, 20286-20292.
14. Nagy, T.; Emami, K.; Fontes, C. M.; Ferreira, L. M.; Humphry, D. R.; Gilbert, H. J. *J. Bacteriol.* **2002,** *184*, 4925-4929.
15. Johnson, K. G.; Harrison, B. A.; Schneider, H.; MacKenzie, C. R.; Fonrana, J. D. *Enzyme Microb. Technol.* **1988,** 10, 403-409.
16. Shao, W.; Obi, S. K. C.; Puls, J.; Wiegel, J. *Appl. Environ. Microbiol.* **1995,** *61*, 1077-1081.
17. Smith, D. C.; Forsberg, C. W. *Appl. Environ. Microbiol.* **1991,** 57, 3552-3557.
18. Wood, T. M.; Wilson, C. A. *Appl. Microbiol. Biotechnol.* **1995,** *43*, 893-900.
19. De Vries, R. P.; Poulsen, C. H.; Madrid, S.; Visser, J. *J. Bacteriol.* **1998,** *180*, 243-249.
20. Van Peij, N. N.; Gielkens, M. M.; de Vries, R. P.; Visser, J.; de Graaff, L. H. *Appl. Environ. Microbiol.* **1998,** *64*, 3615-3619.
21. De Vries, R. P.; Van De Vondervoort, P. J. I.; Hendriks, L.; Van De Belt, M.; Visser, J. *Mol. Genet. Genomics.* **2002,** *268*, 96-102.
22. Shulami, S.; Gat, O.; Sonenshein, A. L.; Shoham, Y. *J. Bacteriol.* **1999,** *181*, 3695-3704.
23. Uchida, H.; Nanri, T.; Kawabata, Y.; Kusakabe, I.; Murakami, K. *Biosci. Biotechnol. Biochem.* **1992,** 56, 1608-1615.
24. Dekker, R. F. H. *Biotechnol. Bioeng.* **1983,** *35*, 1127-1146.
25. Ruile, P.; Winterhalter, C.; Liebl, W. *Mol. Microbiol.* **1997,** *23*, 267-279.
26. Tenkanen, M.; Siika-aho, M. *J. Biotechnol.* **2000,** *78*, 149-161.
27. Siika-aho, M.; Tenkanen, M.; Buchert, J.; Puls, J.; Viikari, L. *Enzyme Microb. Technol.* **1994,** *16*, 813-819.
28. Kawabata, Y.; Kuno, A.; Gama, Y.; Kusakabe, I. *Biosci. Biotechnol. Biochem.* **1996,** *60*, 524-525.
29. Choi, I. D.; Kim, H. Y.; Choi, Y. J. *Biosci. Biotechnol. Biochem.* **2000,** *64*, 2530-2537.
30. Nurizzo, D.; Nagy, T.; Gilbert, H. J.; Davies, G. J. *Structure* **2002,** *10*, 547-556.
31. Henrissat, B.; Davies, G. J. *Curr. Opin. Struct. Biol.* **1997,** *7*, 637-644.
32. De Vries, R. P.; Kester, H. C.; Poulsen, C. H.; Benen, J. A.; Visser, J. *Carbohydr. Res.* **2000,** *327*, 401-410.

Chapter 15

Microbial Feruloyl Esterases

Karl Rumbold[1], Georg Gübitz[2], and Bernard A. Prior[1]

[1]Department of Microbiology, University of Stellenbosch, Private Bag X1, Matieland 7602, South Africa

[2]Department of Environmental Biotechnology, Graz University of Technology, Graz, Austria

Feruloyl esterases (EC 3.1.1.73) are referred to as carbohydrate esterases with high specific activity for hydroxycinnamates. Generally these enzymes are found in supernatants of fungal and bacterial cultures grown on plant material or plant cell wall extracts. For the identification and characterization of feruloyl esterase activity a range of different assay methods have been developed, which can also be used to group enzymes according to their substrate specificity. According to their sequence structure, both fungal and bacterial feruloyl esterases are very diverse. Sequence information of several feruloyl esterases was used to build a phylogenetic tree and to discover homology to other closely related enzyme families.

INTRODUCTION

Lignocellulose is considered to be the most important energy and commodity source for the future (*1*). At present, lignocellulose serves as raw material for the paper, pulp and fiber industry, the food industry, and the animal feed industry. The relative abundance of lignocellulose has resulted in extensive biological and chemical investigations on the material. Lignocellulose from wood and from grasses and cereals is of special interest for biotechnological applications. With the advent of enzyme technology, new methods for the extraction of cellulose, the digestion of forages and the production of dietary fiber were created (*2, 3, 4*). Another key area of investigation is the conversion of plant biomass to bioenergy and biocommodities (*5, 6*). Thus, lignocellulose-degrading microorganisms, their enzyme systems, and the plant cell wall have been extensively investigated with the ultimate goal to apply the knowledge to the numerous existing and prospective industries. However, the recalcitrance and poor solubility of plant cell wall material is a major challenge in the enzymatic degradation of lignocellulose. The extent of wall digestibility by microorganisms may well depend upon the cross-linked nature of wall polysaccharides (*7*). Alkaline treatment of the wall dramatically increases the microbial digestibility of graminaceous forage, enhancing the nutritional value of the forage fiber. Cleavage of alkaline-labile linkages has a major role in the increased digestibility and solubility of lignocellulosic material, suggesting the involvement of ester bonds (*8, 9*). Hydroxycinnamic acid ester cross-linkages between wall components are present in both the primary and the secondary walls. Therefore, hydroxycinnamic acid-polysaccharide complexes are important in the structure and function of walls during plant development as well as the degradation of the walls for extraction of fibers and for bio-energy. These so-called lignin-carbohydrate complexes restrict the extend of enzymatic plant cell wall degradation in several industrial applications (*10, 11*).

Feruloyl esterases have been identified to play a major role in the enzymatic decomposition of cell wall polymers. Both microorganisms and the plant itself produce these enzymes for degradation and cell wall extension purposes respectively. Therefore, feruloyl esterases are diverse in their catalytic properties *(12)*.

PRODUCTION OF FERULOYL ESTERASES BY MICROORGANISMS

The presence of feruloyl esterase activity was first detected in culture filtrates of *Streptomyces olivochromogenes* grown on oats spelt xylan and wheat bran (*13*) and have subsequently been found in a number of bacterial and fungal genera (Table I). Most researchers have found that feruloyl esterases are inducible in microorganisms and that a lignocellulosic substrate is required for induction. Repression occurs when the growth substrate contains glucose, suggesting that feruloyl esterases are catabolically repressed similar to most enzymes involved in lignocellulose degradation. The mechanism of induction of feruloyl esterase activity is still unclear and also appears to vary between different microorganisms. The production rate depends on the type of lignocellulosic substrate, especially the content of esterified phenolic acid and supplemented free phenolic acid. However, a high amount of esterified phenolic acids in the substrate does not always guarantee high activity levels. Feruloyl esterase activity found in culture supernatants from *S. olivochromogenes* grown on oat spelt xylan (*13*) and *Schizophyllum commune* grown on cellulose (*14*), both of which are substrates containing negligible amounts of esterified hydroxycinnamic acids was higher compared to destarched wheat bran that contains 1% esterified ferulic acid. However, *Aspergillus niger* produced a 2.3-fold increase in feruloyl esterase (FAEA) activity in the supernatant when oat spelt xylan was supplemented with free ferulic acid compared to oat spelt xylan alone (*15*). This level of FAEA activity was comparable to the amount of enzyme activity produced by growth on destarched wheat bran. Similarly, de-esterification of the wheat bran prior to incubation led to a 2.4-fold decrease in FAEA activity in the supernatant. Additionally, Smith et al. (*16*) reported that growth of several *A. niger*, *Aspergillus phoenicis* and *Trichoderma reesei* strains on meadow fescue grass, containing considerable esterified hydroxycinnamic acids, led to the induction of feruloyl esterase activity, whereas oat spelt xylan did not induce significant amounts of either activity. The growth substrate determines also the type of feruloyl esterase induced. Feruloyl esterases with different substrate specificities have also been reported to be induced by growth of *A. niger* on oats spelt xylan (*17*) and sugar beet pulp (*18*).

Apparently different bound or unbound phenolic acids act as inducers of feruloyl esterase activity in different organisms. Molecular genetic studies on the expression of a *Butyrivibrio fibrisolvens* feruloyl esterase have revealed that regulation is mediated by feruloylated oligosaccharides (*19*). Cloning of the *cinB* gene, encoding the feruloyl esterase enzyme revealed the presence of another open reading frame *cinR*, coding for a protein with high homology to bacterial repressor proteins. Expression of the *cinB* gene was repressed by binding of the product of the *cinR* gene to a specific tandem repeat sequence occurring in the 170 base pair region between the two open reading frames. Repression of *cinB* expression was abolished by the presence of *O*-{5-*O*-[(E)-feruloyl]-α-L-arabinofuranosyl}-(1,3)-*O*-β-D-xylanopyranosyl-(1,4)-D-xylanopyranose (FAXX) and 5-*O*-(*trans*-feruloyl)-arabinofuranose, both probable xylan degradation products, but not free ferulic or *p*-coumaric acid, xylose, xylobiose or any other monomeric hemicellulose component.

Kroon and co-workers (*20*) identified two putative binding motifs on the *faeB* promotor of *Penicillium funiculosum.* Northern analysis showed that transcription of *faeB* was tightly regulated, being stimulated by growth of the fungus on sugar beet pulp but inhibited by free glucose. The *faeB* promoter sequence contains putative motifs for binding an activator protein, XlnR, and a carbon catabolite repressor protein, CreA.

A study on expression profiling of 26 pectinolytic genes of the fungus *A. niger*, amongst them *faeA* and *faeB*, both encoding feruloyl esterases, revealed that the expression of these genes respond to the presence of ferulic acid (*21*). Neither *faeA* nor *faeB* are expressed in response to the presence of D-galacturonic acid. However, monosaccharides can also act as inducers. For example, *faeA* expression was also observed in the presence of D-xylose and this observation was attributed to a xylanolytic activation factor XylR (*22*).

ENZYME ASSAYS

The determination of feruloyl esterase activity is based on the quantification of hydroxycinnamic acid released from the substrate. A variety of different activity assays have been developed. Separation and quantification of reaction products is achieved by chromatographic methods such as high performance liquid chromatography (*13, 23, 24*), thin layer chromatography (*14*) or gas chromatography of the silylated derivatives of the reaction products (*23*). Additionally, hydroxycinnamic acids can be quantified in a photometric assay exploring the difference in their molar absorbance at 340 nm to the substrate (ferulic acid esters) (*25, 26*). The assay of feruloyl esterase activity was initially performed using native cell wall material in the form of starch-free wheat bran as a substrate (*13*). Alternatively, the methyl and/or ethyl esters of ferulic, caffeic, *p*-coumaric and sinapic acids have been used as substrates (*17, 18*). Carbohydrate oligomers containing esterified hydroxycinnamic acids have been isolated from enzymatic hydrolysates of plant cell wall material and successfully employed as substrates in the assay of feruloyl esterases. Borneman et al. (*23*) isolated *O*-[5-*O*-(*trans*-*p*-coumaryl)-α-L-arabinofuranosyl]-(1,3)-*O*-β-D-xylopyranosyl-(1,4)-D-xylopyranose (PAXX) from Coastal Bermuda grass shoots treated with Driselase, a fungal enzyme preparation containing a mixture of polysaccharide hydrolases. The feruloylated equivalent (FAXX) of the above mentioned oligosaccharide could also be prepared from the same hydrolysate. An assay in order to differentiate feruloyl esterases exhibiting affinity for 5-*O* and 2-*O*-feruloylated α-L-arabinofuranosyl residues was developed by Biely et al. (*26*). The assay uses synthetic *p*-nitrophenyl 5-*O*-*trans*-feruloyl-α-L-arabinofuranoside and *p*-nitrophenyl 2-*O*-*trans*-feruloyl-α-L-arabinofuranoside and is based on coupling the feruloyl esterase-catalyzed formation of *p*-

nitrophenyl α-L-arabinofuranoside with its efficient hydrolysis by α-L-arabinofuranosidase to release *p*-nitrophenol. The new substrates represent convenient tools to differentiate feruloyl esterases on the basis of substrate specificity and avoid the tedious isolation of natural substrate from plant material. A plate screening method for the detection of feruloyl esterase activity by microorganisms has also been developed (*27*). Colonies are grown on agar plates containing the ethyl esters of ferulic or *p*-coumaric acid that are subsequently flooded with a pH indicator, bromocresol green. A yellow zone against a blue background is observed around the colonies that release the hydroxycinnamic acids from their alkyl esters.

SUBSTRATE SPECIFICITY

Several feruloyl esterases have been purified and characterized (Table 1). However, comparison of their properties is difficult as the range of natural and synthetic substrates used to characterize these enzymes is diverse and the enzyme assays are not uniform. The substrates range in size and complexity from small, soluble esters such as feruloylated oligosaccharides isolated from plant cell walls and phenolic acid methyl esters or synthetic feruloylated arabinosides to larger, more complex and often less soluble substrates such as feruloylated polymeric plant cell wall fractions (*28*). The only criterion used in all cases is the release of free ferulic acid or another hydroxycinnamic acid by hydrolysis of an ester bond. Specificity, as defined by the rate of catalysis (k_{cat}) divided by the Michaelis constant (K_m) gives the best indication of preferred substrates. However, hydrolysis of polymeric substrates is more complicated since not all of the esterified substituents are chemically equal, and effects such as decreased solubility and steric hindrance further complicates any results obtained. Therefore, these data should not be extrapolated to obtain kinetic constants.

The use of small, soluble substrates allows the determination of kinetic constants, giving some information on the affinity (from K_m values) and catalytic efficiency (k_{cat} V_{max} / K_m). Substrates in question are effectively two components joined by an ester bond: the phenolic component and the sugar moiety. Specificity for both of these components defines the overall catalytic rate of the reaction. The selectivity for each component gives important information for the classification on feruloyl esterases (*12*).

Phenolic acids esterified to methyl or ethyl groups instead of the natural occurring sugar have been used for the characterization of most feruloyl esterases. They are relatively easy to synthesize, a variety of phenolic acids can be used to examine the enzymes specificity and the assay is reliable and fast, especially useful during purification.

Table I. Purified and characterized fungal and bacterial feruloyl esterases. M_r determined by SDS-PAGE.

Organism	Enzyme	M_r kDa	pH opt.	pI	MFA	Ara_5F	Ara_2F	Ref.
					K_m mM			
Fungi								
Aspergillus. awamori	FAEA	112	5.0	3.7	0.93	ND	ND	(*29*)
Aspergillus niger	FAEA	29.7	5.0	3.3	0.7	0.006	-	(*30*)
A. niger	FAEB	75.8	6.0	4.8	1.3	0.41	0.29	(*18*)
Aspergillus oryzae	FAE	30	4.5-6.0	3.6	ND	ND*	-	(*31*)
Aureobasidium pullulans	FeE	210	6.7	6.5	0.05	0.268	0.23	(*32*)
Fusarium oxysporum	FAE-I	31	7.0	9.5	0.6	1.6	0.06	(*33*)
F. oxysporum	FAE-II	27	7.0	9.9	0.58	1.01	-	(*34*)
Neocallimastix MC-2	FAE-I	69	6.2	4.2	ND	0.032	ND	(*35*)
Neocallimastix MC-2	FAE-II	24	7.0	5.8	ND	0.01	ND	(*35*)
Penicillium expansum	FAE	57	5.6	ND	2.6	ND	ND	(*36*)
Penicillium funiculosum	FAEB	53	ND	6	0.047	0.024	0.14	(*19*)
Penicillium pinophilum	p-CAE/FAE	57	6	4.6	0.14	ND	ND	(*37*)
Piromyces equi	FAEB(EstA)	37	6	ND	ND	0.004	0.075	(*38*)
Bacteria								
C. stercorarium	FeE	33	8.0	ND	0.04	ND	ND	(*39*)
Clostridium thermocellum	$FAECDB_{XYN}$	45	6.0	5.8	ND	5	ND	(*40*)
Butyrivibrio fibrisolvens	CinI (CinA)	30	ND	ND	ND	ND	ND	(*41*)
B. fibrisolvens	CinII (CinB)	35	ND	ND	ND	ND	ND	(*19*)
S. avermitilis CECT	FAE	ND	6.0	ND	ND	0.06	ND	(*42*)
S. avermitilis UAH 30	FeE	ND	6.0	ND	ND	ND	ND	(*43*)
Streptomyces olivochromogenes	FeE	29	5.5	7.9 - 8.5	1.9	0.24	ND	(*44*)
S. viridosporus	CinnAE	ND	9	ND	ND	ND	ND	(*45*)

ND=not determined; ND*=K_m value not determined, but substrate is catalyzed; - =no activity; MFA=methyl ferulate; Ara_5F=feruloyl group on the C5 of arabinose; Ara_2F= feruloyl group on the C2 of arabinose;

The specificity of several fungal feruloyl esterases for the phenolic moiety of various cinnamic acid methyl esters such as methyl ferulate, methyl caffeate, methyl *p*-coumarate and methyl sinapate has been evaluated (*32*). The feruloyl esterase of *Aureobasidium pullulans* has a broad substrate range and can be distinguished from feruloyl esterases from *A. niger*, *P. funiculosum and Penicillium expansum, Aspergillus awamori* and *Penicillium pinophilum* that are more selective. For example, methyl sinapate is a very poor substrate for *A. niger* FAEB and *P. expansum* and *P. funiculosum* feruloyl esterases, whereas *A. niger* FAEA shows no activity on methyl caffeate and methyl *p*-coumarate (*20, 36, 46*). A similar order of activity, methyl *p*-coumarate > methyl caffeate > methyl ferulate > methyl sinapate can be observed for *A. pullulans* FeE, as well as for *A. niger* FAEB, *Fusarium oxysporum* FAE-I, *P. funiculosum* and *Penicillium pinophilum* feruloyl esterase. A different order of activity, methyl sinapate > methyl ferulate > methyl *p*-coumarate > methyl caffeate can be found for *A. niger* FAEA and *F. oxysporum* FAE-II. These results suggest that the active centers of individual feruloyl esterases must be very different, at least for recognition of the phenolic moiety.

The specificity of feruloyl esterases for the sugar moiety of the substrate is more difficult to examine systematically, mainly because only a limited number of substrates are available. Substrates are usually obtained by limited hydrolysis of plant cell walls, or by organic synthesis. Comparison of catalytic properties gives information about the possible existence of catalytic sites on feruloyl esterases for the sugar moiety, the significance of the position of attachment of the feruloyl group to the sugar, and the type of sugar (*12*). A feruloyl group can be attached to the C2 or C5 position of arabinose, depending on the source of the starting material. Only a few feruloyl esterases have been examined from this point of view; however, Table 1 shows a summary of fungal and bacterial enzymes that exhibit preferences for both or one or the other substrate. *A. pullulans* feruloyl esterase and *A. niger* FAEB catalyze 2-feruloylated and 5-feruloylated substrates equally well whereas *Piromyces equi* and *P. funiculosum* feruloyl esterases show a clear preference for 2-feruloylated substrates. *A. niger* FAEA, *A. oryzae* feruloyl esterase and *F. oxysporum* FAE-II show activity exclusively on 5-feruloylated substrates. The reasons for this specificity remain to be elucidated, but differences in the structure of the active center might be a valid explanation. A lack of uniform data does not allow clear prediction on the affiliation of individual enzymes to specific groups. Kinetic data of bacterial feruloyl esterases is even less available than of fungi. Therefore, an examination of all available feruloyl esterases with the same substrates would enable a more comprehensive classification of these enzymes.

Crystal structures of feruloyl esterases from *Clostridium thermocellum* XYNY and XYNZ have been published (*47*). The enzymes display the α/β-hydrolase fold and possess a classical Ser-His-Asp catalytic triad. The active center reveals the binding site for ferulic acid and related compounds. Ferulate binds in a small surface depression that possesses specificity determinants for both the methoxy- and hydroxy-ring substituents of the substrate. There appears to be a lack of specificity for the xylan backbone, which may reflect the intrinsic chemical heterogeneity of the natural substrate (*47*).

SEQUENCE STRUCTURE AND RELATIONSSHIP

Until the beginning of 2003, sequences of nine fungal and seven bacterial feruloyl esterases (EC 3.1.1.73) have been published (Table 2) but the derived amino acid sequences do not appear to be highly conserved. Similar sequences from *Bacillus halodurans, Mesorhizobium loti* and *Agrobacterium tumefaciens* have been discovered in genomic databases by BLAST searches (www.ncbi.nlm.nih.gov/BLAST) but feruloyl esterase activity has not been confirmed experimentally. Therefore, conclusions on feruloyl esterase classification based on amino acid sequences alone should be done with care. All feruloyl esterases, however share one conserved motif (GXSXG) sequence, which is a universal signature sequence common to many esterases and lipases. Coutinho and Henrissat (*48*) classified cell wall-degrading enzymes into families of carbohydrate esterases and their database (http://afmb.cnrs-mrs.fr/CAZY/CE.html) shows that certain feruloyl esterases are in family 1 (including *P. funiculosum* FAEA and FAEB, *Neurospora crassa, Orpinomyces sp., P. equi, C. thermocellum* XYNY and XYNZ and *A. tumefaciens*) whereas the remaining enzymes are not represented in any family (Table 2). However, the alignment of amino acid sequences and using modern phylogenetic analysis tool suggests that distant relationships can be established (Figure 1).

Amongst fungal feruloyl esterases, *Aspergillus awamori* FAEA, *Aspergillus tubingensis* FAEA and *A. niger* FAEA are most closely related, exhibiting more than 90% average identity over the entire amino acid sequence. This group also contains a conserved domain that is found in several lipases of class 3 (triglyceride lipases EC 3.1.1.3), including those of *Thermomyces lanuginosus* (37% identity), *Aspergillus oryzae* (34% identity) and *Penicillium camembertii* (34% identity). The most conserved region in all these proteins is centered around a serine residue. However, Aliwan et al. (*49*) showed, that despite sequence identity (32 %), *A. niger* FAEA does not exhibit lipase activity. Andersen and co-workers (*50*) tested amongst others *T. lanuginosus* lipase for ferulic acid esterase activity. The wild type had no significant ferulic acid esterase activity, but rationally designed structural variants showed significant activity on ferulic acid esters while their lipase activity decreased.

Table II. Fungal and bacterial feruloyl esterases with sequence data available. The active serine moiety is underlined.

Gene	Enzyme	Organism	Putative active site sequence	CE family	Ref.
Fungi					
FaeB	FAEB	*Aspergillus niger*	GC<u>S</u>TG	-	(*51*)
FaeA	FAEA	*A. niger*	GH<u>S</u>LG	-	(*52*)
FaeA	FAEA	*Aspergillus tubingensis*	GH<u>S</u>LG	-	(*52*)
FerA	FAEA	*Aspergillus awamori*	GH<u>S</u>LG	-	(*53*)
FaeB	FAEB	*Neurospora crassa*	GD<u>S</u>LG	1	(*54*)
FaeB	FAEB	*Penicillium funiculosum*	GS<u>S</u>SG	1	(*20*)
FaeA	FAEA	*P. funiculosum*	GC<u>S</u>PG	1	(*55*)
FaeA	FAEA	*Orpinomyces sp.*	GL<u>S</u>MG	1	(*40*)
EstA	FAEB	*Piromyces equi*	GF<u>S</u>MG	1	(*38*)
Bacteria					
CinA	CEH	*Bacillus halodurans*	GC<u>S</u>EG	-	(*56*)
CinA	CEH	*Mesorhizobium loti*	not identified	-	(*57*)
XynY	XYNY	*Clostridium thermocellum*	GF<u>S</u>MG	1	(*58*)
XynZ	XYNZ	*C. thermocellum*	GL<u>S</u>MG	1	(*59*)
CinA	CINI	*Butyrivibrio fibrisolvens*	GH<u>S</u>QG	-	(*41*)
CinB	CINII	*B. fibrisolvens*	GG<u>S</u>QG	-	(*19*)
EstA	CEH	*Agrobacterium tumefaciens*	GY<u>S</u>GG	1	(*60*)

CE family = Carbohydrate esterase family (*48*)
- =not represented

P. funiculosum FAEB and *Neurospora crassa* FAEB show 50% identity and both are closely related to acetyl xylan esterases from *A. awamori* (44% identity; *29*), *Aspergillus ficuum* (44%; *61*) and *Penicillium purpurogenum* (39%; *62*). *A. niger* FAEB is very remotely related to this group (<10%). On the other hand, *P. funiculosum* FAEA, a member of the *Trichocomaceae* family, shows 30% identity to *P. equi* FAEB that belongs to the completely different family of *Neocallimasticaceae*. Both share a conserved domain that is common in enzymes of glycosyl hydrolase family 62. Furthermore, *Orpinomyces* sp. FAEA shows 31 % homology to the bacterial XYNZ of *C. thermocellum*, an endoxylanase from *Ruminococcus flavefaciens* (28%) and a tributyrin esterase from *Streptococcus pneumoniae* (24%). *A. niger* FAEB exhibits significant identity to tannases from *Xanthomonas campestris* (26%) and *A. oryzae* (24%).

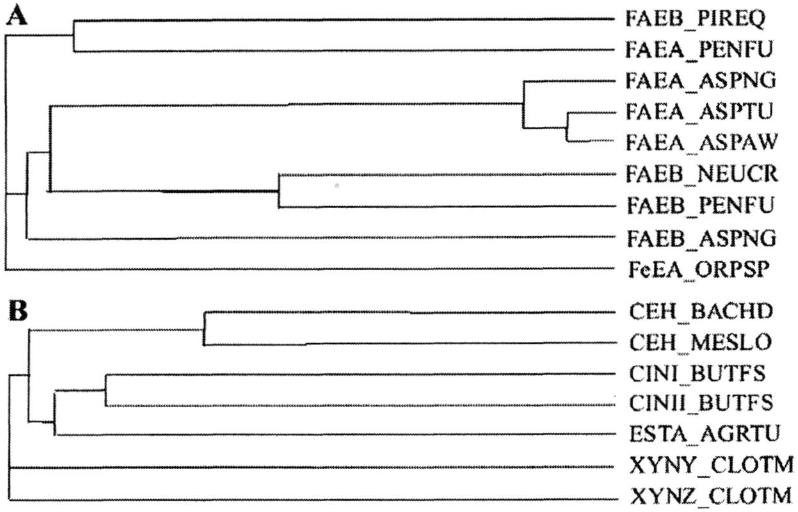

Figure I Phylogenetic tree of fungal (A) and bacterial (B) feruloyl esterases. For multiple sequence alignment BLOSUM62 was used as score matrix, gap penalty existence was 10, extension 0.1. FAEB_PIREQ: Piromyces equi; *FAEA_PENFU:* Penicillium funiculosum *FAEA; FAEA_ASPNG:* Aspergillus niger *FAEA; FAEA FAEA_ASPTU:* Aspergillus tubingensis; *FAEA_ASPAW:* Aspergillus awamori; *FAEB_NEUCR:* Neurospora crassa; *FAEB_PENFU:* P. funiculosum *FAEB; FAEB_ASPNG:* A. niger *FAEB; FeEA_ORPSP:* Orpinomyces sp.; *CEH_BACHD:* Bacillus halodurans; *CEH_MESLO:* Mesorhizobium loti; *CINI_BUTFS:* Butyrivibrio fibrisolvens *CINI; CINII_BUTFS:* B. fibrisolvens *CINII; ESTA_AGRTU:* Agrobacterium tumefaciens; *XYNY_CLOTM:* Clostridium thermocellum *XYNY; XYNZ_CLOTM:* C. thermocellum *XYNZ.*

Homologies between different bacterial feruloyl esterases are generally lower than those found in fungi. Highest identity values can be found between *B. halodurans* CEH and *M. loti* CEH (39%) and 27% between CINI and CINII of *B. fibrisolvens. A. tumefaciens* ESTA is slightly related to the latter (15% identity). XYNY and XYNZ of *C. thermocellum* are not related to other bacterial enzymes but *C. thermocellum* XYNZ is homologous (31% identity) to the fungal *Orpinomyces* sp. FAEA.

When sequence data is compared to kinetic data it is even more difficult to establish relationships between individual and groups of enzymes. This points to the need for more data on properties of these enzymes before the functional properties can be established and before the feruloyl esterases can be organized into discrete families based on ancestral origins. Furthermore, it is apparent that the feruloyl esterases are highly diverse enzymes with little conservation amongst closely related species.

ROLE OF FERULOYL ESTERASE IN XYLAN DEGRADATION

Together with α-arabinofuranosidase, α-glucuronidase and acetyl xylan esterases, feruloyl esterases are referred to as xylan-debranching or hemicellulose accessory enzymes (*12, 63, 64, 65*). Feruloyl esterases remove single ferulic and *p*-coumaric acid residues from the xylan backbone but also di-ester groups interlinking xylan molecules (*66*). Synergy of two or more hemicellulolytic enzymes acting together resulting in a greater rate of xylan degradation than the sum of the rates of degradation when the enzymes act individually, has been extensively reported between most of the accessory enzymes and xylanase. The action of xylanase releases substituted oligosaccharides that are more readily available and have more favorable interactions with the accessory enzymes, while the removal of side-chain branches by the accessory enzymes open up previously unavailable sites for cleavage by xylanases. The release of substituted oligosaccharides by xylanase appears to be of greater importance as many of the xylanolytic accessory enzymes have a preference for oligomeric substrates and show very little activity against polymeric xylan. Simultaneously, the enzymic liberation of acetyl and feruloyl groups from xylan through the action of carbohydrate esterases is essential for the efficient action of xylan degrading enzymes (*66, 67, 68, 69, 70*).

APPLICATION

Feruloyl esterases show intriguing differences in substrate specificity and sequence structure. The main reason for the recent increase in interest in these enzymes is their potential application in the production of ferulic acid. Ferulic acid is an important precursor in the flavor industry (*71*) and has multiple uses as an ultraviolet light protection agent in sun creams and cosmetics (*72*) by suppressing inflammatory responses and skin tumor development (*73*). Feruloyl esterases can also be used to selectively remove ferulic acid from agro-industrial waste products on an industrial scale (*74*). Ferulic acid is thought to be involved in the processing of foods. For example, an important but yet uncertain role for ferulic acid is apparent in the formation of cross-links in the dough during baking (*75*). Furthermore, ferulic acid acts as an antioxidant and is released and taken up in the intestines with the help of feruloyl esterases distributed along the intestines of mammals (*76, 77*). There is no doubt that applications together with a better understanding of the mechanism of action of feruloyl esterases, should increase in the future.

ACKNOWLEDGEMENTS

The authors would like to acknowledge the Austrian Bundesministerium für Bildung, Wissenschaft und Kultur, Postgraduate Stipendium GZ 14.910 and the South African National Research Foundation for financial support.

REFERENCES

1. Berndes, G.; Azar, C.; Kaberger, T.; Abrahamson, D. *Biomass Bioenerg.* **2001,** *20,* 371-383.
2. Breen, A.; Singleton, F.L. *Curr. Opin. Biotechnol.* **1999,** *10,* 252-258.
3. Martin, C.; Galbe, M.; Wahlbom, C.F.; Hahn-Hägerdal, B.; Jönsson, L.J. *Enzyme Microb. Technol.* **2002,** *31,* 274-282.
4. Tengerdy, R.P.; Szakacs, G. *Biochem. Eng. J.* **2003,** *13,* 169-179.
5. Himmel, M.E.; Ruth, M.F.; Wyman, C.E. *Curr. Opin. Biotechnol.* **1999,** *10,* 358-364.
6. Lynd, L.R.; Wyman, C.E.; Gerngross, T.U. *Biotechnol. Prog.* **1999,** *15,* 777-793.
7. Hartley, R.D.; Ford, C.W. In *Plant Cell Wall Polymers, Biogenesis and Biodegradation;* Lewis, N.G.; Paice, M.G.; Eds.; American Chemical Society Symposium Series 399; ACS: Washington, DC, 1989; pp 137-149.
8. Oosterveld, A.; Beldman, G.; Schols, H.A.; Voragen, A.G.J. *Carbohyd. Res.* **2000,** *328,* 185-197.
9. Sun, R.; Sun, X.F.; Wang, S.Q.; Zhu, W.; Wang, X.Y. *Ind. Crop. Prod.* **2002,** *15,* 179-188.
10. Christov, L.P.; Prior, B.A. *Enzyme Microb. Tech.* **1996,** *18,* 244-250.
11. Gübitz, G.M.; Stebbing, D.W.; Johansson, C.I.; Saddler, J.N. *Appl. Microbiol. Biotechnol.* **1998,** *50,* 390-395.
12. Williamson, G.; Faulds, C.B.; Kroon, P.A. *Biochem. Soc. Trans.* **1998,** *26,* 205-209.

13. MacKenzie, C.R.; Bilous, D.; Schneider, H.; Johnson, K.G. *Appl. Environ. Microbiol.* **1987,** *53,* 2835-2839.
14. MacKenzie, C.R.; Bilous, D. *Appl. Environ. Microbiol.* **1988,** *54,* 1170-1173.
15. Faulds, C.B.; deVries, R.P.; Kroon, P.A.; Visser, J.; Williamson, G. *FEMS Microbiol. Lett.* **1997,** *157,* 239-244.
16. Smith, D.C.; Bhat, M.K.; Wood, T.M. *World J. Microbiol. Biotechnol.* **1991,** *7,* 475-484.
17. Faulds, C.B.; Williamson, G. *Microbiol-UK* **1994,** *140,* 779-787.
18. Kroon, P.A.; Faulds, C.B.; Williamson, G. *Biotechnol. Appl. Biochem.* **1996,** *23,* 255-262.
19. Dalrymple, B.P.; Swadling, Y. *Microbiol-UK* **1997,** *43,* 1203-1210.
20. Kroon, P.A.; Williamson, G.; Fish, N.M.; Archer, D.B.; Belshaw, N.J. *Eur. J. Biochem.* **2000,** *267,* 6740-6752.
21. de Vries, R.P.; Jansen, J.; Aguilar, G.; Parenicova, L.; Joosten, V.; Wulfert, F.; Benen, J.A.E.; Visser, J. *FEBS Lett.* **2002,** *530,* 41-47.
22. van Peij, N.N.M.E.; Gielkens, M.M.C.; de Vries, R.P.; Visser, J.; de Graaff, L.H. *Appl. Environ. Microbiol.* **1998,** *64,* 3615-3619.
23. Borneman, W.S.; Hartley, R.D.; Morrison, W.H.; Akin, D.E.; Ljungdahl, L.G. *Appl. Microbiol. Biotechnol.* **1990,** *33,* 345-351.
24. O'Neill, F.H.; Christov, L.P.; Botes, P.J.; Prior, B.A. *World J. Microbiol. Biotechnol.* **1996,** *12,* 239-242.
25. Ralet, M.C.; Faulds, C.B.; Williamson, G.; Thibault, J.F. *Carbohyd. Res.* **1994,** *263,* 257-269.
26. Biely, P.; Mastihubova, M.; van Zyl, W.H.; Prior, B.A. *Anal. Biochem.* **2002,** *311,* 68-75.
27. Donaghy, J.A.; McKay, A.M. *World J. Microbiol. Biotechnol.* **1994,** *10,* 41-44.
28. Williamson, G.; Kroon, P.A.; Faulds, C.B. *Microbiol-UK* **1998,** *144,* 2011-2023.
29. McCrae, S.I.; Leith, K.M.; Gordon, A.H.; Wood, T.M. *Enzyme Microb. Technol.* **1994,** *16,* 826-834.
30. de Vries, R.P.; Michelsen, B.; Poulsen, C.H.; Kroon, P.A.; van den Heuvel, R.H.; Faulds, C.B.; Williamson, G.; van den Hombergh, J.P.; Visser, J. *Appl. Environ. Microbiol.* **1997,** *63,* 4638-4644.
31. Tenkanen, M.; Schuseil, J.; Puls, J.; Poutanen, K. J. Biotechnol. 1991, 18, 69-84.
32. Rumbold, K.; Biely, P.; Mastihubová, M.; Gudelj, M.; Gübitz, G.; Robra, K.-H.; Prior, B.A. Appl. Environ. Microbiol. 2003, 69, 5622-5626.
33. Topakas, E.; Stamatis, H.; Biely, P.; Kekos, D.; Macris, BJ.; Christakopoulos, P. *J. Biotech.* **2003,** *102,* 33-44.

34. Topakas, E.; Stamatis, H.; Mastihobova, M.; Biely, P.; Kekos, D.; Macris, BJ.; Christakopoulos, P. *Enzyme Microb. Technol.* **2003,** *33,* 729-737.
35. Borneman, W.S.; Ljungdahl, L.G.; Hartley, R.D.; Akin, D.E. *Appl. Environ. Microbiol.* **1992,** *58,* 3762-3766.
36. Donaghy, J.; McKay, A.M. *J. Appl. Microbiol.* **1997,** *83,* 718-726.
37. Castanares, A.; McCrae, S.I.; Wood, T.M. *Enzyme Microb. Technol.* **1992,** *14,* 875-884.
38. Fillingham, I.J.; Kroon, P.A.; Williamson, G.; Gilbert, H.J.; Hazlewood, G.P. *Biochem. J.* **1999,** *343,* 215-224.
39. Donaghy, J.; Bronnenmeier, K.; Soto-Kelly, P.F.; McKay, A.M. *J. Appl. Microbiol.* **2000,** *88,* 458-466.
40. Blum, D.L.; Kataeva, I.A.; Li, X.L.; Ljungdahl, L.G. *J. Bacteriol.* **2000,** *182,* 1346-1351.
41. Dalrymple, B.P.; Swadling, Y.; Cybinski, D.H.; Xue, G.P. *FEMS Microbiol. Lett.* **1996,** *143,* 115-120.
42. Garcia, B.L.; Ball, A.S.; Rodriguez, J.; Perez-Leblic, M.I.; Arias, M.E.; Copa-Patino, J.L. *Appl. Microbiol. Biotechnol.* **1998,** *50,* 213-218.
43. Garcia, B.L.; Ball, A.S.; Rodriguez, J.; Perez-Leblic, M.I.; Arias, M.E.; Copa-Patino, J.L. *FEMS Microbiol. Lett.* **1998,** *158,* 95-99.
44. Faulds, C.B.; Williamson, G. *J. Gen. Microbiol.* **1991,** *137,* 2339-2345.
45. Donelly, P.K.; Crawford, D.L. *Appl. Environ. Microbiol.* **1988,** *54,* 2237-2244.
46. Kroon, P.A.; Williamson, G. *Biochem. Soc. Trans.* **1998,** *26,* 167.
47. Prates, J.A.M.; Tarbouriech, N.; Charnock, S.J.; Fontes, C.M.G.A.; Ferreira, L.M.A.; Davies, G.J. *Structure* **2001,** *9,* 1183-1190.
48. Coutinho, P.M.; Henrissat, B. In *Recent Advances in Carbohydrate Bioengineering*; Gilbert H.J.; Davies G.; Henrissat B.; Svensson B.; Eds.; The Royal Society of Chemistry: Cambridge, UK, 1999; pp 3-12.
49. Aliwan, F.O.; Kroon, P.A.; Faulds, C.B.; Pickersgill, R.; Williamson, G. *J. Sci. Food Agric.* **1999,** *79,* 457-459.
50. Andersen, A.; Svendsen, A.; Vind, J.; Lassen, S.F.; Hjort, C.; Borch, K.; Patkar, S.A. *Colloid. Surface. B* **2002,** *26,* 47-55.
51. de Vries, R.P.; vanKuyk, P.A.; Kester, H.C.; Visser, J. *Biochem. J.* **2002,** *363,* 377-386.
52. de Vries, R.P.; Michelsen, B.; Poulsen, C.H.; Kroon, P.A.; van den Heuvel, R.H.; Faulds, C.B.; Williamson, G.; van den Hombergh, J.P.; Visser, J. *Appl. Environ. Microbiol.* **1997,** *63,* 4638-4644.
53. Koseki, T.; Nishimori, C.; Iefuji, H.; Iwano, K.; Matsuzawa, H. **1999,** unpublished, GenBank accession number AB032760.
54. Crepin, V.F.; Faulds, C.B.; Connerton, I.F. *Biochem. J.* **2003,** *370,* 417-427.
55. Furniss, C.; Williamson, G.; Fish, N.; Kroon, P. **2001,** unpublished, GenBank accession number AJ312296.

56. Takami, H.; Nakasone, K.; Takaki, Y.; Maeno, G.; Sasaki, R.; Masui, N.; Fuji, F.; Hirama, C.; Nakamura, Y.; Ogasawara, N.; Kuhara, S.; Horikoshi, K. *Nucleic Acids Res.* **2000,** *28,* 4317-4331.
57. Kaneko, S.; Kuno, A.; Muramatsu, M.; Iwamatsu, S.; Kusakabe, I.; Hayashi, K. *Biosci. Biotechnol. Biochem.* **2000,** *64,* 447-451.
58. Fontes, C.M.G.A.; Hazlewood, G.P.; Morag, E.; Hall, J.; Hirst, B.H.; Gilbert, H.J. *Biochem. J.* **1995,** *307,* 151-158.
59. Grepinet, O.; Chebrou, M.C.; Beguin, P. *J. Bacteriol.* **1988,** *170,* 4582-4588.
60. Goodner, B.; Hinkle, G.; Gattung, S.; Miller, N.; Blanchard, M.; Qurollo, B.; Goldman, B.S.; Cao, Y.; Askenazi, M.; Halling, C.; Mullin, L.; Houmiel, K.; Gordon, J.; Vaudin, M.; Iartchouk, O.; Epp, A.; Liu, F.; Wollam, C.; Allinger, M.; Doughty, D.; Scott, C.; Lappas, C.; Markelz, B.; Flanagan, C.; Crowell, C.; Gurson, J.; Lomo, C.; Sear, C.; Strub, G.; Cielo, C.; Slater, S. *Science* **2001,** *294,* 2323-2328.
61. Chung, H.J.; Park, S.M.; Kim, H.R.; Yang, M.S.; Kim, D.H. *Enzyme Microb. Technol.* **2002,** *31,* 384-391.
62. Gutierrez, R.; Cederlund, E.; Hjelmqvist, L.; Peirano, A.; Herrera, F.; Ghosh, D.; Duax, W.; Jornvall, H.; Eyzaguirre, J. *FEBS Lett.* **1998,** *423,* 35-38.
63. Jeffries, T.W. *Biodegradation* **1990,** *1,* 163-176.
64. Eriksson, K.E.L. *Wood Sci. Technol.* **1990,** *24,* 79-101.
65. Thomson, J.A. *FEMS Microbiol. Rev.* **1993,** *104,* 65-82.
66. Faulds, C.B.; Sancho, AI; Bartolome, B, *Appl. Microbiol. Biotechnol.* **2002,** *60,* 489-493.
67. Biely, P.; Kluepfel, D.; Morosoli, R.; Shareck, F. *Biochim. Biophys. Acta* **1993,** *1162,* 246-254.
68. Bartolome, B.; Faulds, C.B.; Tuohy, M.; Hazlewood, G.P.; Gilbert, H.J.; Williamson, G. *Biotechnol. Appl. Biochem.* **1995,** *22,* 65-73.
69. Tenkanen, M. *Biotechnol. Appl. Bioc.* **1998,** *27,* 19-24.
70. de Vries, R.P.; Kester, H.C.; Poulsen, C.H.; Benen, J.A.; Visser, J. *Carbohyd. Res.* **2000,** *327,* 401-410.
71. Bonnin, E.; Brunel, M.; Gouy, Y.; Lesage-Meessen, L.; Asther, M.; Thibault, J.-F. *Enzyme Microb. Technol.* **2001,** *28,* 70-80.
72. Saija, A.; Tomaino, A.; Lo, C.; Trombetta, D.; Proteggente, A.; De Pasquale, A.; Uccella, N.; Bonina, F. *J. Sci. Food Agric.* **1999,** *79,* 476-480.
73. Murakami, A.; Nakamura, Y.; Koshimizu, K.; Takahashi, D.; Matsumoto, K.; Hagihara, K.; Taniguchi, H.; Nomura, E.; Hosoda, A.; Tsuno, T. *Cancer Lett.* **2002,** *180,* 121-129.
74. Thibault, J.; Micard, V.; Renard, C.; Asther, M.; Delattre, M.; Lesage-Meessen, L.; Faulds, C.; Kroon, P.; Williamson, G.; Duarte, J. *Lebensm. Wiss. Technol.* **1998,** *31,* 530-536.

75. Schooneveld-Bergmans, M.E.F.; Dignum, M.J.W.; Grabber, J.H.; Beldman, G.; Voragen, A.G.J. *Carbohyd. Polym.* **1999,** *38,* 309-317.
76. Andreasen, M.F.; Christensen, L.P.; Meyer, A.S.; Hansen, A. *J. Sci. Food Agric.* **1999,** *79,* 411-413.
77. Chesson, A.; Provan, G.J.; Russell, W.R.; Scobbie, L.; Richardson, A.J.; Stewart, C. *J. Sci. Food Agric.* **1999,** *79,* 373-378.

Chapter 16

Cellobiose Dehydrogenase: An Extracellular Flavocytochrome from the Phytopathogenic Basidiomycete *Sclerotium (Athelia) rolfsii*

Dietmar Haltrich[1], Roland Ludwig[1], and Marcel Zámocky[1,2]

[1]Division of Biochemical Engineering, Institute of Food Technology, BOKU - University of Natural Resources and Applied Life Sciences at Vienna (Universität für Bodenkultur Wien), Muthgasse 18, A–1190 Wien, Austria
[2]Institute of Molecular Biology, Slovak Academy of Sciences, 845 51 Bratislava, Slovak Republic

Cellobiose dehydrogenase (CDH) is an extracellular flavocytochrome that is produced by wood-degrading and plant pathogenic fungi, both basidiomycetes and ascomycetes. Typically, CDH is a monomeric protein with a bipartite domain organization, consisting of an N-terminal heme domain containing a cytochrome *b* type heme, and a C-terminal flavin domain with a noncovalently bound FAD. These are linked by a protease-sensitive linker region. Based on the currently known sequences CDH can be divided in two distinct classes, class-1 CDH comprising five known sequences from basidiomycetes, and class-2 CDH represented by three sequences from ascomycetes. CDH oxidizes cellobiose and higher cellodextrins efficiently to their corresponding lactones. Concurrently a wide spectrum of different electron acceptors including various quinones, organic radical species and metal ion complexes are reduced. This report reviews recent progress made in understanding the physiology, structure and function of CDH.

Sclerotium rolfsii (teleomorph: *Athelia rolfsii*) was first described by Peter Henry Rolfs more than 100 years ago as a plant pathogen causing tomato blight. *S. rolfsii* causes disease (Southern blight or Rolfs Disease) in over 500 plant species, many of which are important crop plants such as tomatoes, peppers, sunflowers, peanuts, etc., and causes enormous economic damage. During its attack on plant material *S. rolfsii* produces large amounts of different enzymes which rapidly destroy host tissue and cell walls, thus enabling it to enter the host organism. Additionally, it secretes copious amounts of oxalic acid which binds calcium ions from calcium pectate of the host cell wall which enables together with acidification the polygalacturonase to hydrolyze pectates of the middle lamella.

S. rolfsii is known to produce cellulolytic and hemicellulolytic enzymes in high amounts *(1-3)*, as a matter of fact it has been recognized as one of the few producers of these hydrolases that are of industrial interest *(4)*. *S. rolfsii* secretes a complete cellulose degrading enzyme system consisting of endoglucanases *(5, 6)*, cellobiohydrolases *(7)*, and β-D-glucosidases *(8)* which have been isolated and characterized extensively in the past. In addition, cellobiose dehydrogenase was reported as part of the cellulolytic enzyme system of *S. rolfsii (9, 10)*.

Cellobiose-oxidizing enzymes were first described 30 years ago in cellulolytic cultures of the white rot fungi *Phanerochaete chrysosporium* and *Trametes versicolor (11)*. In the scientific literature two types of these enzymes are described: the hemoflavoprotein cellobiose dehydrogenase (CDH, EC 1.1.99.18; formerly cellobiose oxidase) and the flavoprotein cellobiose:quinone oxidoreductase (CBQ, EC 1.1.5.1), which for a long time was believed to be a separate enzyme. However, it was shown that the intact hemoflavoprotein can be cleaved by proteases, either by endogenous ones formed by the organism itself *(12, 13)* or by proteases such as papain *(14, 15)*, into a catalytically active flavin domain, which is similar to CBQ, and an inactive heme domain. Recently, it was shown that the flavin fragment originates from the *cdh* gene and that the primary RNA transcript is not differently spliced. This implies that this heme-devoid fragment is formed at the protein level by limited proteolysis *(16)*.

Characterization of Cellobiose Dehydrogenase

Cellobiose oxidoreductases, both intact CDH and/or the flavin-only fragment CBQ, were purified and characterized from a number of fungi, including *P. chrysosporium* (*Sporotrichum pulverulentum*) *(17, 18)*, *Chrysonilia (Monilia) sitophila (19)*, *Coniophora puteana (20)*, *Heterobasidion annosum (Fomes annosus) (21)*, *Humicola insolens (22)*, *Irpex lacteus (23)*, *Pycnoporus cinnabarinus (24, 25)*, *Schizophyllum commune (26)*, *S. rolfsii (10)*, *Thielavia heterothallica (Sporotrichum thermophile) (27, 28)*, *T. versicolor (29)*, *Trametes pubescens* and *Trametes villosa (30)*. Typically, CDH is a monomeric enzyme consisting of the N-terminal heme domain with one heme *b*, which is hexacoordinate *(31)*, and the flavin domain containing one noncovalently bound

FAD (or alternatively one 6-hydroxy-FAD *(32)*). These two domains are connected by a protease-sensitive linker region that is rich in serine and threonine.

In accordance with reports on other CDH enzymes, intact CDH from *S. rolfsii* contains both the heme and the flavin as prosthetic groups. The heme cofactor was identified as protoheme IX (heme *b*) and estimated as one heme per CDH molecule. Similarly, FAD was quantified as one nucleotide per molecule by spectrophotometric analysis. Both the oxidized and the reduced spectrum of CDH are shown in Figure 1. The major peak of the oxidized spectrum at 421 nm can be attributed to the heme cofactor, whereas the broad absorbance shoulder between 450 and 500 nm can be mainly attributed to the FAD group. The extinction coefficients for the oxidized state of the enzyme at 421, 460, 531 and 563 nm were 105, 24, 12 and 9 $mM^{-1}\cdot cm^{-1}$, respectively. Upon reduction of CDH at pH 4 by its substrate cellobiose or reducing agents such as sodium dithionite, strong peaks appeared at 429, 533 and 564 nm, representing the Soret, β and α peaks of a typical heme protein. Absorption between 450 and 500 nm decreased drastically, presumably representing the reduced form of FAD. The extinction coefficients for 429, 460, 533 and 564 nm were determined to be 140, 10, 17 and 28 $mM^{-1}\cdot cm^{-1}$, respectively, for the reduced enzyme.

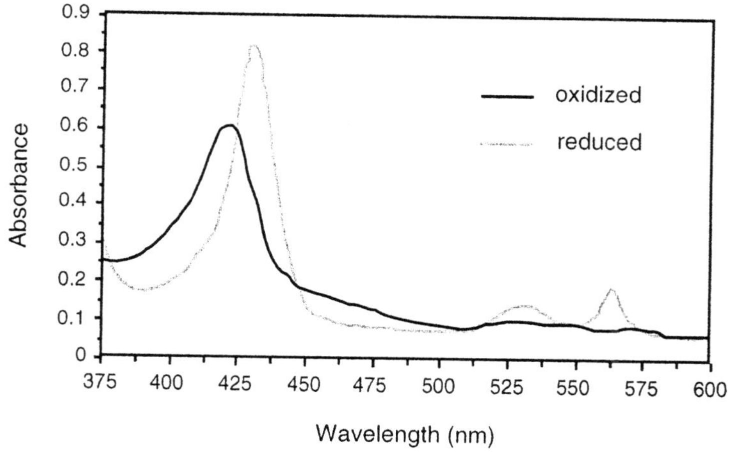

Figure 1. Absorption spectra of CDH from S. rolfsii *in the oxidized and reduced state. The reduction was performed with cellobiose.*

Table I lists some properties of CDH enzymes isolated from different organisms. Typically, CDH is a monomeric glycoprotein with a molecular mass of approximately 90–100 kDa and an acidic pI-value. Cellobiose oxido-reductases oxidize various sugars at their anomeric carbon atoms to the

Table I. Comparison of Cellobiose Dehydrogenases Isolated from Different Fungal Organisms

Source	M_r	*pI-value*	*pH optimum*[a]	K_m*-value (µM)* cellobiose	DCIP[b]	cyt c[c]	k_{cat}/K_m $(mM^{-1}\cdot s^{-1})$ cellobiose	DCIP	cyt c	Reference
Sclerotium rolfsii	101,000	4.2	3.2–4.8	120	15	0.30	225	2,000	113,0	*(10)*
Coniophora puteana	192,000 (dimer)	3.9	4.0	84	14	nr[d]	110	1,300	nr	*(62, 80)*
Humicola insolens	92,000	4.0	6.5–7.0	11	26	93	1,30	650	290	*(22)*
Irpex lacteus	97,000	nr	3.0–5.0[e]	34	13	4.0	430	1,200	4,4	*(23)*
Thielavia heterothallica	95,000	4.1	4.5–5.0	3.4	2.4	0.67	1,80	3,100	11,0	*(27, 28)*
Phanerochaete chrysosporium	90,000	4.2	4.5	25	3.6	1.2	960	9,200	17,0	*(40)*
Pycnoporus cinnabarinus	101,000 92,000	3.8	4.5	110	23	31	630	nr	nr	*(24, 25)*
Schizophyllum commune	102,000	nr	4.5	30	10	2.8	430	6,300	11,0	*(26)*
Trametes pubescens	90,000	4.2	4.5–5.0	210	4.9	0.69	380	18,900	153,0	*(30)*
Trametes versicolor	97,000	4.2	4.5–5.0	120	7.8	7.8	51	620	1,3	*(29)*
Trametes villosa	89,000	4.4	4.5–5.0	210	9.9	1.8	400	10,400	64,2	*(30)*

[a] pH optimum for the reduction of 2,6-dichlor-indophenol or 3,5-di-*tert*-butyl-1,2-benzoquinone
[b] 2,6-dichlorindophenol
[c] cytochrome *c*
[d] nr, not reported
[e] determined with cyt *c* as substrate

corresponding lactones. These sugar substrates typically are β-1,4-linked di- or oligosaccharides and include cellobiose (glc-β-1,4-glc), which in fact is the preferred substrate of CDH *(33)*, cello-oligosaccharides, lactose (gal-β-1,4-glc), and even cellulose *(34, 35)*. Some other electron donors that are accepted by different cellobiose dehydrogenases as substrates, albeit with sometimes only very low activity and / or affinity, include glucose, mannose, mannobiose (man-β-1,4-man), xylobiose (xyl-β-1,4-xyl) and maltose (glc-α-1,4-glc) *(10, 22, 36)*. The substrate specificity of *S. rolfsii* CDH for some of its electron donors (sugar substrates) is shown in Table II. There are certain differences with respect to sugar specificity between CDH from various sources. While CDH from *P. chrysosporium* oxidizes mannobiose and is not active on xylobiose, the enzyme from *S. rolfsii* oxidizes the latter disaccharide and is inactive with mannobiose.

Table II. Apparent Kinetic Constants of Cellobiose Dehydrogenase from *Sclerotium rolfsii* for Different Electron Donors

Substrate	K_m (mM)	k_{cat} (s^{-1})	k_{cat}/K_m ($mM^{-1} \cdot s^{-1}$)
Glucose	1250	1.5	1.2×10^{-3}
Cellobiose	0.12	27	225
Cellotriose	0.49	26	53.1
Cellotetraose	0.60	24	40.0
Cellopentaose	0.54	24	44.4
Lactose	2.4	26	10.8
Xylobiose	5.4	27	5.0
Maltose	240	0.8	3.3×10^{-3}

Data are from reference 10.

Based on kinetic data Henriksson *et al. (36)* suggested two hexose-binding subsites in the active center of the enzyme, the catalytic subsite (site C) and a second subsite involved in substrate binding (site B). Based on the crystal structure of *P. chrysosporium* CDH and molecular docking of the substrate cellobiose in the active site of the enzyme it was proposed that the enzyme discriminates against glucose by favoring interactions with the nonreducing end of cellobiose. This confirms the existence of two glucosyl-binding sites, the C subsite adjacent to the flavin ring and the B subsite close to the tunnel entrance *(37)*. These two subsites seem also to be valid for CDH from *S. rolfsii* as is suggested from the data for the Michaelis constant K_m determined for glucose and a series of cello-oligosaccharides. The K_m-value in the cello-oligosaccharide series of G2 to G5 (number of glucose units) is lowest for cellobiose and then increases for the higher cello-oligosaccharides (Table II), while the K_m of glucose is 10,000 times higher than that of cellobiose.

In contrast to this relatively high specificity of cellobiose oxidoreductases for β-1,4-linked di- or oligosaccharides they are highly non-specific with respect to the electron acceptor and can reduce a wide range of quinones, quinone-analogue dyes such as dichloroindophenol, organic radical species, and variously complexed metal ions *(29, 33)*. Some of these electron acceptors of *S. rolfsii* CDH together with their kinetic constants are listed in Table III.

Table III. Apparent Kinetic Constants of Cellobiose Dehydrogenase from *S. rolfsii* for Several Electron Acceptors

Electron acceptor	K_m (μM)	k_{cat} (s^{-1})	k_{cat}/K_m ($mM^{-1} \cdot s^{-1}$)
2,6-Dichloroindophenol	15	30	2,000
1,4-Benzoquinone	25	30	1,200
2,6-Dimethyl-1,4-benzoquinone	550	15	27
3,5-Di-*tert*-butyl-1,2-benzoquinone (TBBQ)	53	23	434
Methylene Blue	4.4	18	4,100
Methylene Green	4.3	18	4,200
ABTS cation radical	0.4	27	67,500
Cytochrome *c* (equine)	0.3	34	113,000
$K_3Fe(CN)_6$	20	37	1,850

Data are from reference 10; nd, not determined.

An obvious difference between the hemoflavoprotein CDH and the non-heme protein CBQ is their ability to reduce one-electron acceptors. Cyt *c* is efficiently reduced only by CDH, while the heme-devoid fragment exerts negligible activity with this substrate *(38)*. Whereas the flavin fragments from *P. chrysosporium* and *T. versicolor* reduce other one-electron acceptors such as complexed metal ions [$Fe(CN)_6^{3-}$, Fe(III), Cu(II)] much more slowly *(29, 38)*, the *S. rolfsii* CDH fragment reduces at least ferricyanide as efficiently as the intact enzyme *(10)*. In addition, cellobiose oxidoreductases can transfer electrons to oxygen, although this activity is only very low compared to the above mentioned electron acceptors. Both hydrogen peroxide and superoxide anion have been suggested as primary reaction products of molecular oxygen *(35, 39)*. It has been shown that H_2O_2 is formed stoichiometrically from cellobiose and oxygen in the absence of alternative electron acceptors *(40, 41)*. Recently, an overall mechanism for oxygen reduction by CDH was reported, in which electrons are primarily transferred to the flavin, while the heme group is reoxidized by the reactive oxygen species superoxide or hydrogen peroxide formed in this primary reaction *(42)*.

Structure and Phylogenetic Relationship

The primary structure of CDH was determined for different organisms by sequencing the cDNA and/or genomic DNA. These organisms include *S. rolfsii* (gb:AY187232), *P. chrysosporium (43, 44)*, *T. versicolor (45)*, *Py. cinnabarinus (46)*, *Th. heterothallica (28)*, *H. insolens (47)*, and *Grifola frondosa (48)*. At least in *P. chrysosporium* CDH is encoded by two allelic variants, the nucleotide sequences of which exhibit 97% similarity. The translation products of these two variants, however, have identical amino acid sequences *(49)*. Based on the currently known sequences CDH can be divided in two distinct classes: class-1 CDH which is represented by five known sequences from *Basidiomycetes*, and class-2 CDH represented by three known sequences (one putative from *Aspergillus fumigatus*) from *Ascomycetes (50)*. The typical features of class-1 CDHs are a rather longer and flexible linker region and no obvious cellulose binding motif. Class-2 CDHs possess a shorter linker region probably stabilized by disulfide bridges and a separate cellulose-binding domain on their C-terminus, the latter however has been only shown in *Th. heterothallica (28)*. The overall sequence homology is higher within each class than between the two classes. However, the typical features of CDHs represented by two heme ligands, one intra-domain disulfide bridge in the heme domain, conserved flavin binding and catalytic site, and the presence of the Rossman fold at the N-terminus of the flavin domain are completely conserved among all known CDH representatives. These two classes of cellobiose dehydrogenases can also be seen from the phylogenetic relationship of eight known CDH sequences, in which the ascomycetous sequences are clearly separated from those of the basidiomycetes (Figure 2).

To date the crystal structure of intact CDH has not been elucidated. Apparently crystallization of the complete flavocytochrome is hampered by the extensive interdomain peptide linker. The group of Dr. Christina Divne in Stockholm has instead determined the crystal structures of the heme and the flavin domain of PcCDH separately. The heme domain was shown to have a β-sandwich fold. The previously proposed ligation *(31)* of the heme iron in this *b*-type cytochrome by Met65 and His163 was confirmed. This Met/His ligation is quite uncommon in *b*-type cytochromes, in fact, CDH is the first cyt *b*-containing enzyme with such a ligation. Furthermore, the heme propionate groups are surface exposed, which could indicate a role in the association between the heme and the flavin domain as well as in the interdomain electron transfer *(51)*. The crystal structure of the flavin domain, which was determined at 1.5 Å resolution, shows an overall PHBH (*p*-hydroxybenzoate hydroxylase) fold *(52)*. It can be partitioned into an FAD-binding and a substrate-binding subdomain. Based on docking studies of the heme and the flavin domain it was proposed that the heme domain covers the active-site entrance of the flavin domain and that the resulting distance between the two prosthetic groups of less than 15 Å is within acceptable limits for interdomain electron transfer *(37)*.

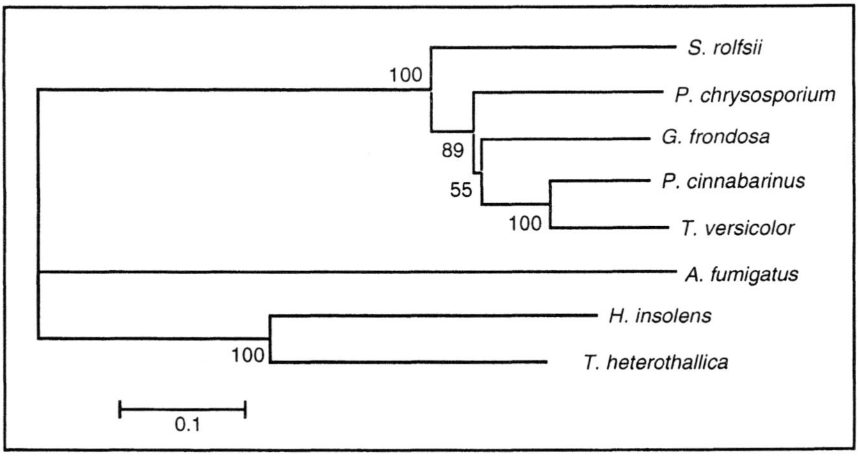

Figure 2: Unrooted phylogenetic tree for eight protein sequences of cellobiose dehydrogenases. The numbers in nodes represent bootstrap values for 100 replicates. The scale bar indicates the branch length corresponding to 0.1 amino acid substitutions per site.

Microbial Production of Cellobiose Dehydrogenase

Cellobiose oxidoreductases are inducible enzymes that are formed during growth of the fungal organisms on cellulose *(19, 53, 54)* and with some fungi also on cellobiose *(24, 28, 29)*. This induction of CDH synthesis was further confirmed by Northern blot analysis of total RNA from cellulose-, cellobiose-, and glucose-grown *P. chrysosporium* which indicated that CDH mRNA is produced only in cellulose-grown cells *(43)*. Similar results were also obtained for *Py. cinnabarinus (46)*. Further, CDH synthesis is repressed in the presence of easily metabolizable carbon sources such as glucose and maltose in the culture medium *(48, 55)*. It has been proposed that this repression of CDH production is mediated at the level of gene transcription. The supplementation of the cellulose medium with lignin seems to enhance CDH formation at least for *T. versicolor*, but this was not further investigated *(29)*. A similar enhancing effect of lignosulfonate has been suggested for *H. annosum* but not studied in any detail *(21)*. It has been postulated that the initial translation product is the hemoflavoenzyme CDH which is then processed to the flavin-only fragment CBQ by proteolytic cleavage *(45)*. Only a few studies have been performed in which the simultaneous formation of CDH and CBQ under laboratory conditions has been investigated *(53, 54, 56, 57)*. Bao *et al. (53)* reported a strong influence of the pH value and the buffer system used on CDH/CBQ production by *P. chrysosporium* with lower pH values being deleterious for enzyme production. A correlation between the ratio of CDH to CBQ formed with the protease activity, which was also measured for these cultivations, was not possible. However, the proteolytic cleavage of CDH also strongly depends on the pH *(13)*. Habu et al. *(57)* reported that the addition of bovine calf serum significantly increased CDH production by *P. chrysosporium*. This was explained in part by a nutritional effect, mainly resulting from the albumin fraction of the serum, as well as by an additional factor, probably a protease inhibitor. Yet, the addition of serum, and hence also of the putative protease inhibitor, to the culture medium only slightly changed the ratio of CDH to CBQ formed by the organism.

We found that the nitrogen (N) source has a significant effect on CDH production by *S. rolfsii* as well as on the ratio of intact to truncated enzyme formed, with both the type and the concentration of the N source being important *(54)*. Complex organic N sources such as peptone from meat were found to be optimal for CDH production. When concentrations of 60 $g{\cdot}l^{-1}$ or even higher were used, approximately 14,000 units of CDH activity per liter were formed; most of this activity could be attributed to the intact form of CDH with only a minor fraction occurring as the truncated flavin fragment. Reducing the concentration of peptone resulted in the formation of less CDH and relatively more of the flavin-only fragment *(54)*. When these complex nutrients were replaced by inorganic N sources CDH formation was negligible. The expensive peptone could partly be replaced by certain amino acids which supported excellent CDH production of up to 15,000 $U{\cdot}l^{-1}$, corresponding to 225 $mg{\cdot}l^{-1}$ (Table IV). The nature of the amino acid used in the growth medium significantly affected the ratio of cytochrome *c* (cyt *c*) to dichloroindophenol (DCIP) activity obtained. Intact flavoheme CDH is active with cyt *c* while the flavin-only fragment is not, hence the assay with cyt *c* is specific for CDH. DCIP is reduced by both enzyme species, so that this assay detects both intact and truncated CDH. Hence these two activities can be used as a measure for the ratio of intact to truncated CDH. While with some of the hydrophobic amino acids, e.g., leucine, this ratio was approximately 1.25, indicating that most of the cellobiose oxidizing activity can be ascribed to intact CDH, it was as low as 0.23 when using glutamine. This indicates that up to 80% of the total CDH activity formed is transformed to the flavin fragment. Even though protease activity varies to some extent when using the different media, there is no clear correlation between extracellular protease activity and the fraction of CDH cleaved. It was shown however that peptone as well as certain amino acids including valine or leucine inhibit the proteases secreted by *S. rolfsii*, thereby preventing the cleavage of CDH during the cultivation *(54)*.

Recombinant CDH was expressed both homologously and heterologously. Homologous expression of *P. chrysosporium* CDH in cultures supplemented with glucose, i.e., under conditions when no endogenous CDH is formed, was achieved by placing the *cdh-1* gene under control of the D-glyceraldehyde-3-

phosphate dehydrogenase promoter. Physical, spectral, and kinetic characteristics of the homologously expressed rCDH were similar to those of wtCDH *(58)*. Heterologously expressed *P. chrysosporium* CDH was achieved in the yeast *Pichia pastoris* using the expression vector pPIC9K containing the alcohol oxidase promoter. The expression level of 1,800 $U \cdot l^{-1}$ (79 $mg \cdot l^{-1}$) is considerably higher than that of the wild-type fungus of 750 $U \cdot l^{-1}$ (33 $mg \cdot l^{-1}$) *(57)*. rCDH had a slightly increased molecular mass and similar spectral and kinetic properties as compared to wtCDH. in addition, the binding affinities for cellulose were similar *(59)*. Recombinant CDH from *H. insolens* was expressed in *Fusarium venenatum* and *Aspergillus oryzae* as hosts but no details were given *(47)*.

Table IV. Effect of Various Amino Acids on the Production of CDH Activities by *S. rolfsii.*

Amino acid	*DCIP activity ($U \cdot ml^{-1}$)*	*cyt c activity ($U \cdot ml^{-1}$)*	*Ratio cyt c to DCIP activity*	*Protease activity ($\Delta A_{330}/h$)*
Blank	1.9	0.80	0.33	0.64
Valine	5.7	5.7	1.01	0.45
Leucine	11.0	13.6	1.24	0.31
Isoleucine	7.2	7.9	1.10	0.33
Glutamic acid	5.8	2.2	0.38	0.37
Aspartic acid	4.5	1.6	0.35	0.40
Glutamine	6.0	1.4	0.23	0.48
Asparagine	6.2	4.2	0.67	0.35

Amino acids were added in equal concentrations (30 $g \cdot l^{-1}$) to the basal cellulose-based medium containing 20 $g \cdot l^{-1}$ peptone from meat. CDH activity was measured with dichloroindophenol (DCIP) which measures both intact CDH and the catalytically active flavin domain, as well as with cytochrome *c* (cyt*c*) which selectively determines activity of intact CDH. Adapted from reference 54.

Biological Function of Cellobiose Dehydrogenase

Cellobiose oxidoreductases have been proposed to be involved in both cellulose and lignin biodegradation *(33)*. Bao and Renganathan *(60)* reported that the addition of CDH to *Trichoderma reesei* cellulases significantly increased the degradation of microcrystalline cellulose and suggested that the physiological function of CDH is cellulose degradation. A participation of CDH in cellulose degradation is also strongly suggested from recent work on CDH from *T. versicolor*. CDH-deficient strains of this basidiomycete were unable to grow on agar plates containing highly crystalline cellulose as sole carbon source, while the reference strain producing CDH grew efficiently on these plates. In contrast, both the parent strain and the CDH-deficient strain were able to grow on amorphous cellulose (CM-cellulose). These mutants also had a greatly decreased ability to colonize and degrade native white birch wood, a natural substrate of *T. versicolor*. These data suggest that CDH plays an important role in the degradation of native, substantially crystalline cellulose and thus is essential for wood invasion and lignocellulose degradation by *T. versicolor (61)*. Another proposed degradative function of CDH, which however is less specific, is via the production of Fenton's reagent. This is in agreement with the fact that CDH can continuously reduce Fe(III) to Fe(II) in the presence of cellobiose (which results from the degradation of cellulose by cellulases) and also possibly produces H_2O_2 — although hydrogen peroxide can also be formed by additional fungal oxidases or result from the reaction of Fe(II) with molecular oxygen *(39, 62)*. The hydroxyl radicals thus formed can then degrade not only cellulose, but also xylan and lignin *(63)*. Cellobiose oxidoreductases have also been suggested to be part of the lignin degrading enzyme system of fungi. Its function could be to prevent the repolymerization of aromatic radicals that are formed by the action of lignin peroxidases *(38, 64, 65)*. CDH could work in co-operation with manganese peroxidase, making Mn(II) available by reducing $Mn(IV)O_2$, which is found during delignification as insoluble deposits in wood, and simultaneously forming suitable complexing agents, i.e., cellobionic acid, for the reactive Mn(III) ion *(66)*.

An entirely different physiological function of cellobiose dehydrogenase has been proposed by Morpeth *(67)* who suggested that its main role is as a fungal defense mechanism. Plants form quinones that are highly toxic for microorganisms and serve as one of their main antimicrobial systems. Often these are formed in response to fungal attack *(68)*. Toxic quinones are also formed during fungal degradation of lignocellulose. By reducing these quinones to the corresponding hydroquinones, one of the reactions catalyzed by cellobiose oxidoreductases, both phytopathogenic and wood-rotting fungi should have a considerable advantage. Hence, CDH could also be involved in pathogenesis. This is supported by the fact that different isolates of phytopathogenic *Sclerotium* spp. screened for the formation of CDH activity formed considerable amounts of CDH at least under the selected *in vitro* conditions, accounting for up to 2.2% of total extracellular protein *(69)*. In accordance to the above described degradative function and the proposed involvement of CDH in the initial degradation of highly crystalline cellulose CDH could also play a comparable role in plant pathogenesis by loosening tightly packed, rigid lignocellulose, thus making it susceptible to further attack by high molecular weight enzymes. A possible mechanism for this could be via hydroxyl radicals formed through the CDH-catalyzed reduction of ferric ions in the presence of hydrogen peroxide.

Applications of Cellobiose Dehydrogenase

Cellobiose dehydrogenases have several proposed applications which include their use in biosensors for the detection of soluble cello-oligosaccharides *(70)*, for the continuous monitoring of cellulase activity *(71)*, or for the detection of diphenols *(72)*. In addition, their application in the pulp and paper industry has been suggested *(73)*, e.g., for modifying the degree of polymerization of pulp *(74)*. CDH has been used for several bioremediation applications which in principle are based on the formation of the hydroxyl radical. It has been shown that CDH can depolymerize polyacrylate *(75)*, degrade environmentally harmful explosives such as 2,4,6-trinitrotoluene (TNT) or hexahydro-1,3,5-trinitro-1,3,5-triazine *(76)*, or indirectly degrade many more chemicals *(77)*. Recently, we have reported a biocatalytic application of CDH, using the enzyme as biocatalyst for the formation of various aldonic acids *(78)* that have potential as building blocks in new glycopolymers with possible biomedical applications *(79)*.

Acknowledgement

Our research was supported by grant P14537-B11 from the Austrian Science Fund FWF (Fonds zur Förderung der wissenschaftlichen Forschung).

References

1. Lachke, A. H.; Deshpande, M. V. *FEMS Microbiol. Rev.* **1988,** *54*, 177–194.
2. Haltrich, D.; Laussamayer, B.; Steiner, W.; Nidetzky, B.; Kulbe, K. D. *Bioresource Technol.* **1994,** *50*, 43–50.
3. Sachslehner, A.; Haltrich, D.; Nidetzky, B.; Kulbe, K. D. *Appl. Biochem. Biotechnol.* **1997,** *63–65*, 189–201.
4. Goyal, A.; Ghosh, B.; Eveleigh, D. *Bioresource Technol.* **1991,** *36*, 37–50.
5. Lindner, W. A.; Dennison, C.; Berry, R. K. *Biochim. Biophys. Acta.* **1983,** *746*, 160–167.
6. Sadana, J. C.; Lachke, A. H.; Patil, R. V. *Carbohydr. Res.* **1984,** *133*, 297–312.
7. Patil, R. V.; Sadana, J. C. *Canadian Journal of Biochemistry and Cell Biology.* **1984,** *62*, 920–926.
8. Sadana, J. C.; Shewale, J. G.; Patil, R. V. *Carbohydr. Res.* **1983,** *118*, 205–214.
9. Sadana, J. C.; Patil, R. V. *J. Gen. Microbiol.* **1985,** *131*, 1917–1923.
10. Baminger, U.; Subramaniam, S. S.; Renganathan, V.; Haltrich, D. *Appl. Environ. Microbiol.* **2001,** *67*, 1766–1774.
11. Westermark, U.; Eriksson, K.-E. *Acta Chem. Scand.* **1974,** *B 28*, 209–214.

12. Eggert, C.; Habu, N.; Temp, U.; Eriksson, K.-E. L. In *Biotechnology in the Pulp and Paper Industry.;* Srebotnik, E.; Messner, K., Eds.; Facultas-Universitätsverlag: Vienna, 1996; pp 551–554.
13. Habu, N.; Samejima, M.; Dean, J. F. D.; Eriksson, K.-E. L. *FEBS Lett.* **1993,** *327*, 161–164.
14. Henriksson, G.; Pettersson, G.; Johansson, G.; Ruiz, A.; Uzcategui, E. *Eur. J. Biochem.* **1991,** *196*, 101–106.
15. Wood, J. D.; Wood, P. M. *Biochim. Biophys. Acta.* **1992,** *1119*, 90–96.
16. Raíces, M.; Montesino, R.; Cremata, J.; García, B.; Perdomo, W.; Szabó, I.; Henriksson, G.; Hallberg, B. M.; Pettersson, G.; Johansson, G. *Biochim. Biophys. Acta.* **2002,** *1576*, 15–22.
17. Ayers, A. R.; Ayers, S. B.; Eriksson, K.-E. *Eur. J. Biochem.* **1978,** *90*, 171–181.
18. Bao, W.; Usha, S. N.; Renganathan, V. In *Biotechnology in Pulp and Paper Industry;* Kuwahara, M.; Shimada, M., Eds.; Uni Publishers: Tokyo, 1992; pp 377–382.
19. Dekker, R. F. H. *J. Gen. Microbiol.* **1980,** *120*, 309–316.
20. Schmidhalter, D. R.; Canevascini, G. *Arch. Biochem. Biophys.* **1993,** *300*, 559–563.
21. Hüttermann, A.; Noelle, A. *Holzforschung.* **1982,** *36*, 283–286.
22. Schou, C.; Christensen, M. H.; Schülein, M. *Biochem. J.* **1998,** *330*, 565–571.
23. Hai, P. Q.; Nozaki, K.; Amano, Y.; Kanda, T. *J. Appl. Glycosci.* **2000,** *47*, 311–318.
24. Temp, U.; Eggert, C. *Appl. Environ. Microbiol.* **1999,** *65*, 389–395.
25. Sigoillot, C.; Lomascolo, A.; Record, E.; Robert, J. L.; Asther, M.; Sigoillot, J. C. *Enzyme Microb. Technol.* **2002,** *31*, 876–883.
26. Fang, J.; Liu, W.; Gao, P. J. *Arch. Biochem. Biophys.* **1998,** *353*, 37–46.
27. Canevascini, G.; Borer, P.; Dreyer, J.-L. *Eur. J. Biochem.* **1991,** *198*, 43–52.
28. Subramaniam, S. S.; Nagalla, S. R.; Renganathan, V. *Arch. Biochem. Biophys.* **1999,** *365*, 223–230.
29. Roy, B. P.; Dumonceaux, T.; Koukoulas, A. A.; Archibald, F. S. *Appl. Environ. Microbiol.* **1996,** *62*, 4417–4427.
30. Ludwig, R.; Varga, J.; Zámocky, M.; Haltrich, D. submitted for publication.
31. Cox, M. C.; Rogers, M. S.; Cheesman, M.; Jones, G. D.; Thomson, A. J.; Wilson, M. T.; Moore, G. R. *FEBS Lett.* **1992,** *307*, 233–236.
32. Igarashi, K.; Verhagen, M. F.; Samejima, M.; Schülein, M.; Eriksson, K. E.; Nishino, T. *J. Biol. Chem.* **1999,** *274*, 3338–3344.
33. Henriksson, G.; Johansson, G.; Pettersson, G. *J. Biotechnol.* **2000,** *78*, 93–113.
34. Kremer, S. M.; Wood, P. M. *FEMS Microbiol. Lett.* **1992,** *92*, 187–192.
35. Morpeth, F. F. *Biochem. J.* **1985,** *228*, 557–564.
36. Henriksson, G.; Sild, V.; Szabo, I. J.; Pettersson, G.; Johansson, G. *Biochim. Biophys. Acta.* **1998,** *1383*, 48–54.

37. Hallberg, B. M.; Henriksson, G.; Pettersson, G.; Divne, C. *J. Mol. Biol.* **2002,** *315*, 421–434.
38. Henriksson, G.; Johansson, G.; Pettersson, G. *Biochim. Biophys. Acta.* **1993,** *1144*, 184–190.
39. Kremer, S. M.; Wood, P. M. *Eur. J. Biochem.* **1992,** *208*, 807–814.
40. Bao, W.; Usha, S. N.; Renganathan, V. *Arch. Biochem. Biophys.* **1993,** *300*, 705–713.
41. Nutt, A.; Salumets, A.; Henriksson, G.; Sild, V.; Johansson, G. *Biotechnol. Lett.* **1997,** *19*, 379–383.
42. Mason, M. G.; Wilson, M. T.; Ball, A.; Nicholls, P. *FEBS Lett.* **2002,** *518*, 29–32.
43. Li, B.; Nagalla, S. R.; Renganathan, V. *Appl. Environ. Microbiol.* **1996,** *62*, 1329–1335.
44. Raíces, M.; Paifer, E.; Cremata, J.; Montesino, R.; Ståhlberg, J.; Divne, C.; Szabó, I. J.; Henriksson, G.; Johansson, G.; Pettersson, G. *FEBS Lett.* **1995,** *369*, 233–238.
45. Dumonceaux, T. J.; Bartholomew, K. A.; Charles, T. C.; Moukha, S. M.; Archibald, F. S. *Gene.* **1998,** *210*, 211–219.
46. Moukha, S. M.; Dumonceaux, T. J.; Record, E.; Archibald, F. S. *Gene.* **1999,** *234*, 23–33.
47. Xu, F.; Golightly, E. J.; Duke, K. R.; Lassen, S. F.; Knusen, B.; Christensen, S.; Brown, K. M.; Brown, S. H.; Schülein, M. *Enzyme Microb. Technol.* **2001,** *28*, 744–753.
48. Yoshida, M.; Ohira, T.; Igarashi, K.; Nagasawa, H.; Samejima, M. *FEMS Microbiol. Lett.* **2002,** *217*, 225–230.
49. Li, B.; Nagalla, S. R.; Renganathan, V. *Appl. Environ. Microbiol.* **1997,** *63*, 796–799.
50. Zámocky, M.; Hallberg, M.; Ludwig, R.; Divne, C.; Haltrich, D. *Gene,* submitted for publication.
51. Hallberg, B. M.; Bergfors, T.; Bäckbro, K.; Pettersson, G.; Henriksson, G.; Divne, C. *Structure.* **2000,** *8*, 79–88.
52. Mattevi, A. *Biophys. Chem.* **1998,** *70*, 217–222.
53. Bao, W.; Lymar, E.; Renganathan, V. *Appl. Microbiol. Biotechnol.* **1994,** *42*, 642–646.
54. Ludwig, R.; Haltrich, D. *Appl. Microbiol. Biotechnol.* **2003,** *61*, 32–39.
55. Stapleton, P. C.; Dobson, A. D. W. *FEMS Microbiol. Lett.* **2003,** *221*, 167–172.
56. Costa-Ferreira, M.; Ander, P.; Duarte, J. *Enzyme Microb. Technol.* **1994,** *16*, 771–776.
57. Habu, N.; Igarashi, K.; Samejima, M.; Pettersson, B.; Eriksson, K.-E. L. *Biotechnol. Appl. Biochem.* **1997,** *26*, 97–102.
58. Li, B.; Rotsaert, F. A.; Gold, M. H.; Renganathan, V. *Biochem. Biophys. Res. Commun.* **2000,** *270*, 141–146.

59. Yoshida, M.; Ohira, T.; Igarashi, K.; Nagasawa, H.; Aida, K.; Hallberg, B. M.; Divne, C.; Nishino, T.; Samejima, M. *Biosci. Biotech. Biochem.* **2001,** *65*, 2050–2057.
60. Bao, W.; Renganathan, V. *FEBS Lett.* **1992,** *302*, 77–80.
61. Dumonceaux, T.; Bartholomew, K.; Valeanu, L.; Charles, T.; Archibald, F. *Enzyme Microb. Technol.* **2001,** *29*, 478–489.
62. Hyde, S. M.; Wood, P. M. *FEMS Microbiol. Lett.* **1996,** *145*, 439–444.
63. Henriksson, G.; Ander, P.; Pettersson, B.; Pettersson, G. *Appl. Microbiol. Biotechnol.* **1995,** *42*, 790–796.
64. Ander, P.; Mishra, C.; Farrell, R. L.; Eriksson, K.-E. *J. Biotechnol.* **1990,** *13*, 189–198.
65. Ander, P. *FEMS Microbiol. Rev.* **1994,** *13*, 297–312.
66. Roy, B. P.; Paice, M. G.; Archibald, F. S.; Misra, S. K.; Misiak, L. E. *J. Biol. Chem.* **1994,** *269*, 19745–19750.
67. Morpeth, F. In *Chemistry and Biochemistry of Flavoenzymes;* Müller, F., Ed.; CRC Press: Boca Raton, 1991; Vol. 1, pp 337–348.
68. Daniel, M. In *Handbook of Phytoalexin Metabolism and Action;* Daniel, M.; Purkayastha, R. P., Eds.; Marcel Dekker: New York, 1995; pp 533–553.
69. Ludwig, R.; Haltrich, D. *Lett. Appl. Microbiol.* **2002,** *35*, 261-266.
70. Nordling, M.; Elmgren, M.; Ståhlberg, J.; Pettersson, G.; Lindquist, S.-E. *Anal. Biochem.* **1993,** *214*, 389–396.
71. Kremer, S. M.; Wood, P. M. *Appl. Microbiol. Biotechnol.* **1992,** *37*, 750–755.
72. Lindgren, A.; Stoica, L.; Ruzgas, T.; Ciucu, A.; Gorton, L. *Analyst.* **1999,** *124*, 527–532.
73. Duarte, J. C.; Costa-Ferreira, M.; Sena-Martins, G. *Bioresource Technol.* **1999,** *68*, 43–48.
74. Mansfield, S. D.; de Jong, E.; Saddler, J. N. *Appl. Environ. Microbiol.* **1997,** *63*, 3804–3809.
75. Cameron, M. D.; Aust, S. D. *Arch. Biochem. Biophys.* **1999,** *367*, 115–121.
76. Stahl, J. D.; van Aken, B.; Cameron, M. D.; Aust, S. D. *Biorem. J.* **2001,** *5*, 13–25.
77. Cameron, M. D.; Timofeevski, S.; Aust, S. D. *Appl. Microbiol. Biotechnol.* **2000,** *54*, 751–758.
78. Baminger, U.; Ludwig, R.; Galhaup, C.; Leitner, C.; Kulbe, K. D.; Haltrich, D. *J. Mol. Catal. B. Enzymatic.* **2001,** *11*, 541–550.
79. Casali, M.; Riva, S.; Ferruti, P. *J. Bioact. Comp. Polym.* **2001,** *16*, 479–491.
80. Schmidhalter, D. R.; Canevascini, G. *Appl. Microbiol. Biotechnol.* **1992,** *37*, 431–436.

Chapter 17

Improvement of the Enzyme Character for Lignocellulose Degradation by Gene Manipulation

Kiyoshi Hayashi, Mamoru Nishimoto, Farooqahmed S. Kittur, and Motomitsu Kitaoka

Enzyme Laboratory, National Food Research Institute, Tsukuba, Japan

One of the powerful method to alter enzyme character is manipulating the enzyme genes. One example in successful result in increasing the catalytic activity of xylanese by fusing carbohydrate binding module and the other in shifting optimum pH for the catalytic reaction by shuffling the two xylanase genes have been described in detail.

Searching for useful enzymes possessing desirable characteristics is carried out in many laboratories commonly by the two methods; conventional screening and gene manipulation. Important factor in using a conventional screening method successfully is to establish an appropriate method which helps to concentrate enzymes with desirable characteristics.

Other than looking for the enzymes which exist in nature, biotechnological methods enable us to modify the character of natural enzymes. One method is to use site-directed mutagenesis to modify a single amino acid residue among the few hundreds comprising the enzyme molecule. However, in this case, a drastic change in the enzyme character will not usually occur. Another method for DNA manipulation is gene shuffling and prepartaion of chimeric genes which tends to drastically alter the enzyme character (1).

By preparing chimeric xylanase, enzymes with increased catalytic activity and altered pH optimum were successfully obtained as described below.

Preparation of Chimeric Genes by Overlapping PCR

The construction of the chimeric enzyme genes was carried out by shuffling the parental genes at the selected borders using a self-priming polymerase chain reaction (2). The chimeric sites were mostly targeted at the homologous regions of the amino acid sequences of the two enzymes, as previously reported (3). High fidelity KOD-plus-DNA polymerase (TOYOBO Biochemicals, Japan) was used in the three step PCR reactions in order to overcome the incorporation of the undesired errors during the multiple PCR steps.

The first PCR step amplified the selected N- and C-terminal regions of the two genes, as shown in Fig. 1. The primer 2 (primer 3) consisted with the 20 bases of the gene 1 (gene 2) and 10 bases of the additional DNA sequence of the gene 2 (gene 1). The denaturation and annealing steps were performed at 98°C for 1 min and 55°C for 1 min respectively, and primer extension was carried out at 68°C for 1-2 min, repeating for 20 cycles. These conditions were also employed in the second and the third PCR. The amplified PCR products were separated by agarose gel electrophoresis and purified.

The purified PCR products were used as template DNA for the second stage-overlapping PCR without primers. In this process, strands having the same base pairs (20 bases) overlapped and acted as primers for each other.

In the third PCR, the combined fragments were amplified using forward and reverse primers (primer 1 and 4).

Enhancing the Catalytic Activity of Xylanase by Fusing the Carbohydrate Binding Module

In order to enhance the catalytic activity of the thermostable xylanase (XynB) from *Thermotoga maritima*, fusion of carbohydrate binding module (CBM) was conducted. *Streptomyces thermoviolaceus* STX-II has a modular architecture composed of two distinct domains (4). Residues 1-234 correspond to a xylanase catalytic domain belonging to glycoside hydrolase family 11. Downstream of the catalytic domain (residues 246-335) is a family 2b CBM. The interdomain region is a typical polyglycine linker sequence. *T. maritima* XynB, on the other hand, is a single domain enzyme with a catalytic domain belonging to glycosyl hydrolase family 10 (5).

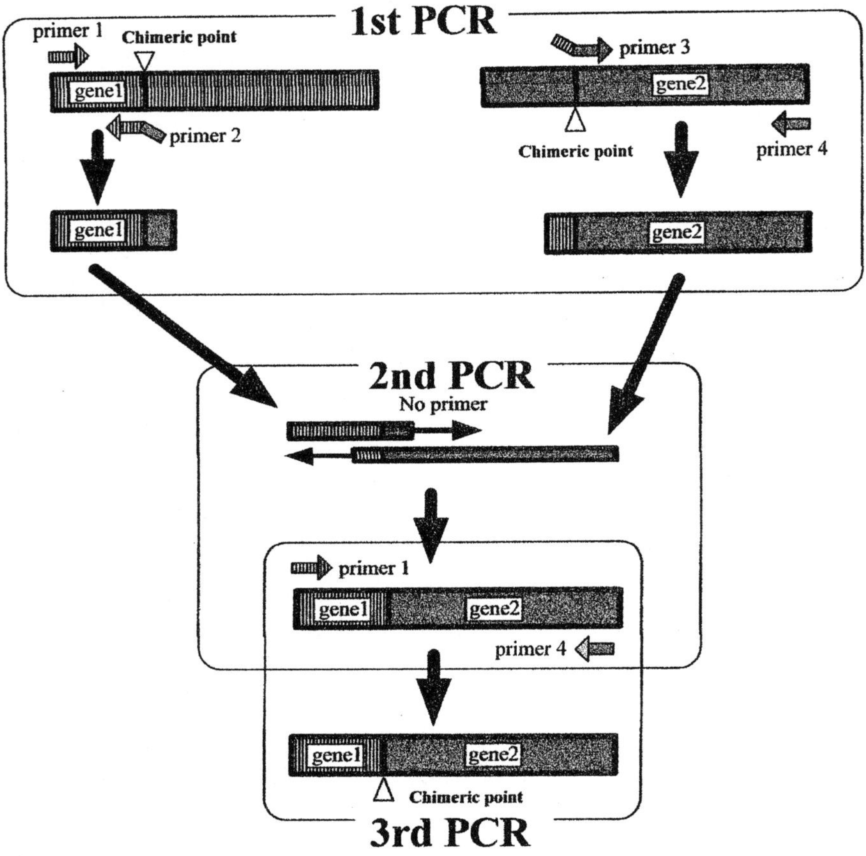

Figure 1. Shuffling genes at any designed site by overlapping PCR.

The CBM of *S. thermoviolaceus* was fused at the C-terminus of XynB (XynB-CBM2b) as shown in Fig. 2. The chimeric gene was cloned and expressed in *E. coli* BL-21CodonPlus-RIL under the control of the inducible phage T7 promoter of vector pET-28b(+). The over expressed protein displayed xylanase activity, indicating that fusion of the XynB with the binding domain did not result in misfolding.

When the substrate of solble burchwood xylan is used, chimeric xylanase displayed an optimum pH of 6.2 at 50 °C, and it was stable in the pH range 5.0-11.0 at 70 °C. Fig. 3 shows the temperature optimum and thermal stability profiles of chimeric and the parental enzyme. XynB-CBM2b displayed the identical temperature activity profile as that of XynB; a temperature optimum at 90 °C and stability up to 90 °C. These results indicate that fusion of the CBM2b to XynB has no effect on the structure and function of the catalytic domain.

The function of CBMs in xylanases is ambiguous, although they might attach/secure the enzymes on cellulose in plant cell walls, resulting in the increase of substrate (xylan) concentrations around the enzyme. Since the family 2b CBMs have affinity for soluble xylan chains, the binding ability and the hydrolytic activities of XynB-CBM2b and XynB toward soluble and insoluble substrates was investigated.

The ability of XynB-CBM2b to bind insoluble xylan was investigated by incubating enzymes with insoluble birch wood and oat spelt xylan at 0°C for 1 hr. XynB-CBM2b bound to the insoluble xylan, whereas XynB was unable to adhere to these polymers (Table 1). It also shows difference in the relative adsorption values for birch wood and oat spelt xylan, adsorption being higher in the case of oat spelt xylan than birch wood. The difference in the binding affinity is probably due to the differences in the composition of the two polysaccharides. In the case of birch wood xylan, the backbone is decorated with more 4-O-methyl-α-D-glucuronopyranosyl residues (acidic) (6). α-L-arabinofuranosyl residues (neutral) are common side chains in oat spelt xylan.

Though the binding ability of XynB-CBM2b to insoluble xylan at 50 °C was not directly confirmed because the insoluble xylan was readily hydrolyzed in the presence of the catalyst, the CBM2b was considered to be intact because the binding ability of XynB-CBM2b at 0 °C did not decrease after 30 min of incubation at 50 °C.

When initial rates of reaction were measured at 50 °C in various concentrations of birch wood xylan, the *k*cat values of XynB-CBM2b was greater than that of XynB (Table 2). On the other hand, when insoluble birch wood xylan was used as substrate, no significant difference in hydrolytic activities was observed, suggesting that the family 2b CBM has no role in the hydrolysis of insoluble xylan.

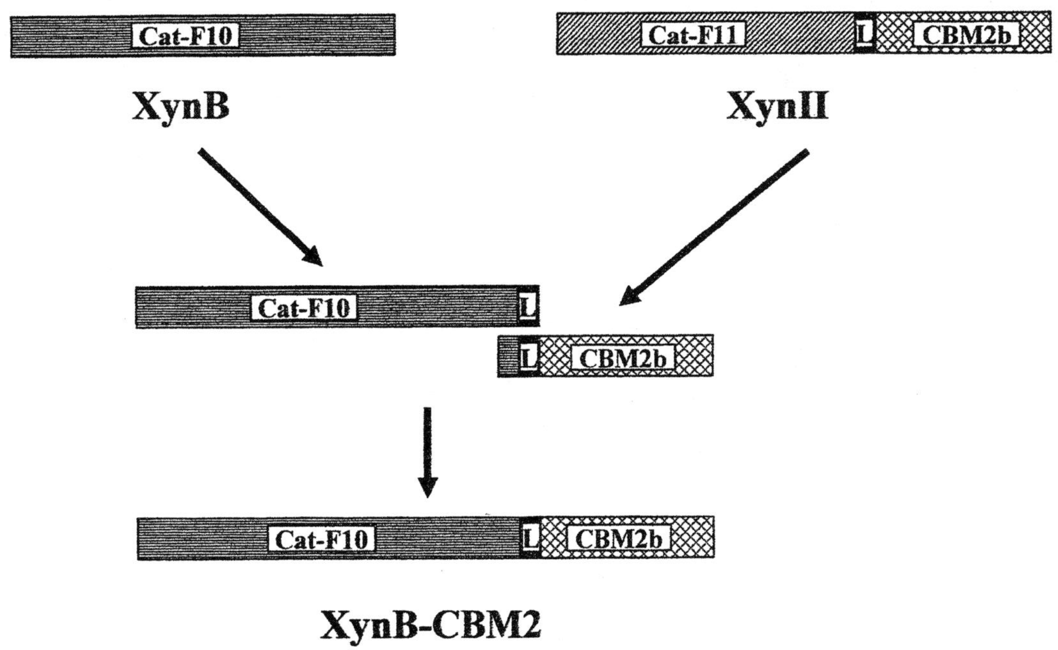

Figure 2. Strategy for the construction of the chimeric enzyme (XynB-CBM2b).
Cat-F10, family 10 catalytic domain of XynB;
Cat-F11, family 11 catalytic domain of STX-II;
L, linker region of STX-II;
CBM2b, family 2b CBM of STX-II.

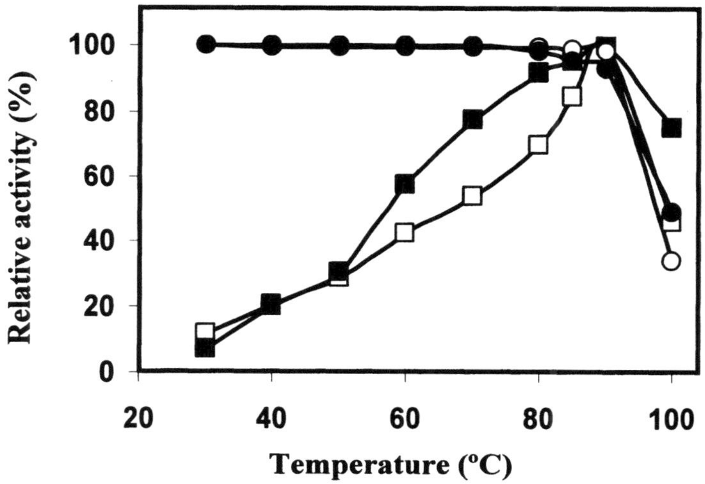

Figure 3. Temperature optimum and thermal stability of XynB and XynB-CBM2b.
Temperature optimum; □, XynB; ■, XynB-CBM2b
Thermal stability; ○, XynB; ●, XynB-CBM2b

Table I. Adsorption of XynB and XynB-CBM2b to Insoluble Xylans

	XynB	XynB-CBM2b
Birch wood xylan	2%	31%
Oat spelt xylan	2%	71%

Table II. Activities of XynB and XynB-CBM2b on the Hydrolysis of Various substrates

	XynB	XynB-CBM2b
Soluble birch wood xylan		
Km (%)	0.21	0.28
*k*cat (s^{-1})	33	57
Insoluble birch wood xylan		
v at 5 g/l	0.97	1.05
p-Nitrophenyl-β-xylopyranoside		
Km (mM)	9.4	9.9
*k*cat (s^{-1})	2.8	2.2

The kinetic parameters for *p*-nitrophenyl-ß-xylopyranoside, a substrate for which family 2b CBMs have no affinity was also investigated as shown in Table 2. No significant difference in *k*cat was observed.

Members of families 2b, 6, 13, and 22, though, bind soluble chains, but they enhance the catalytic activity only toward insoluble substrates. The increase in the catalytic activity of XynB toward soluble xylan upon fusion with a CBM is not clear at present.

Grafting of CBM has also been shown to alter the activity and thermal stability of the catalytic domain in some cases. Fusion of CBM4 to the feruloyl acetyl esterase domain (FAE) of *Clostridium stercorarium* XynZ was shown to result in a drastic decrease in the thermal stability of the FAE domain (7). On the other hand, Karita et al. (8) reported improvement in the thermal stability of *R. albus* endoglucanase upon fusion with a family 6 CBM.

Changing pH Profile of Xylanase by Preparing Chimeric Gene

Alkaline xylanases are highlighted since they are useful for pulp bleaching in the paper industry and in the production of xylooligosaccharides from xylan (9). The family 10 xylanase (XynA) produced by the alkaliphilic *Bacillus halodurans* C-125 displays full hydrolytic activity toward *p*-nitrophenyl-ß-D-xylobioside (pNPX2) over the pH range from 5.8 to 8.8 (10, 11).

While xylanase B (XynB) from *Clostridium stercorarium* F9 belongs to the same family (12), it displays 51% identity in the amino acid sequence to the XynA. The range of the pH optimum of XynB is between pH 5.3 and 6.6 (13), meaning that the acidic range of the pH optimum is similar to that observed for XynA, whereas the basic range is significantly narrower than that of XynA. Thus, XynB was selected as a shuffling partner for XynA to construct chimeric xylanases.

The reaction mechanism of family 10 xylanases has been studied in detail: Glu127 is acting as a proton donor and Glu233 is acting as the catalytic nucleophile (numbering are taken from the Cex from *Cellulomonas fimi*) (14). In general, the typical bell-shaped pH-activity profile that shows loss of activity in the direction of increasing acidity is attributed to protonation of the nucleophile, while loss of activity in the direction of increasing basicity is thought to be due to the dissociation of a proton from the proton donor (15). Therefore, it is assumed that the proton donor of XynA (Glu195) has a higher pK_a value than that of a general xylanase.

The XynA and XynB were subdivided into four fragments at highly homologous regions as shown in Fig. 4: an amino-terminal region (A or a; regions originating from XynA are denoted by uppercase letters and those from XynB by lowercase letters), a region containing the putative proton donor (P or p), a region containing the putative catalytic nucleophile (N or n), and a

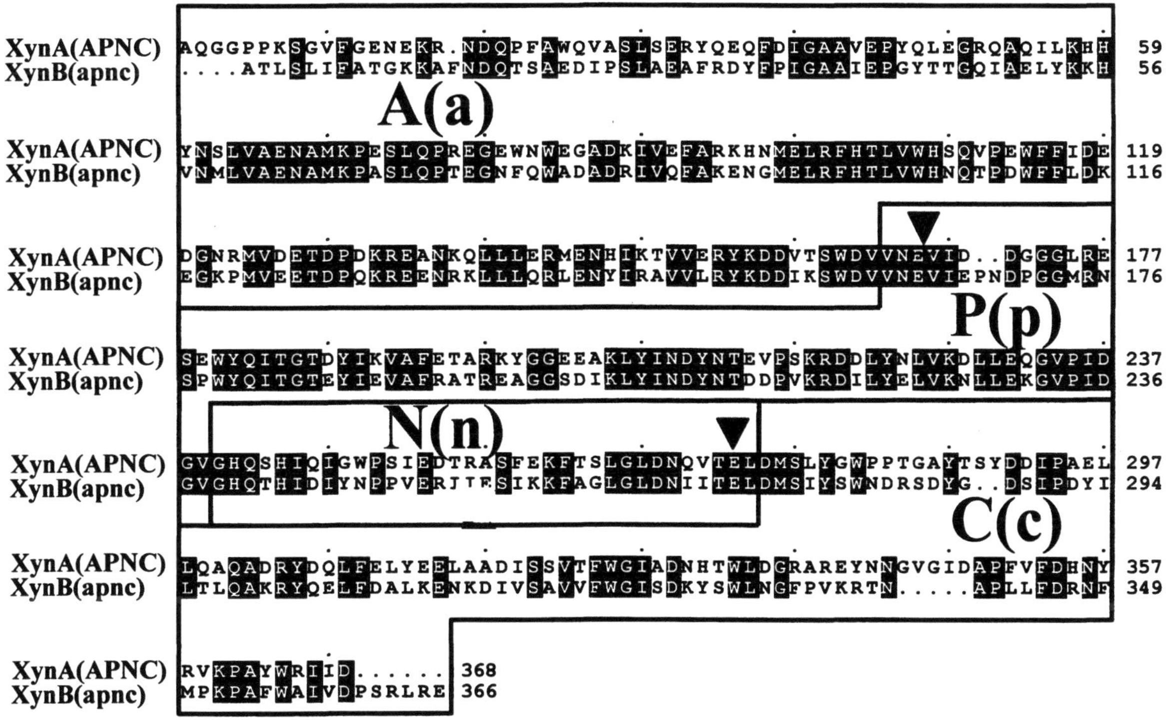

Figure 4. Alignment of the parental xylanases of XynA and XynB.
The putative proton donor and catalytic nucleophile are shown with triangle.

carboxyl-terminal region (C or c). Six chimeric xylanases (APNc, APnc, Apnc, aPNC, apNC and apnC) were constructed by the selective substitution of the four fragments using an overlapping PCR technique. Two of the six xylanases, APnc and Apnc, were obtained as catalyticaly active form in *E. coli* while the other four xylanases were obtained as inactive form of inclusion bodies. The APnc and Apnc chimeric xylanases were purified and characterized. The two chimeric enzymes are as active as parental enzymes at pH 6.6 (Table 3).

The respective pH and temperature stabilities of the purified enzymes were observed from pH 5.6 to 11.6 and up to 45 °C, for APnc and Apnc from pH 5.6 to 11.2 and up to 45 °C. Thus, these enzymes were slightly less stable than the parental xylanases (Fig. 5). This relative instability could be due to the loss of some interaction in the protein molecule such as a hydrogen bond, caused by changing the region of the enzyme.

An assessment of the pH-activity relationships for the chimeric xylanases employed *p*-nitrophenyl-ß-D-xylobioside as the substrate. The pK_{a1} values for the APnc and Apnc chimeric enzymes were 4.3 and 4.2, respectively, which were almost identical to those for the parental xylanases. In contrast, the pK_{a2} values obtained for APnc and Apnc were 9.1 and 8.5, respectively; these values fall between those for the parental xylanases, XynA (9.4) and XynB (7.8) as shown in Table 4 and Fig. 6. These results indicate that the main regions necessary to maintain the high pK_{a2} value of XynA locate in the A and P fragments. It is often reported that chimeric enzymes show properties intermediate of those of the parental enzymes (16, 17). In this study, the specific activities and pK_{a2} values of APnc and Apnc were intermediate of those observed for the parental enzymes.

The pK_{a2} value of XynB (apnc) increased from 7.8 to 8.5 by substitution of the N-terminal region with that of XynA (Apnc). This result indicates that the N-terminal fragment contains some region that affects the pK_{a2} of the proton donor. Moreover, the pK_{a2} value of Apnc increased to 9.1 by the substitution of the p fragment of Apnc to give the APnc chimeric enzyme, a value very close to that observed for XynA (9.4). The contributions of A and P to the increase in the pK_{a2} were estimated to be equal because exchange of A and P caused similar increases in pK_{a2}, 0.7 and 0.6, respectively. This result indicates that the section containing the proton donor Glu195 also contains the main region necessary to maintain a high pK_{a2} value. However, the pK_{a2} value obtained for APnc was still slightly less than that obtained for XynA, suggesting that the N and/or C parts may weakly affect on the pK_{a2} value.

Chimeric enzyme whose pK_{a2} is modified were successfully constructed. Roberge *et al.* reported that the pK_a values of the two catalytic residues on XlnA, a family 10 xylanase A from *Streptomyces lividans*, were reduced from 4.9 and 9.4 to 4.1 and 9.0 by a single amino acid mutation (Asn127Asp) (18). Similarly, Joshi *et al.* observed that a mutation of Asn35Asp of BCX, a family 11 xylanase

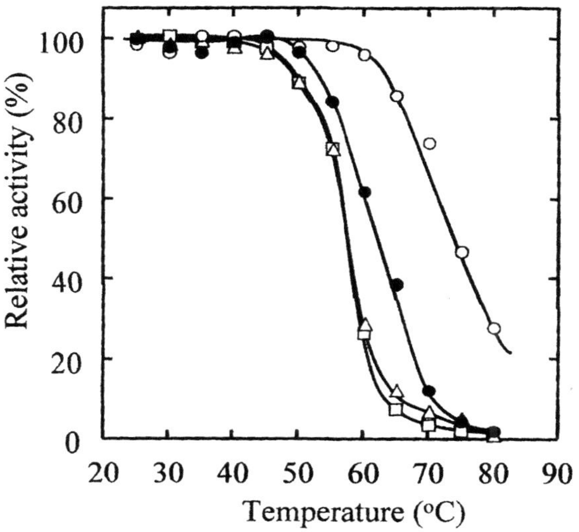

Figure 5. Thermal stability of the chimeric and parental xylanases.
●, XynA; ○, XynB; □, Apnc; △, APnc.

Table III. Kinetic Parameters of the Parental and Chimeric Xylanases at pH6.6

	*K*m (mM)	*k*cat (s^{-1})	*k*cat/*K*m ($mM^{-1}s^{-1}$)
APNC (XynA)	0.196	32.4	165
APnc	0.074	40.3	544
Apnc	0.063	38.2	611
apnc (XynB)	0.086	58.1	676

Table IV. pK_a Values of the Parental and Chimeric Xylanases

	pK_{a1}	pK_{a2}
APNC (XynA)	4.1	9.4
APnc	4.3	9.1
Apnc	4.2	8.5
apnc (XynB)	3.9	7.8

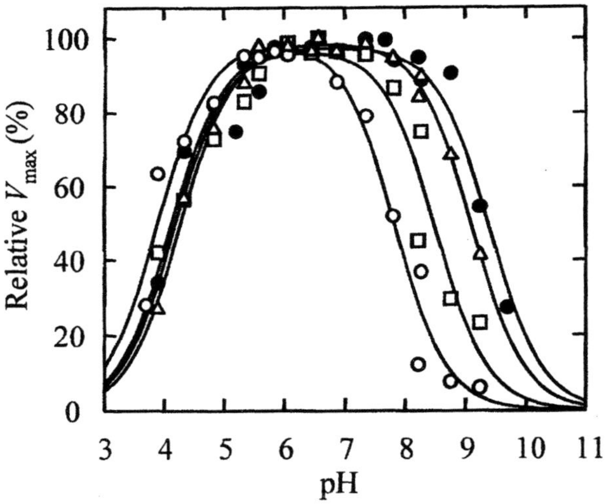

Figure 6. Comparison of the pH-activity (Vmax).
●, XynA; ○, XynB; □, Apnc; △, APnc.

from *Bacillus circulans*, decreased both the optimum pH (from 5.7 to 4.6) and the pK_a values of the enzyme (from 4.6 and 6.7 to 3.7 and 5.7, respectively) (19). These results strongly suggest that pH optimum of the enzyme can be designed by changing amino acid residues.

Out Look of Preparation of Chimeric Enzymes

One of the advantages in preparing chimeric enzymes is that it is not necessary to know the three dimensional structure. As long as the amino acid sequence of the target enzyme is available, it is not difficult to prepare chimeric enzymes, though it is desirable to have the three dimensional structure for designing the shuffling position.

The other advantages of chimeric enzymes is that the character of the chiemeric enzymes can be predicted to some extent. In general, they becomes admixture of the parental enzymes. This can be quite an advantage especially as the prediction of enzyme character based on the three dimensional structure is almost impossible at present.

Preparation of chimeric enzymes will provide a relatively accessible way for altering enzyme character.

References

1. Stemmer, W. P. *Nature* **1994,** *370*, 389-391.
2. *Genetic Engineering with PCR*; Horton R. M.; Tait R. C., Ed.: Horizon Scientific Press: Norfolk, 1998; pp 97-106.
3. Goyal, K.; Kim, B. J.; Kim J. D.; Kim, Y. K.; Kitaoka, M.; Hayashi, K. *Arch. Biochem. Biopys.* **2002,** *407*, 125-134.
4. Tsujibo, H.; Ohtsuki, T.; Iio, T.; Yamazaki, I.; Miyamoto, K.; Sugiyama, M.; Inamori, Y. *Appl. Environ. Microbiol.* **1997,** *63*, 661-664.
5. Zhengqiang, J.; Kobayashi, A.; Ahsan, M. M.; Lite, L.; Kitaoka, M.; Hayashi, K. *J. Biosci. Bioeng.* **2001,** *92*, 423-428.
6. Timell. T. E. *Adv. Carbohydr. Chem.* **1964,** *19*, 247-302.
7. Kataeva, I. A.; Blum, D. L.; Li, X. L.; Ljungdahl, L. *Protein Eng.* **2001,** *14*, 167-172.
8. Karita, S.; Sakka, K.; Ohmiya, K. *J. Ferment. Bioeng.* **1996,** *81*, 553-556.
9. Bissoon, S.; Christov, L.; Singh, S. Process Biochem. 2002, 37, 567-572.
10. Takami, H.; Nakasone, K.; Takaki, V.; Maeno, G.; Sasaki, R.; Masui, N.; Fuji, F.; Hirama, C.; Nakamura, V.; Ogasawara, N.; Kuhara, S.; Horikoshi, K. *Nucleic Acids Res.* **2000,** *28*, 4317-4331.
11. Nishimoto, M.; Honda, V.; Kitaoka, M.; Hayashi, K. J. Biosci. Bioeng. **2002,** *93*, 428-430.
12. Fukumura, M.; Sakka, K.; Shimada, K.; Ohmiya, K. *Biosci. Biotechnol. Biochem.* **1995,** *59*, 40-46.

13. Honda, Y.; Kitaoka, M.; Sakka, K.; Ohmiya, K.; Hayashi, K. *J. Biosci. Bioeng.* **2002,** *93,* 313-317.
14. Notenboom, V.; Williams, S. J.; Hoos, R.; Withers, S. G.; Rose, D. R. *Biochemistry* **2000,** *39,* 11553-11563.
15. Lawson, S. L.; Wakarchuk, W. W.; Withers, S. G. *Biochemistry* **1997,** *36,* 2257-2265.
16. Kaneko, S.; Iwamatsu, S.; Kuno, A.; Fujimoto, Z.; Sato, V.; Vura, K.; Go, M.; Mizuno, H.; Taira, K.; Hasegawa, T.; Kusakabe, I.; Hayashi, K. *Protein Eng.* **2000,** *13,* 873-879.
17 Hayashi, K.; Ying, L.; Singh, S.; Kaneko, S.; Nirasawa, S.; Shimonishi, T.; Kawata, Y.; Imoto, T.; Kitaoka, M. *J. Mol. Cat. B Enzymatic* **2001,** *11,* 811-816.
18. Roberge, M.; Dupont, C.; Morosoli, R.; Shareck, F.; Kluepfel, D. *Protein Eng.* **1997,** *10,* 399-403.
19. Joshi, M. D.; Sidhu, G.; Pot, I.; Brayer, G. D.; Withers, S. G.; McIntosh, L. P. *J. Mol. Biol.* **2000,** *299,* 255-279.

Applications

Chapter 18

Biotechnological Production of Xylitol from Lignocellulosic Materials

Maria G. A. Felipe

Department of Biotechnology, Faculty of Chemical Engineering of Lorena, Rod. Itajubá-Lorena, km 74, 5, P.O. Box 116, 12600–970 Lorena, SP, Brazil (mgafelipe@debiq.faenquil.br)

Abstract

The conversion of xylose into xylitol by microorganisms (mainly yeasts) cultivated in hydrolysates obtained by diluted-acid hydrolysis of lignocellulosic materials has been extensively investigated. As the hydrolytic process generates a variety of toxic compounds that inhibit the xylose metabolism of the microorganisms (e.g. acetic acid, furfural, hydroxymethylfurfural and phenolic compounds), different detoxification methods have been used to remove these inhibitors from the hydrolysates, with a view to improving their fermentability and increasing the xylitol production. This paper reviews several studies dealing with some of these detoxification methods and also with the major factors in the xylose-xylitol bioconversion.

Xylitol

Physicochemical properties and natural occurrences of Xylitol

Xylitol is a five-carbon sugar alcohol (molecular weight=152.15) as sweet as sucrose and sweeter than sorbitol. Owing to its high negative heat of solution, this white and crystalline powder produces a cooling and fresh sensation in the mouth. Xylitol is chemically more stable than sucrose, and when present in some product containing aminoacids, it does not participate in the Maillard reactions when this product is heated. This sugar can be found in various fruits, cereals, vegetables, and even in some microorganisms, but because it always occurs in very low quantities, its extraction on a commercial scale is economically unviable. In mammals xylitol is also a normal metabolic intermediate in carbohydrate metabolism (*1*).

Physiological properties and applications of xylitol

Xylitol is anticariogenic and cannot be fermented by oral bacteria (*2*, *3*, *4*, *5*). Studies performed with the oral bacteria *Streptococcus mutans* showed that xylitol also has bacteriostatic effect (*6*). Besides, it promotes mineral absorption and consequently prevents demineralization of the tooth enamel (*7*). Experiments with people that used xylitol-containing chewing gum indicated that the effects of this sweetener on the dental plaque persisted for several weeks or even for years (*4*). Xylitol does not depend on insulin to enter the glycogenolytic pathways and so it is appropriate to replace glucose in diets of diabetics. It can also be consumed by obese people and is apparently absorbed much more slowly than glucose (*8*).When injected intravenously, xylitol has caused only minimal insulin release in comparison with other carbohydrates, as well as a reduction in the levels of free fatty acid in the blood of both diabetic and healthy persons (*1*). Diseases such as erythrocytic glucose-6-phosphate dehydrogenase deficiencies (*9*), osteoporosis (*10*), airway infections (*11*) and otitis (*12*) can be treated with xylitol. This sugar can be utilized either alone or in conjunction with other sweeteners in confectionery, food, pharmaceutical or oral care industries (*1*).

Industrial production of xylitol

The industrial production of xylitol is based on the chemical reduction of xylose found in hemicellulosic hydrolysates. By this process, xylitol is recovered from pentose-rich solutions obtained preferably from xylan-containing raw materials. This chemical process comprises hydrolysis of the raw material, purification of the hydrolysate by ion exclusion techniques for color removal, and chromatographic fractionation of the purified solution for increasing the level of xylose for further chemical catalysis. Pure xylitol is then recovered from the reaction mixture by chromatographic fractionation. This process has the advantage of providing a xylose solution of high purity that ensures hydrogenation (*13*). On the other hand, purifying xylitol of other polyols and sugars increases the production cost (*14*, *15*).

Xylitol production by microorganisms

Certain microorganisms are known to have the ability to ferment D-xylose to xylitol. Most researches on the biotechnological production of xylitol employ yeasts (*16, 56*) although some have employed bacteria (*57, 58*) or even the fungus *Petromyces albertensis* (59). In bacteria, D-xylose is initially converted to D- xylulose by xylose isomerase in a single step (*60)*, while in yeasts and filamentous fungi, D-xylose is metabolized mostly via a two-stage oxidative-reductive pathway (*60, 61*) consisting in reduction of xylose to xylitol by an NADPH/NADH-linked xylose reductase and oxidation of xylitol to xylulose by an NAD-linked xylitol dehydrogenase (*61*). Xylulose is then phosphorylated by xylulose kinase to D-xylulose-5-phosphate, which can be catabolized by pentose phosphate, Embden-Meyerhof-Parnas, or by phosphoketolase pathways (*62*). An imbalance of the NAD^+:NADH redox system in D-xylose metabolism of yeasts is avoided by reducing xylose to xylitol with NADH (*61*). In xylose-fermenting microorganisms, xylose reductase generally uses both NADPH and NADH, while xylitol dehydrogenase is mainly NAD^+dependent (*21, 22, 28, 63, 64)*. The oxygen availability greatly influences the cofactors requirements of these enzymes. Anaerobic or oxygen-limited conditions cause a redox imbalance which interferes with the production of xylitol (*14, 22, 61*) and of the by-products of this metabolism (ethanol and/or glycerol) (*14, 22*). The activities of xylose reductase and xylitol dehydrogenase are believed to be influenced by other carbohydrates (*65, 66*).

Lignocellulosic Materials for Xylitol Production by Microorganisms

Hydrolysis of lignocellulosic materials

Forest and agricultural residues are lignocellulosic materials utilized in the biotechnological production of xylitol from xylose. Despite the variations in their chemical composition, these materials roughly contain about 45% cellulose , 30% hemicellulose, 25% lignin and 10% extractives. Cellulose is a linear homopolymer of repeating units of beta-D-glucose, while hemicellulose is a heterogeneous polymer mainly composed of pentoses (xylose, arabinose) and hexoses (glucose, manose, galactose). Some kinds of hemicellulose also contain uronic acid. The most common hemicelluloses are xylans, arabinoxylan, glucomannan and galacto-glucomannan. Lignin is a complex heteropolymer of irregular structure consisting in an aromatic system formed from phenylpropane units. Extractives are represented by a great diversity of materials such as aromatic compounds (phenolics), terpenes, aliphatic acids, alcohols and inorganic substances (*67*).

Hydrolysates obtained from lignocellulosic residues are suitable for xylose-xylitol conversion by microorganisms, because their hemicellulosic fraction contain high quantities of xylose (80% in average, in the case of agricultural residues) (68). The literature describes various methods of hydroysis of lignocellulosic materials, diluted-acid hydrolysis standing out as the one most frequently employed in experiments for xylitol production. This process, which requires high pressures and temperatures (*54*, *69*), generates, besides sugars, a range of compounds that are toxic to the microorganisms (70). These compounds include sugar degradation products such as furfural and hydroxymethylfurfural (HMF) (derived from pentosans and hexosans, respectively), formic acid (formed when furfural and HMF are broken down), levulinic acid (resulting from HMF degradation), acetic acid and phenolic compounds. Phenolic compounds originate from a partial breakdown of lignin or from carbohydrate degradation and include vanillic, syringic and palmitic acids. The types and amounts of compounds in the hydrolysates vary according to the lignocellulosic materials from which they are obtained (*70*). The principal drawback of any chemical process utilizing lignocellulosic residues on a technical scale or on a laboratory scale is the impossibility of separating cellulose, hemicellulose and lignin without changing their chemical structures (*71*). For this reason, it is necessary to establish hydrolysis conditions able to provide high concentrations of sugars with low concentrations of toxic compounds, in order that high yields and productivity rates can be achieved. Pessoa Junior *et. al.* (*69*) reported that the temperature, acid concentration and residence time used in the diluted-acid hydrolysis of sugar cane bagasse influenced both the xylose recovery yield and the quantity of toxic compounds. According to this author, 140°C, 20min and 100mg sulfuric acid /g of bagasse (dry weight) provided the highest xylose recovery yield (83%), a xylose production of 18.5g/l and low concentrations of acetic acid (3.9g/l), furfural (2.0 g/l) and HMF (0.08 g/l) (*69*). Employing the same conditions also for hydrolysis of bagasse, Rodrigues et al. (*72*) detected the presence of phenolic compounds in the hydrolysate (p-hydroxybenzoic, vanillic, syringic, p-coumaric and ferulic acids), and inorganic constituents such as calcium, chrome, iron and potassium. The highest concentrations of syringic, vanillic and p-hydroxybenzoic acids were (g/L) 0.005, 0.0042 and 0.002, respectively (*72*). Acid hydrolysis of *Eucalyptus grandis* chips employing 140°C, 10min and 0.2 % of sulfuric acid resulted in the highest xylose concentration (21.18g/l) and in 6.14g/l acetic acid (*33*). Roberto et al. (*54*) found that the best conditions for hydrolysis of rice straw were 1% sulfuric acid, 27min reaction time and 121°C. Under these conditions, 77% xylose was recovered and the concentration of acetic acid was low (2.0g/l). Increasing the reaction time and the acid concentration caused the levels of furfural and HMF to increase.

The concentrations of arabinose (3.5g/l) was 50% higher than those reported for bagasse (*72*) and eucalyptus hydrolysates (*33*).

Detoxification of hydrolysates

The literature on the methods for removal of toxic compounds from hemicellulosic hydrolysates obtained by acid hydrolysis is quite extensive. The hydrolysates utilized for xylitol production by microorganisms are usually pretreated by overlime (*33,34,36,73,74,75*), neutralization (*33,72,73,76,77*), adsorption on activated charcoal (*34,72,76,77,78*) and ion-exchange resins (*76,79,80*), but other detoxification methods such as molecular sieve (*74*), sulfitation (*78*) and extraction with solvents (*81*) have also been reported. Overlime consists in increasing the initial pH of the hydrolysate with bases ($CaCO_3$, $Ca(OH)_2$), and then decreasing it with acids (mainly H_2SO_4 and H_3PO_4).When in association with activated charcoal, this method provides an effective reduction in toxic compounds and improves the xylitol production. In experiments with sugar cane bagasse hydrolysate for xylitol production by *Candida guilliermondii* yeast, Alves et al. (*34*) attained maximum values of xyose-xylitol bioconversion efficiency (86%), xylitol yield (0.79 g/g) and productivity (0.47g/l.h^{-1}) increasing the pH of the hydrolysate from 0.5 to 7.0 with CaO and reducing it to 5.5 with H_3PO_4 before using activated charcoal (2.4%) for adsorption. In another study, the same authors (*75*) reported that treating bagasse hydrolysate with overliming combined with activated charcoal also had a positive effect on the enzyme activity of *C. guilliermondii*. This pretreated hydrolysate attained the high xylose reductase/xylitol dehydrogenase ratio of 4.5. The combination of overliming and activated charcoal was also used to detoxify rice straw hydrolysate before fermentation by *C. guilliermondii*. In this case, a hydrolysate:charcoal ratio of 40 :1 (g/g) resulted in a removal of 27% of phenolic compounds, and in high values of xylitol yield factor (0.72g/g) and volumetric productivity (0.61g/l.h $^{-1}$) (*48*). Besides reducing the concentrations of phenolic compounds in the hydrolysates (*48*,55,*72,76,77*), the treatment with charcoal also reduces the concentrations of acetic acid (55,*72,76,77*), furfural and HMF (55,*72,76*). However, undesirable decreases in the sugar contents have also been detected (*72,77*). Overliming combined with diethyl ether extraction was capable of improving the xylose-xylitol bioconversion by *Debaryomyces hansenii* in eucalyptus wood hydrolysate, providing a xylitol yield of 0.70g/g and a volumetric productivity of 1.11g/l.h $^{-1}$ (*81*). Nilvebrant et. al. (79) reports that one of the most efficient methods for removing inhibiting compounds prior to fermentation is treatment of the hydrolysate with ion-exchange resins. According these authors the anion exchanger effectively removed phenols, furan aldehydes and aliphatic acids from softwood hydrolysate at pH 10.0 (79). In another experiments, the treatment of the sugar cane bagasse hydrolysate with the anion exchange resin alone removed 84% of the acetic acid, while cation exchange resin removed only 27% of acid. When the two resins were used in combination, the percentage of D-xylose removal was lower than when cation-exchange resin was used alone (*80*). Neutralizing the initial pH of hydrolysates with bases was not sufficient to reduce the amount of toxic substances, and so the rates of xylose consumption and xylitol production were low (*33,72,76,78,81*). The pH and temperature utilized during vacuum evaporation, a procedure employed for increasing the xylose concentration in the hydrolysate, can also promote detoxification, because the concentration of volatile compounds decreases as a result of evaporation (*72,78*).

Another way to overcome the inhibiting effect of toxic compounds on the xylose-xylitol bioconversion is adapting the microorganism to the hydrolysate. A study by Felipe et al. (*25*) reports that adapting *C. guilliermondii* cells to sugar cane bagasse hydrolysate favored the consumption of xylose and acetic acid by the yeast provided a productivity of 0.65 $g/l.h^{-1}$, against 0.22$g/l.h^{-1}$ observed for non-adapted cells. In order to obtain adapted cells of *C. guilliermondii*, Sene et al. (35) cultivated the cells in diluted sugar cane bagasse hydrolysate and repeatedly transferred them to more concentrated hydrolysates. These authors observed that the necessity for cell adaptation is related to the concentration of inhibitors in the hydrolysate, and that the higher the adaptation, the higher the xylitol yield and volumetric productivity. According to them, this method resulted in a xylitol productivity increase of about 15% (*35*). In rice straw hemicellulosic hydrolysate, the xylose consumption by *C. guilliermondii* increased from 52 to 83% after adaptation (*46*). Isolating *Candida sp.* by continuous adaptation enhanced the tolerance of this yeast to untreated neutralized sugar cane bagasse hydrolysate and increased the xylitol yield to over 85% (*17*). The production of xylitol from various hemicellulosic hydrolysates by yeasts is presented in Table I.

Major factors in the xylose-xylitol bioconversion

Xylose concentration

The xylose transport across cell membrane is the initial step of the xylose utilization by the yeasts. Pentose-fermenting yeasts generally posses multiple xylose uptake systems, which can be classified as low- or high-affinity systems. The types of transport system vary between the yeasts and depend on nutritional conditions (82,*83*,*84*,*85*). In batch processes, the increase in sugar concentration

Table I. Lignocellulosic materials utilized in the biotechnological. production of xylitol by yeasts

Hemicellulosic hydrolysates	*Yeasts*	*Yp/s*	*Qp*	*References*
Sugar cane bagasse	*Candida guilliermondi FTI 20037*	0.74	0.75	Felipe et al. (26)
Sugar cane bagasse	*Candida sp. B-22*	0.89	1.0	Chen, L- F; Gong, C-S (17)
Sugar cane bagasse	*Candida sp. 11-2*	-	0.205	Domingues et al. (76)
Rice straw	*Candida guilliermondi FTI 20037*	0.72	0.61	Mussato et al. (48)
Rice straw	*Candida mogii NRRL Y-17032*	0.54	0.43	Mayerhoff et al. (88)
Eucal.yptus grandis	*Candida guilliermondi FTI 20037*	0.54	0.62	Silva et al. (33)
Eucal.yptus globulus	*Debaryomyces hansenii NRRL Y-7426*	0.70	1.11	Parajó et al. (81)
Aspenwood	*Candida parapsilosis ATCC 28474*	0.50	0.44	Preziosi - Belloy et. al.. (91)
Corn Cob	*C. parapsilosis KFCC 10875*	0.72	0.61	Kim et al. (36)

Yp/s, xylitol yield factor on D-xylose consumed (g/g)
Qp, xylitol volumetric productivity (g of xylitol/[L.h])

in the medium usually leads to increased yields and productions rates, if the sugar concentration and the osmotic pressure are not above the threshold of tolerance of the microorganism (*21*). Several studies (*14*, *21*, *30*, *45*, *86*, *87*) show that increasing the xylose concentration in synthetic culture medium has a beneficial effect on the xylose-xylitol conversion by yeasts, improving the xylitol production rates. As for hemicellulosic hydrolysates, the yeast capacity to tolerate non-volatile toxic compounds is a major determinant of the xylose concentration, since the evaporation necessary to concentrate this sugar causes the concentrations of these compounds to increase as well (*72*). In experiments with *Candida sp.* grown in sugar cane bagasse hemicellulosic hydrolysate, the xylitol production rate increased when the initial xylose concentration was increased from 30 to 50 g/l, reaching a maximum of 21.35 g/l after 72h, when the xylose concentration was 60g/l the ethanol production was the highest (*76*). Felipe et al. (*26*) evaluated the capacity of *C. guilliermondii* to produce xylitol in concentrated sugar cane bagasse hemicellulosic hydrolysate and attained a xylitol yield of 0.74g/g and a volumetric productivity of 0.75g/l.h^{-1} with a xylose concentration of 54.5 g/l. This productivity value was 44% higher than the one obtained with 37.6 g/l xylose, but when the xylose concentration increased to 74.2 g/l, the xylitol productivity decreased by 24% (*26*). For *C. guilliermondii* grown in rice straw hemicellulosic hydrolysate, the optimum xylose concentration was 82 g/l, which gave a xylitol yield of 0.65 g/g (*47*). When the same yeast was grown in eucalyptus hemicellulosic hydrolysate containing 60.0 g/l xylose, the xylitol yield and productivity were only 0.2 g/g and 0.1 g/l.h^{-1} respectively, probably due to the high levels of toxic compounds in the hydrolysate (*44*). In rice straw hydrolysate with a xylose concentration of 49.3 g/l, the xylitol yield after 36h of fermentation by *C. mogii* was 0.59 g/l, which corresponded to 82% xylose consumption. With 69.8 g/l xylose, the xylitol yield reached a peak (0.65 g/g) after 48h, but only 37% of xylose was consumed (*88*).

Presence of glucose

The bioconversion of xylose into xylitol can also be inhibited by glucose present in the hydrolysate. The preference of xylose-fermenting yeasts for glucose rather than xylose has been attributed to the catabolic repression exerted by glucose on the induction of D-xylose enzymes (*66,89,90,91*). In experiments with *Candida parapisilosis* grown in aspenwood hemicellulosic hydrolysate, this preference was attributed to the low enzyme activities observed during glucose consumption. However, the glucose repression of the pentose uptake did not affect the xylose conversion to xylitol, since the xylitol yield obtained from the mixture of xylose and glucose (0.53g/g) was similar to the one obtained from xylose (0.51 g/g) and the volumetric productivity was the same for both cases

(0.69 g/l.h^{-1}) (*91*). In synthetic medium, xylose and glucose were simultaneously consumed by *Candida shehatae*, but the cells were previously cultivated in medium containing xylose (*90*). With *D. hansenii*, when a glucose/xylose feeding ratio of 10% was used, the xylitol yield increased by more than 30%, whereas higher glucose concentrations brought about the formation of ethanol and glycerol (*38*). The addition of glucose to synthetic medium during xylose fermentation alleviated the blockage in xylose metabolism occurring under anaerobiosis (22).

Nitrogen supplementation and inoculum

The supplementation of aspenwood hemicellulosic hydrolysate with yeast extract improved the hydrolysate utilization by *C. parapsilosis* for growth and the xylitol yield, probably because this nitrogen source also minimizes the inhibitory action of the toxic compounds present in the hydrolysate (91). Ammonium sulfate proved to be a better nitrogen source than urea for growing *C. guilliermondii* in rice straw hemicellulosic hydrolysate, since it caused the xylitol yield and productivity to increase (*92*). In another experiment with the same yeast, the xylitol yield (0.65 g/g) obtained from rice straw hydrolysate without supplementation was similar to the one obtained from hydrolysate supplemented with nitrogen source (*47*). In several studies (23-27) employing *C. guilliermondii* for xylitol production, ammonium sulfate was added to sugar cane bagasse hydrolysate together with rice bran extract solution. The optimal initial inoculum concentration for xylitol production by *C. guilliermondii* cultivated in sugar cane bagasse (*26*) and rice straw (*92*) hydrolysates was dependent on the xylose contained in the hydrolysate. In sugar cane bagasse hydrolysate, a xylitol yield of 0.74 g/g and a productivity of 0.75 g/l.h^{-1} were achieved with 54.5 g/l xylose and 3.0 g/l inoculum, while for 37.6 g/l xylose, an inoculum concentration of only 0.1g/l was sufficient to attain a xylitol productivity of 0.52 g/l.h^{-1} (26). In rice straw hydrolysate, a similar xylitol yield and a productivity of 0.56 g/l.h^{-1} were attained with approximately 62 g/l of xylose and 0.90 g/l inoculum (*92*). In both cases a 24-h-old inoculum was proved more appropriate to produce xylitol. Also in experiments with *D. hansenii* and wood hydrolysate, the xylitol production was enhanced by increasing the initial inoculum concentration in the medium (78).

Temperature and pH

Temperatures around 30°C have been utilized for xylitol production yeasts cultivated in synthetic medium or in hemicellulosic hydrolysates produced from different raw materials. In experiments with *C. guilliermondii*, Felipe et al. (*27*) used a temperature of 30°C and obtained maximum values of xylitol yield (0.75g/g) and volumetric productivity (0.57g/l.h^{-1}) from sugar cane bagasse hydrolysate containing 40.0g/l xylose and 4.3 g/l acetic acid. However, pH values lower than 4.5 interfered with the consumption of glucose, xylose and arabinose, as well as with the production of cells and xylitol (*27*). An identical xylitol yield (0.75g/g) and a productivity of 0.95g/l.h $^{-1}$ were attained by Sene et al. (*41*) when working with the same yeast and the same kind of hydrolysate at a higher temperature (35°C). In this case, 71% acetic acid was consumed. At 25 °C the acetic acid consumption reached 73% and the specific activity of NADPH-linked xylose reductase was the highest. Besides temperature, the pH also had influence on the results (*41*). In a study (*92*) with *C. guilliermondii* grown in rice straw hydrolysate, maximum values of xylitol yield (0.68g/g) and productivity (0.51g/l.h^{-1}) were achieved with pH 5.3. When the same yeast was cultivated in sugar cane bagasse hydrolysate containing 60g/l xylose and 4.7g/l acetic acid, increasing the pH from 3.5 to 5.5 caused the fermentation time to decrease from 116 h to 48 h and the xylitol volumetric productivity to be as high as 0.76 g/l.h^{-1}.The yeast started to use xylose only when the acetic acid was almost completely exhausted (55). The ability of the yeast to assimilate acetic acid suggests that the yeast cells act as detoxification agents (*23,40*). The inhibitory effect of the acetic acid is basically due to the fact that, in its undissociated form, this acid diffuses into the cell cytoplasm and acidifies the cell interior, diminishing the growth and the xylose metabolism of the cells in order to keep the intracellular pH constant (*93, 94, 95, 96*). The toxic effect of the acetic acid can be potentiated by phenolic compounds (*91*), but low concentrations of this acid can favor the xylose bioconversion by *C. guilliermondii* (*23*).

Oxygen level

Oxygen is one of the most important factors in the xylose-xylitol bioconversion by pentose-fermenting yeasts. The fact that oxygen is required for xylose metabolism is frequently ascribed to the apparent redox imbalance occurring under anaerobic conditions, due to the difference in the cofactors required by xylose reductase and xylitol dehydrogenase, two key enzymes in the initial phase of the xylose metabolism (61). The level of oxygen required by the yeasts depends on the yeast strain and mainly on the conditions of the culture medium such as the presence of glucose and toxic compounds liberated from the lignocellulosic materials during hydrolysis. A number of studies have been published on the effect of the aeration on xylitol production, but only a few describe the effect of this parameter on hemicellulosic hydrolysates. In an experiment with sugar cane bagasse hemicellulosic hydrolysate containing 54.0

g/l xylose and 3.75 g/l acetic acid, the increase in air flow rate from 0.4 to 0.8 vvm resulted in increased rates of both xylose consumption and xylitol production by *C. guilliermondii*, but with the depletion of xylose, the yeast started to assimilate xylitol. The xylitol volumetric productivity at 0.8vvm was low (0.39 $g/l.h^{-1}$), but after adaptation of *C. guilliermondii* to the hydrolysate, it increased to 0.65 $g/l.h^{-1}$ with 0.6vvm (25). The most suitable aeration rate for xylitol volumetric productivity was not the same one that provided the highest xylose reductase and xylitol dehydrogenase activities (42). Using a continuous stirred-tank reactor and an oxygen transfer volumetric coefficient (K_La) of $20h^{-1}$ to grow the same yeast in bagasse hydrolysate resulted in a high xylitol volumetric productivity (0.70 $g/l.h^{-1}$), but when the k_La was raised to $30h^{-1}$ the yeast produced less xylitol and consumed more arabinose and acetic acid (*37*). The oxygen is not only an important factor to optimize the xylitol production but it is also an essential component for xylose assimilation (*22*). According to Granström et al. (*97*) a surplus of NADH during a limited oxygen supply inhibited the activity of xylitol dehydrogenase of *C. guilliermondii*, causing the yeast to accumulate xylitol.

Conclusions

The properties of xylitol, which attract the attention of pharmaceutical, odontological and food industries, along with the discovery of several xylose-fermenting microorganisms and the great availability of xylan-rich lignocellulosic residues, have given an impulse to the researches into new technologies for obtaining xylitol by biotechnological process. Efforts have been made to find optimal conditions for hydrolysis of lignocellulosic materials, as well as to reduce the toxicity of the hydrolysates and to establish the best fermentation parameters for xylitol recovery.

Nevertheless, the xylitol yields, productivities and concentrations so far obtained have not yet been sufficiently high (*98*) to make the biotechnological process able to compete with the chemical process currently used to produce xylitol (*13*). Further investigations are necessary to better understand the xylose metabolism into xylitol, the effect of the oxygen availability on the bioconversion, and the synergistic action of the microbial inhibitors present in the hydrolysates. Researches into the utilization of immobilized cells (*50, 51, 99*), mixed cultures (*100*), enzymatic technology (*101*) and genetic engineering techniques (*102*) have also been carried out with a view to finding new ways of improving this bioprocess.

Acknowledgment
The author acknowledge financial support of FAPESP and CNPq and thank Ms. Maria Eunice M. Coelho for assistance in writing this paper.

References

1. MANZ, U.; VANNINEN, E.; VOIROL, F. Xylitol - Its Properties and Use as a Sugar Substitute in Foods. In: FOOD. A. Symp. Sugar and Sugar Replacements, London, 10 oct., 1973.
2. MAKINEN, K.K. The Futurist, **1976**, june, p. 135-139.
3. ASSEV, S.; WALER, S.M.; RÖLLA, G. *Acta Pathol. Microbio. Immunol.* Section B, **1983**, v.91, p. 261-265.
4. WÄLER, S.M.; ASSEV, S.; RÖLLA, G. *J. Dental Research*, **1992**, v.100, p. 319-321.
5. ALANEN, P.; ISOKANGAS, P.; GUTMAN, K. *Communit. Dent. Oral Epidem.*, **2000**, v. 28, p. 218-224.
6. ASSEV, S.; RÖLLA, G. *Acta Pathol. Microbio. Immunol.* Scandinavica Section B, **1984**, v. 92, n. 2, p. 89-92.
7. HÄMÄLÄINEN, M.M.; MAKINEN, K.K. *Miner. Eletr. Metabol.*, **1989**, v. 15, p. 346-352.
8. YLIKAHRI, R. Advances in Food Research, **1979**, v. 25, p. 159-180.
9. EMODI, A. *Food Technol.*, **1978**, v. 32, n. 1, p. 28-32.
10. MATILA, P.T.; KNUUTTILA, M.L.E.; SVANBERG, M.J. *Metabolism*, **1998**, v. 47, p. 578-583.
11. ZABNER, J.; SEILER, M.P.; LAUNSPACH, J.L.; KARP, P.H.; KEARNEY, W.R.; LOOK, D.C.; SMITH, J.J.; WELSH, M.J. *Proc. Natl. Acad. Sci. USA*, **2000**, v. 97, n. 21, p. 11614-11619.
12. UHARI, M.; TAPAINEN, T.; KONTIOKARI, T. *Vacine*, **2001**, v. 19, p. 5144-5147.
13. MELAJA, A.J.; HÄMÄLÄINEN, L. U S N. 4.008.285. 1977.
14. NOLLEAU, V.; PREZIOSI-BELLOY, L. DELGENES, J.P.; NAVARRO, J.M. *Current Microbiol.*, **1993**, v. 27, p. 191-197.
15. DE FAVERI, D.; PEREGO, P. CONVERTI, A.; DEL BORGHI, M. *Chemical Engineer. J.*, **2002**, v.4101, p.1-8.
16. ONISH, H.; SUZUKI, T. *Agric. Biotech. Chemist.*, **1966**, v. 30, n. 11, p. 1139-1144.
17. CHEN, L-F; GONG, C-S. *J. Food Sci.*, **1985**, v. 50, p. 226-228.
18. BARBOSA, M.F.S.; MEDEIROS, M.B.; MANCILHA I.M.; SCHNEIDER, H.; LEE, H. *J. Ind. Microbiol.*, **1988**, v. 3, n. 4, p. 241-251.
19. du PREEZ, J.C.; DRIESSEL, van B.; PRIOR, B.A. *Arch. Microbiol.*, **1989**, v.152, p. 143-147.

20. ROSEIRO, J.C.; PEITO, M.A.; GIRIO, F.M.; AMARAL-COLLAÇO, M.T. *Archiv. Microbiol.*, **1991**, v.165, p. 484-490.
21. HORITSU, H.; YAHASHI, Y.; TAKAMIZAWA, K.; KAWAI, K.; SUZUKI, T.; WATANABE, N. *Biotecnol. Bioengineer.*, **1992**, v.40, p.1085-1091.
22. NOLLEAU, V.; PREZIOSI-BELLOY, L.; NAVARRO, J.M. *Biotechnol. Letters*, **1995**, v. 4, p. 417-422.
23. FELIPE, M.G.A.; VIEIRA, D.C.; VITOLO, M.; SILVA, S.S.; ROBERTO, I.C.; MANCILHA, I.M. *J. Basic Microb.*, **1995**, v.35 (3), p. 171-177.
24. FELIPE, M.G.A.; ALVES, L.A.; SILVA, S.S.; ROBERTO, I.C.; MANCILHA, I.M.; SILVA, J.B.A. *Bioresour. Technol.*, **1996**, v. 56, n. 2-3, p. 281-283.
25. FELIPE, M.G.A.; VITOLO, M.; MANCILHA, I.M. *Acta Biotecnol.*, **1996**, v. 16, n.1, p. 73-79.
26. FELIPE, M.G.A.; VITOLO, M.; MANCILHA, I.M.; SILVA, S.S.; *J. Indust. Microb. Biotech.*, **1997**, v. 18, p. 251-254.
27. FELIPE, M.G.A.; VITOLO, M.; MANCILHA, I.M.; SILVA, S.S. *Biomass Bioenerg.*, **1997**, v.13, n. 1/2, p.11-14.
28. SILVA, S.S.; VITOLO, M.; PESSOA Jr., A.; FELIPE, M.G.A. *J. Basic Microb.*, **1996**, v. 36, n. 3, p. 187-191.
29. KERN, M.; HALTRICH, D.; NIDETZKY, B.; KULBE, K.D. *Microb. Letters*, **1997**, v.149, p. 31-37.
30. ROSA, S.M.A.; FELIPE, M.G.A.; SILVA, S.S.; VITOLO, M. *Appl. Biochem. Biotechnol.*, **1998**, v. 70-72, p. 127-135.
31. ROBERTO, I.C.; SATO, S.; MANCILHA, I.M. *J. Industr. Microbiol.*, **1996**, v.16, p. 348-350.
32. RODRIGUES, D.C.G.A.; SILVA, S.S.; FELIPE, M.G.A. *J. Biotechnol.* , **1998**, v. 62, n. 1, p. 73-77.
33. SILVA, S.S.; FELIPE, M.G.A.; SILVA, J.B.A.; PRATA, A.M.R. Biochem., **1998**, v.33, n.1, p. 63-67.
34. ALVES, L.A.; FELIPE, M.G.A.; SILVA, J.B.A.; SILVA, S.S.; PRATA, A.M.R. *Appl. Biochem. Biotechnol.*, **1998**, v. 70-72, p. 89-98.
35. SENE, L.; FELIPE, M.G.A.; VITOLO, M.; SILVA, S.S.; MANCILHA, I.M. *J. Basic Microb.*, **1998**, v. 38, n. 1, p. 61-69.
36. KIM, S-Y.; OH, D-K. KIM, J-H. *Microb. Letters*, **1999**, v. 21, p. 891-895.
37. MARTINEZ, E.A.; SILVA, S.S.; FELIPE, M.G.A. *Appl. Biochem. Biotechnol.,* **2000**, v.84-86, p.633-641.
38. TAVARES, J.M.; DUARTE, L. C.; AMARAL-COLLACO, M.T.; GÍRIO, F.M. *Enzyme Microb. Technol.*, **2000**, v. 26, p.743-747.
39. AZUMA, M.; IKEUCHI, T.; KIRITANI, R.; KATO, J.; OOSHIMA, H. *Biomass Bioenergy*, **2000**, v. 19. p. 129-135.

40. MORITA, T.A.; SILVA, S.S.; FELIPE, M.G.A. *Appl. Biochem. Biotechnol.*, **2000**, v. 84-86, p. 751-759.
41. SENE, L.; VITOLO, M.; FELIPE, M.G.A.; SILVA, S.S. *Appl. Microbiol. Biotechnol.*, **2000**, v. 84-86, p.371-380.
42. SENE, L.; FELIPE, M.G.A.; SILVA, S.S.; VITOLO, M. *Appl. Microbiol. Biotechnol.*, **2001**, v. 91-93, p. 671-680.
43. CORTEZ, E.V.; FELIPE, M.G.A.; ROBERTO, I.C.; PESSOA JR., A.; VITOLO, M. *Appl. Biochem. Biotechnol.*, **2001**, n. 91-93, p. 753-759.
44. CANETTIERI, E.V.; ALMEIDA e SILVA, J.B.; FELIPE, M.G.A. *Appl. Biochem. Biotechnol.*, **2001**, v.94, p. 159168.
45. WALTHERS, T.; HENSIRISAK, P.; AGBLEVOR, A.F. *Appl. Biochem. Biotechnol.*, **2001**, v.91-93, p. 423-435.
46. SILVA, C.J.S.M.; ROBERTO, I.C. *Lett. Appl. Microbiol.*, **2001**, v. 32, p. 248-252.
47. SILVA, C.J.S.M.; ROBERTO, I.C. *Process. Biotech.*, **2001**, v.36, p. 119-1124.
48. MUSSATTO, S.I.; ROBERTO, I.C. *Biotechnol. Letters*, **2001**, v. 23, p. 1681-1684.
49. AGUIAR Jr., W.B.; FARIA, L.F.F.; COUTO, M.A.P.G.; ARAUJO, O.Q.F.; PEREIRA Jr., N. *Biochem. Engineer.* Journal, **2002**, v. 12, p. 49-59.
50. CARVALHO, W.; SILVA, S.S.; CONVERTI, A.; VITOLO, M.; *Appl. Biochem. Bioengineer.*, **2002**, v. 79, n. 2, p. 489-496.
51. CARVALHO, W.; SILVA, S.S.; CONVERTI, A.; VITOLO, M.; FELIPE, M.G.A.; ROBERTO, I.C.; SILVA, M.B.; MANCILHA, I. *Appl. Biochem. Biotechnol.*, **2002**, v. 98-100, p. 489-496.
52. RODRIGUES, R.C.L.B.; FELIPE, M.G.A.; ALMEIDA, J.B.A.; VITOLO, M. *Proces. Biochem.* **2003**, v. 38, p. 1231-1237.
53. CARVALHO, W.; SILVA, S.S.; SANTOS, J.C.; CONVERTI, A. *Enzyme Microb. Technol.*, **2003**, v. 32, p. 553–559.
54. ROBERTO, I.C.; MUSSATO, S.J.; RODRIGUES, R.C.L.B. *Ind. Crops Produc.*, **2003**, v. 17, p. 171-176.
55. RODRIGUES, R.C.L.B.; FELIPE, M.G.A.; ROBERTO, I.C.; VITOLO, M. *Biop. Biosyst. Engineer.*, **2003** (impress)
56. ALMEIDA e SILVA, J.B.; CANILHA, L. CANETTIERI, E. V.; FELIPE, M.G.A.; SOLENZAL, A.I.N. *J. Chem. Tech. Biotech.*, **2003**, v.58, p.945-948.
57. YOSHITAKE, J.; SHIMAMURA, M.; IMAI, T. *Agricult. Biolog. Chem.*, **1973**, v.37, p. 2251-2259.
58. YOSHITAKE, J.; SHIMAMURA, M.; ISHIZAK, H.; IRIE, Y. *Agricult. Biolog. Chemistry*, **1976**, v.8, p.1493-1503,.
59. DAHIYA, J.S. *Can. J. Microbiol*, **1991**, *v. 37, p. 14-18*

60. JEFFRIES, T.W. *Advanc. Biochem. Engineer.*, **1983**, v. 27, p. 1-32.
61. BRUINEMBERG, P.; BOT, P.H.M.; VAN DIJKEN, J.P.; SCHEFFERS, W.A. *Europ. J. Appl. Microbiol. Biotechnol.*, **1983**, v.18, p.287-292.
62. EVANS, *C.T.;* RATLEDG, C. *Arch. Microbiol.* **1984**, *v. 139, p. 48-52*
63. WILSON, D.K.; KAVANAGH, K.L.; KLIMACEK, M.; NIDETZKY B. *Chem. Biol. Interact*, **2003**, v. 143-144, p. 515-521.
64. VANDESKA, E.; KUZMANOVA, S.; JEFFRIES, T.W. *J. Ferment. Bioengin.*, **1995**, v. 80, n. 5, p. 513-516.
65. LEE, H.; SOPHER, C.R.; YAU, K.Y.F. *J. Chem. Tech. Biotechnol.* **1996**, v.66, 375-379.
66. SUGAI, J.K.; DELGENES, J.P. *J. Ind. Microbiol.*, **1995**, v. 14, n. 1, p. 46-51.
67. FENGEL, D.; WEGENER, G. *Wood Chemistry, Ultraestructure, Reactions*. Berlin: Walter de Gruyter, 1989.
68. KRULL, L.H.; INGLETT, G.E. *J. Agricult. Food Chemis.*, **1980**, 28: 917-919.
69. PESSOA JR.; A.; MANCILHA, I.M.; SATO, S. *Brazil. J. Chem. Engineer.*, **1997**, v. 14, n. 3, p. 309-312.
70. PALMQVIST, E.; HÄGERDAL, B.H. *Biores. Technol.*, **2000**, v. 74, p. 25-33
71. NIMZ, H.H.; CASTEN, R. *Hols als roh-und werstoff*, **1986**, v. 44, p. 207-212.
72. RODRIGUES, R.C.L.B.; FELIPE, M.G.A.; SILVA, J.B.A.; VITOLO, M.; GÓMEZ, P.V. *Brazil. J. Chem. Engineer.*, **2001**, v. 18, n. 3, p. 299-311.
73. ROBERTO, I.C.; FELIPE, M.G.A.; LACIS, L.S.; SILVA, S.S.; MANCILHA, I.M. *Bioresour. Technol.*, **1991**, v.36, n.3, p. 271-275.
74. CONVERTI, A.; DEL BORGHI, M. *Acta Biotechnol.*, **1996**, v. 16, n. 2-3, p. 133-144.
75. ALVES, L.A.; VITOLO, M.; FELIPE, M.G.A.; ALMEIDA e SILVA, J.B. *Appl. Biochem. Biotechnol.*, **2002**, v.98-100, p. 403-413.
76. DOMINGUEZ, J.M.; GONG, C.S.; TSAO G.T. *Appl. Biochem. Biotechnol.*, **1996**, v. 57-58, p. 49-56.
77. PARAJÓ, J.C.; DOMINGUEZ, H.; DOMINGUEZ, J.M. *Biotechnol. Letters.*, **1996**, v.18, n.5, p.593-598.
78. PARAJÓ, J.C.; DOMINGUEZ, H.; DOMINGUEZ, J.M. *Microb. Technol.*, **1997**, v.21, p.18-24.
79. NILVEBRANT, N.; REIMANN, A.; LARSSON, S.; JÖNSSON, L. *Appl. Biochem. Biotechnol.*, **2001**, v.91, p. 35-49.
80. van ZYL, C.; PRIOR, B.A.; du PREEZ, J.C. *Enzyme Microb. Technol.*, **1991**, v. 13, p. 82-86.
81. PARAJÓ, J.C.; DOMINGUES, H. DOMINGUES, J.M. *Proces. Biochem.*, **1997**, v.32, p.599-604.

82. SPENCER-MARTINS, I.; van UDEN, N. *Biochim. Biophys. Acta,* **1985**, v.812, p. 168-172.
83. LUCAS, C.; UDEN, N. *Appl. Microbiol. Biotechnol.*, **1986**, v. 23, p. 491-495.
84. KILIAN, S.G.; UDEN, N. *Appl. Microbiol. Biotechnol.*, **1988**, v. 27, p. 545-548.
85. WEBB, S.R.; LEE, H. *Biotechnol. Adv.*, **1990**, v.8, p.685-697.
86. SPENCER-MARTINS, I. *Biores. Technol.*, **1994**, v.50, p.51-57.
87. DELLWEG, H.; RIZZI, M.; METER, H. *Biotech. Letters*, **1984**, v.6, n. 6., p. 384-340.
88. MAYERHOFF, Z.D.V.L.; ROBERTO, I.C.; FRANCO, T.T. *Appl. Biochem. Biotechnol.*, **2001**, v. 91-93, P. 729-737.
89. BICHO, P.A.; RUNNALS, P.L.; CUNNINGHAM, J.D.; LEE, H. *Appl.Environ. Microbiol.*, **1988**, v. 54, n. 1, p. 50-54.
90. KASTNER, J.R.; ROBERTS, R.S. *Biotechnol. Letters*, **1990**, v. 12, n. 1, p. 57-60.
91. PREZIOSI-BELLOY, L.; NOLLEAU, V.; NAVARRO, J.M. *Enzyme Microb. Technol.*, **1997**, v.21, p. 124-129.
92. ROBERTO, I.C.; SILVA, S.S.; FELIPE, M.G.A.; MANCILHA, I.M.; SATO, S. *Appl. Biochem. Biotechnol.* **1996**, v. 57-58, p. 339-346
93. PAMPULHA, M.E.; LOUREIRO-DIAS, M.C. *Appl. Microbiol. Biotechnol.*, **1989**, v.31, p. 547-550.
94. HERRERO, A.A.; GOMES, R.F.; SNEDECOR, B.; TOLMAN, C.J.; ROBERTS, M.F. *Appl. Microbiol. Biotechnol.*, **1985**, v.22, p. 53-62.
95. FERRARI, M.D.; NEIROTTI, E.; ALBORNOZ, C.; SAUCEDO, E. *Biotechnol. Bioeng.*, **1992**, v.40, p. 753-759.
96. LAWFORD, H.G.; ROUSSEAU, J.D. *Appl. Bioch. Biotechnol.*, **1994**, v.45/46, p.473-448.
97. GRANSTRÖM, T.; OJAMO, H.; LEISOLA, M. *Appl. Microb. Biotechnol.* **2001**, v.55, p.36-42.
98. GURGEL, P.V.; MANCILHA, I.M.; PECANHA, R.P. *Biores. Technol.* **1995**, v. 52, p. 219-223.
99. SANTOS, J.C.; CARVALHO, W.; SILVA, S.S.; CONVERTI, A. *Biotechnol. Progr.*, **2003**, v. 19, p. 1210-1215.
100. CHUNG, Y-S.; KIM, M-D.; LEE, W-J.; RYU, Y-W.; KIM, J-H.; SEO, J-H. *Enzyme Microb. Technol.*, **2002**, v. 30, p. 809-816.
101. DELGENES, J.P.; ESCARE, M.C.; LAPLACE, J.M.; MOLETA, R.; NAVARRO, J.M. *Indust. Crops Prod.,* **1998**, v.7, p.101-111.
102. NIDETZKY, B.; WILFRIED, N.; HALTRICH, D.; KULBE, D.K. *Biotechnol. Bioengineer.,* **1996**, v. 52, p.387-396.

Chapter 19

Biobleaching Efficiency of *Aspergillus oryzae* Xylanases Produced by Solid Substrate Fermentation

Judit Szendefy[1], George Szakacs[1], and Lew Christopher[2,*]

[1]Department of Agricultural Chemical Technology, Technical University of Budapest, 1111 Budapest, Gellért tér 4, Hungary
[2]Sappi Biotechnology Laboratory, Department of Microbial, Biochemical and Food Biotechnology, University of the Free State, P.O. Box 339, Bloemfontein 9300, South Africa and Sappi Forest Products Technology Centre, Sappi Management Services, P.O. Box 3252, Springs 1560, South Africa

The direct use of substrate specific enzymes produced by solid substrate fermentation (SSF) without a downstream processing represents an innovative method for modification of pulp and paper properties and reduction of chemical consumption during bleaching that can bring about positive economical and environmental benefits to the industry. Screening of a number of *Aspergillus oryzae* isolates on oxygen delignified soda-aq pulp of *Eucalyptus grandis* as carbon source and enzyme inducer yielded xylanase activities for the best six strains in the range of 500 – 4300 IU/g dry material (DM) with low levels (<1.0 FPU/g DM) of simultaneously produced cellulase. The *in situ* produced enzymes (whole SSF culture) were used in pulp bleaching without a prior downstream processing of xylanase. The brightness gain over control was enzyme dose dependent and varied between 0.7 and 1.4 points. The optimum moisture content (83%), initial pH (8.3) and time course (4 d) of SSF were determined for the best-performing strain in the bleaching studies (*A. oryzae* NRRL 3485). Addition of corn steep liquor, an inexpensive nitrogen source, enhanced the xylanase yield considerably. The pH and temperature optima of the xylanase complex of *A. oryzae* NRRL 3485 was found to be 6.5 and 65°C, respectively. An improved thermal stability was observed at 50°C and pH 6.5 in the presence of eucalyptus pulp substrate. When used at equal enzyme costs and charges in biobleaching of eucalyptus soda-aq pulp, the *A. oryzae* enzyme outperformed a commercial xylanase control in its efficiency to improve pulp brightness.

Biobleaching of cellulose pulps with xylanase enzymes is a viable technology proven in a number of mills worldwide (1). The use of xylanases to enhance pulp bleachability was first reported in 1986 by Viikari *et al* (2). Pretreatment of pulp with xylanases was shown to reduce the amount of chlorine used in the bleaching process, affording a positive environmental impact. This observation has initiated numerous research investigations to improve the method of biobleaching, explore and characterize new strains which could be suitable for this purpose. The requirements for these microorganisms are to produce high levels of xylanase concomitantly with no or low levels of cellulase. Furthermore, the xylanase produced by these strains should have a high pH and temperature optimum.

The pulp and paper industry uses liquid xylanase preparations (enzyme concentrates) for biobleaching. These enzymes are produced by a conventional submerged fermentation (SF) followed by a downstream processing (concentration and partial purification). It has been recently demonstrated that *in situ* produced solid substrate fermentation (SSF) materials can be used directly in pulp bleaching without a downstream processing of the enzyme (3). SSF offers numerous advantages over submerged fermentation, including high volumetric productivity, high concentration of the products, less effluent generation and simple fermentation equipment (4,5). It has been proven that many enzymes and other biochemicals can be produced by SSF at a fraction of the cost of SF production (6,7). The possibility that raw SSF enzymes (mixture of substrate residue, fungal biomass and different enzymes) can be applied for pretreatment of pulp without a prior downstream processing makes this new biobleaching approach competitive to the use of commercial enzymes. However, as a fermentation method for enzyme production, the SSF has some disadvantages over the SF such as incomplete heat removal (temperature gradient inside the bed) and difficulties in monitoring and regulation of some basic parameters (pH, temperature, moisture content).

Fungal xylanases including *A. oryzae* xylanase have been reviewed recently by Haltrich *et al* (8). Xylanases produced by *A. oryzae* have been already studied from different aspects. Induction of xylanase production by a synthetic xylobiose analog, ß-methyl-D-xyloside, was demonstrated by Bailey *et al* (9). Purification and characterization of xylanase components was studied in detail (10,11). Xylanase production in a 10-l laboratory fermentor was also performed (12).

Binding of xylanase components onto insoluble xylan and cellulose has also been described (13). Evaluation of the bleach-enhancing abilities of cellulase-free xylanases of *A. oryzae,* produced in shake flasks by SF, has been carried out on sulfite pulps (8). Although xylanases from *A. oryzae* has been produced exclusively in SF (4,8,14,31), a SSF alternative has been demonstrated most recently (15). However, no reports are found in the literature on the use of xylanase enzymes produced by this fungus in SSF for biobleaching of cellulose pulps.

To avoid health risks and minimize environmental hazards, safe microorganisms should be used. For example, *Aspergillus oryzae* is a food-grade fungus that is widely used for production of fermented food products and amylolytic enzymes in the Oriental countries. In this paper, the production of SSF xylanases using food-grade strains of *A. oryzae* is presented together with the application of the *in situ* enzymes (whole SSF material) in biobleaching of chemical pulp.

Materials and Methods

Cultures

Aspergillus oryzae strains (14 in total) have been purchased from the following culture collections: Institute for Fermentation, Osaka, Japan (IFO); University of Sciences Jozsef Attila, Szeged, Hungary (JATE); Merck Sharp Dohme Research Laboratories, Rahway, NJ (MSD); Northern Regional Research Center, USDA, Peoria, IL (NRRL); Russian Culture Collection, Moscow (VKM). The fungi were grown and maintained on potato dextrose agar (PDA) Petri plates and slants. The total viable spore number in PDA Petri plate culture (6 d old) was determined by counting colonies following serial dilution and incubation.

Substrate

Oxygen-delignified pulp from *Eucalyptus grandis* was obtained from a soda-aq pulp producing mill in South Africa. Prior to use, pulp was thoroughly washed with distilled water until neutral pH of wash waters was attained, thereafter pulp was air-dried and stored at 4°C until use.

Solid Substrate Fermentation

Non-optimized SSF was carried out in 500-ml cotton plugged Erlenmeyer flasks. Five gram pulp substrate (dry weight equivalent) was wetted with 25 ml salt solution setting the moisture content to 83%. The *A. oryzae* strains were screened on two SSF media. Composition of medium I was as follows (in g/l): NH_4NO_3, 5; corn steep liquor (50 % dry weight) 2; KH_2PO_4, 5; NaCl, 1; $MgSO_4.7H_2O$, 1; and (in mg/l): $CoCl_2.6H_2O$, 2; $MnSO_4$, 1.6; $ZnSO_4.7H_2O$, 3.45; $FeSO_4.7H_2O$, 5. Medium II contained (in g/l)): $(NH_4)_2HPO_4$, 4; corn steep liquor (50% dry weight), 2; urea, 1; KH_2PO_4, 3; NaCl, 1; $MgSO_4.7H_2O$, 1; and (in mg/l): $CoCl_2.6H_2O$, 2; $MnSO_4$, 1.6; $ZnSO_4.7H_2O$, 3.45; $FeSO_4.7H_2O$, 5. The pH of both media was adjusted before sterilization to pH 9.0 or pH 7.0. These values changed after sterilization to pH 8.3 and pH 7.0, respectively. Sterilization was carried out in an autoclave at 121°C for 30 min. After cooling, the medium was inoculated with a spore suspension of the fungus to a final concentration of $2.5x10^6$ spores g^{-1} dry matter (DM) substrate. The spore suspension was prepared by washing the spores from the surface of 6 day-old sporulating PDA plates with water containing 0.1 % Tween-80. The inoculated flasks were incubated at 30°C for 4 days. Thereafter, the whole SSF material in the flask was used for enzyme extraction. All SSF experiments were carried out in duplicate and results were presented as average values.

The effect of easily metabolized sugars and mineral salts on xylanase production was determined by adding glucose, lactose or xylose in a 10 g/l concentration to the above salt solution (medium I). $CaCl_2.2H_2O$ and $FeSO_4.7H_2O$ were added to the basal medium in 1.0 and 0.2 g/l concentrations, respectively.

Ammonium sulfate, ammonium nitrate, potassium nitrate, corn steep liquor (50 % dry weight) and defatted soybean meal were compared as nitrogen sources. Their concentrations were calculated as to provide 0.009 g nitrogen per gram carbon source (0.9%). The pH of the medium was adjusted to pH 9.0 before sterilization.

To determine the optimum moisture level of the optimized medium, different moisture levels (67-86%) were adjusted with the following optimized salt solution (g/l): KH_2PO_4, 5; corn steep liquor (50 % dry weight), 75; NaCl, 1; $MgSO_4.7H_2O$, 1; and (in mg/l): $CoCl_2.6H_2O$, 2; $MnSO_4$, 1.6; $ZnSO_4.7H_2O$, 3.45; $FeSO_4.7H_2O$, 5; and water.

In the time course experiment, 5 g pulp was supplemented to 83 % moisture content with the above described optimized and non-optimized medium. The composition of the non-optimized medium was identical to medium I described above. Before sterilization, the pH of both media was adjusted to pH 9.0.

Enzyme Extraction

The enzyme activity was determined from the culture extract of SSF samples: 5 g (dry weight) fermented substrate was extracted with 100 ml water containing 0.1 % Tween-80 by shaking for 1 h at room temperature (25°C). At the end of the extraction, the suspension was centrifuged (3000 rpm, 10 min) and the supernatants were stored at 4°C until use for enzyme activity determinations.

Enzyme Assays

Xylanase activity was determined by the method of Bailey *et al.* (*16*). The substrate solution contained 1% birchwood xylan (Sigma X-0502) solubilized in 0.05 M citrate-phosphate buffer (pH 6.0). All necessary dilutions of the supernatants were carried out using the same buffer. The reaction mixture consisted of 1.8 ml substrate solution and 0.2 ml properly diluted enzyme. After 5 min incubation at 50°C, the liberated reducing sugars (xylose equivalents) were estimated by the dinitrosalicylic acid method (17). One unit (IU) of xylanase was defined as the amount of enzyme releasing 1 μmol xylose equivalents per minute under the assay conditions. Filter paper activity (FPA) was assayed as described by Ghose (18). β-Xylosidase was determined by measuring the release of *p*-nitrophenol from *p*-nitrophenyl-β-D-xylopyranoside according to Her *et al* (19). Laccase was determined as described by Wolfenden *et al* (20). 2,2'-azino-bis(3-ethylbenzthiazoline-6- sulfonic acid) (ABTS; Sigma A-1888) was used as substrate.

Enzyme Characterization

The pH and temperature optima of xylanase were determined using pH and temperature ranges of 4.0-8.0 and 40-80°C, respectively. For pH adjustment, the following buffers have been used: 0.05M citrate (pH 4.0-5.5), 0.05 M citrate-phosphate (5.5-6.5) and 0.05 M phosphate (pH 6.5-8.0). The thermal stability was determined by incubating the enzyme at the desired temperature for different periods (5, 10, 30 and 60 min) in a 0.05M citrate-phosphate buffer, pH 6.5.

Experimental Design

A 2 factors on 3*4 level design leading to 12 sets of experiments, performed in duplicate, was used to verify the effect of total nitrogen content in the salt solution and its composition on xylanase activity. The variables were coded according to Eq. 1:

$$x_i = X_i\text{-}X_0/\Delta X_i \qquad (1)$$

where x_i is the coded value of a variable, X_i is the real value of the variable, X_0 is the real value of a variable at the center point, and ΔX_i is the step change value. The significance level of each variable effect and interactions is determined by Student's test:

$$t(18) = \frac{b_j}{s_{b_j}}$$

where b_j is the regression coefficient calculated as follows:

$$b_j = \frac{\sum_i y_i x_{ji}}{\sum_i x_{ji}^2}, \qquad b_0 = \frac{\sum_i y_i}{\sum_i x_{ji}^2}, \qquad \sum_i^N x_{ij}^2 = N$$

y_i - xylanase activity, x_{ji} - transformed factors, b_0 – intercept, N - number of trials

$s_{b_j}^2$ is the variance of regression coefficient: $s_{b_j}^2 = \frac{s_y^2}{N}$

The range and the levels of the variables investigated in this study are shown in Table I. The xylanase activity was taken as the dependent variable or response, y_i. Response surface was fitted on experimental results. The quadratic model for predicting the optimal point was expressed according to Eq. 2:

$$Y = b_0 + \Sigma b_i x_i + \Sigma b_{ii} x_i^2 + \Sigma b_{ij} x_i x_j \qquad (2)$$

where Y is the response variable, b the regression coefficients, and x_i the coded levels of the variable. The significance level of each variable effect, and interactions is determined by Student's t test. $t_{\alpha/2} = 2.052$ when $\alpha = 0.05$ and degrees of freedom is 18. Statistica for Windows 2000 (StatSoft, Inc.) was used for the regression analysis of the experimental data obtained.

Table I. Range of values for the 2 factors on 3*4 level design

Variables	*Levels*			
	-1	**-0.33**	**+0.33**	**+1**
X_1: Total nitrogen content (g/l)	1	2	3	4
	Levels			
	-1	**0**	**+1**	
X_2: CSL/$(NH_4)_2SO_4$	0 (0% CSL/100% $(NH_4)_2SO_4$)	0.5 (50% CSL/50% $(NH_4)_2SO_4$)	1 (100% CSL/0% $(NH_4)_2SO_4$)	

CSL: corn steep liquor (SIGMA)

Enzyme Treatment of Pulp

Pretreatment of eucalyptus oxygen-delignified soda-aq pulp with the intact SSF material was carried out under standardized conditions (60°C, pH 6.5 , 10 % pulp consistency, 3 h). As reference, pretreatment of pulp with a commercial enzyme product was carried out at optimum conditions as described above using a charge of 5 IU xylanase/g pulp or a charge corresponding to a cost of USD1/t pulp. The latter was based on a price supplied by the enzyme manufacturer, in the case of the commercial enzyme, and a cost for enzyme production of USD 150 per ton pulp, in the case of the SSF enzyme (7).

Chemical Bleaching of Pulp

The enzyme pretreated pulp was subsequently bleached in a DED sequence. Controls were prepared under the same conditions but omitting the enzyme. To evaluate the bleaching efficiency of the *in situ* SSF xylanase following chemical bleaching, pulp samples were washed, brightness sheets were prepared and analysed for brightness using a ColorTouch 2 brightness machine (Technidyne Corp., New Albany, IN, USA) according to the Standard Methods of the Technical Association of the Pulp and Paper Industry (Tappi, Atlanta, GA, USA).

Results and Discussion

Screening of *Aspergillus oryzae* Isolates

Fourteen *Aspergillus oryzae* isolates were evaluated for xylanase and cellulase production on eucalyptus pulp in SSF on two media at two initial pH values. Xylanase production varied between 561 and 4131 IU/g dry matter (DM) of initial substrate (data not shown). Isolate NRRL 447 showed the lowest productivity while strain NRRL 6270 produced the highest results. The xylanase production for the best six strains is shown in Fig.1. Most isolates produced more xylanase when the initial pH of the substrate was higher (pH 8.3). The only exception was isolate 6270 (initial pH of the substrate was pH 7.0). For the best enzyme producer (NRRL 6270), the medium composition had a significant effect on xylanase activity.

Six strains that produced xylanase activity of 500 IU/g DM or higher and cellulase activity of less than 1.0 FPU/g DM were selected for further studies. In the next step, the SSF *in situ* enzymes (the whole SSF material) of these isolates were evaluated for their biobleaching efficiency on eucalyptus soda-aq pulp. This substrate was used to induce the formation of substrate specific enzyme complexes with enhanced bleaching efficiency. The SSF enzyme of NRRL 3485 showed the best bleaching ability at equal enzyme costs and equal enzyme charges (Fig.2). Interestingly, the best xylanase producer (NRRL 6270) was not the most efficient in in the bleaching experiments. One explanation might be that xylanase is not the only enzyme that takes part in the biobleaching process. De Vries *et al* studied the synergy between enzymes from *Aspergillus niger* (*21*). Arabinofuranosidase, arabinofuranohydrolase, feruloyl esterase and α-glucuronidase can assist in the degradation of the xylan backbone by endo-xylanase and ß-xylosidase (*21*). Similar findings have been described for the action of *Trichoderma reesei* and *A. oryzae* esterases in the deacetylation of hemicellulose (*31*). Based on the bleaching efficiency of the *A. oryzae* SSF enzymes, isolate NRRL 3485 was selected for further optimization studies.

Evaluation of Optimal Culture Conditions

Effect of Easily Metabolized Sugars and Mineral Salts

The effect of three simple sugars as supplements on xylanase production was studied by using a basal SSF medium (Fig. 3). Addition of glucose (10 g/l) to the wetting solution (0.25 g glucose/5 g eucalyptus pulp) practically did not affect xylanase production while lactose and xylose slightly decreased the enzyme yield (Fig.3). The influence of $CaCl_2$ and $FeSO_4$ on xylanase production of *A. oryzae* NRRL 3485 in SSF was also studied. Although significant effects

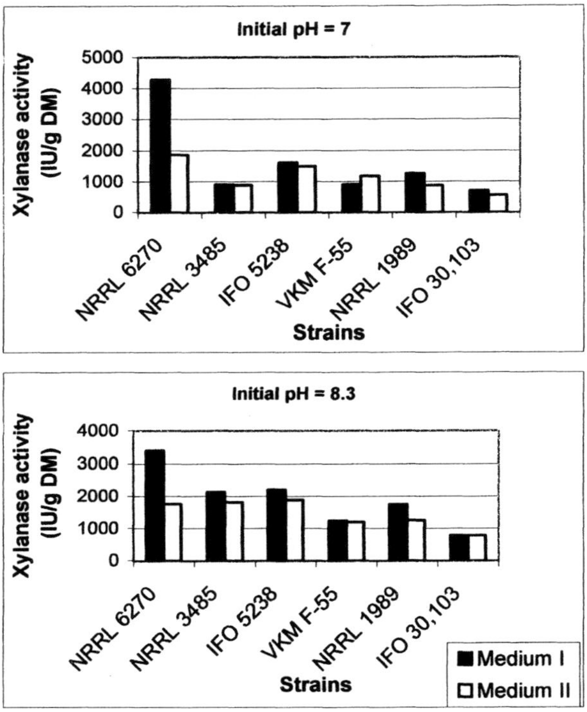

Figure 1. *Effect of medium composition and initial pH on xylanase production by Aspergillus oryzae strains in SSF on eucalyptus pulp (T, 30°C; initial moisture content, 83%;, initial pH, 8.3, fermentation time, 4 d). Medium I (g/l): NH_4NO_3, 5; Corn steep liquor (50% dry weight), 2; KH_2PO_4, 5; NaCl, 1; $MgSO_4$, 1; trace element solution I., II. 1 ml/l. Medium II (g/l): $(NH_4)_2HPO_4$, 4; Corn steep liquor (50% dry weight), 2; urea, 1; KH_2PO_4, 3; NaCl, 1; $MgSO_4$, 1; trace element solution I and II, 1 ml/l.*

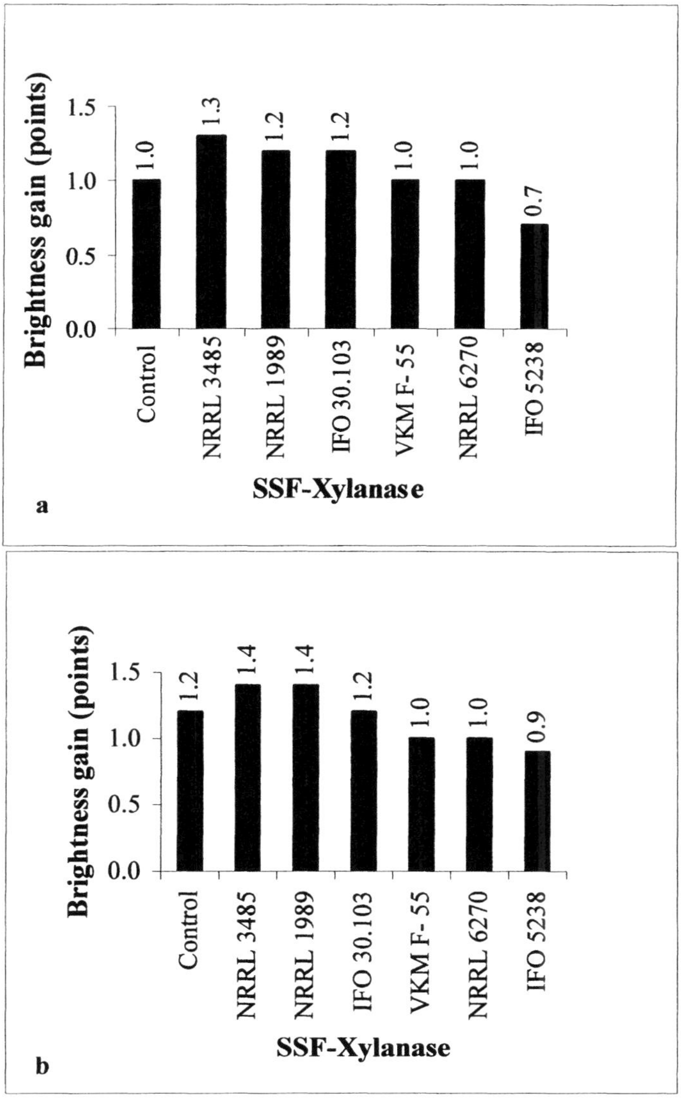

Figure 2. *Comparison of the bleaching efficiency of the SSF Aspergillus oryzae xylanase vs commercial xylanase (control) at equal enzyme costs (a: USD1/t pulp) and equal enzyme charges (b: 5 U/g pulp).*

on xylanase production of some other *Aspergillus* strains by $CaCl_2$ (*23*) and Fe^{2+} (*28*) were reported, none of these salts resulted in any changes in xylanase production in our experiment (Fig.3).

The expression of xylanases in many fungi is subject to regulation by catabolite repression. In submerged cultures, the accumulation of reducing sugars has been reported to have a negative effect in the production of xylanases (*14*). Solid state systems that have been described as resistant to catabolite repression, were so far developed using only wheat bran as substrate (*22*). Our results indicate that cellulose pulps of *Eucalyptus grandis* can also serve as a substrate for xylanase production which is insensitive towards catabolite repression.

Nitrogen Requirements

It was shown earlier (*3*) that the nitrogen content of eucalyptus pulp is extremely low and not enough to support any growth and enzyme production. Therefore the impact of different organic and inorganic nitrogen sources/additives on xylanase production was studied. The ammonium nitrate of the basal medium (see Materials and methods) was replaced by various inorganic and organic ingredients with a total nitrogen content fixed at 1.75 g/l in the wetting solution. Corn steep liquor (CSL) was found to be the best nitrogen source for xylanase production. Interestingly, the growth and enzyme production was very scarce if KNO_3 was used as sole N-source (Fig.4.). Similar nitrogen sources were evaluated for xylanase production in SSF by *Aspergillus terreus* on wheat straw (*23*). CSL was found to be efficient in the fermentation process, however, the use of ammonium salts such as ammonium sulfate and ammonium chloride resulted in the highest xylanase production.

Although the initial pH for all five media tested was the same (pH 8.3), the final pH values of SSF differed. For example, final pH values of 6.8, 6.7, 5.5 and 4.4 were measured for CSL, ammonium nitrate, soybean meal and ammonium sulfate containing media, respectively. The final pH of the SSF fermentation can have a major impact on xylanase yield (*3*). The ammonium sulfate containing media had the lowest pH whereas the use of CSL produced the highest final pH value. Te combination of these two components in the SSF medium seem to impact on the final pH and on the regulation of xylanase production. The design of this experiment and data obtained are given in Table II. The amount of the total nitrogen content in the wetting solution was studied at four levels, namely 1, 2, 3 and 4 g/l total N concentration. Three ratios for CSL and ammonium sulfate were set: 0 (ammonium sulfate alone), 0.5 (ammonium sulfate and CSL in a 1:1 ratio, based on their total N content) and 1 (CSL alone as nitrogen source). Regression analysis was performed to fit the response function with the experimental data. The regression coefficients *t*- and *p* are presented in Table III.

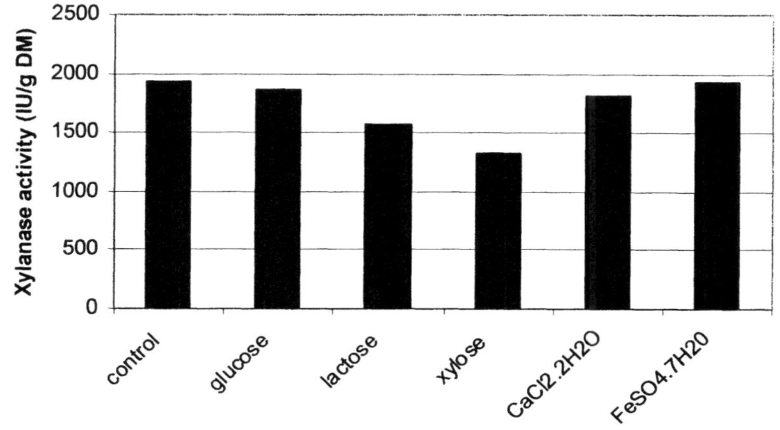

Figure 3. *Effect of easily metabolized sugars and mineral salts on xylanase production of Aspergillus oryzae NRRL 3485. Control wetting solution consisted of (g/l): NH_4NO_3, 5; corn steep liquor (50 % dry weight), 2; KH_2PO_4, 5; NaCl, 1; $MgSO_4$, 1; trace element solutions I and II, 1 ml/l. Sugars were added to control salt solution in 10 g/l concentration,$CaCl_2$ and $FeSO_4$ in 1 and 0.2 g/l concentrations, respectively (T, 30°C; initial moisture content, 83%;, initial pH, 8.3, fermentation time,4 d).*

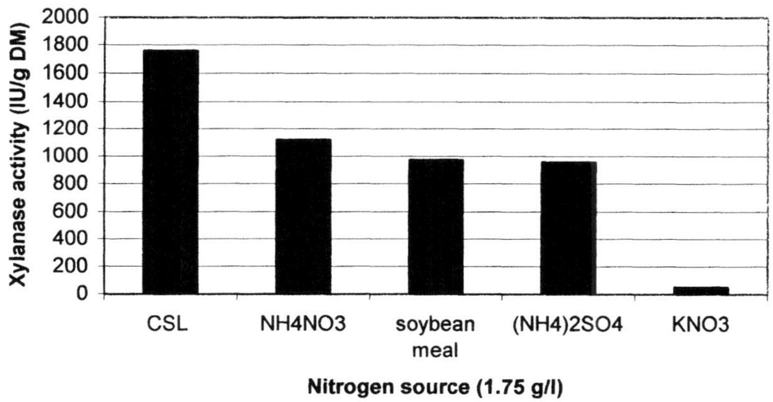

Figure 4. *Effect of different nitrogen sources on xylanase production by Aspergillus oryzae NRRL 3485 in SSF on eucalyptus pulp (T, 30°C; initial moisture content, 83%;, initial pH, 8.3, fermentation time, 4 d). CSL, corn steep liquor (50% dry weight).*

The fit of the polynomial model (R^2) was calculated to be 0.92, indicating that 92% of the variability in the response could be explained by the model. The response equation that represented a suitable model for xylanase production was obtained as follows:

$$Y = 2559 + 301\,x_1 - 561\,x_1^2 + 864\,x_2 - 607\,x_2^2 + 317\,x_1\,x_2$$

The three-dimensional graphs obtained from the calculated response surface are presented in Fig.5. The fitted model predicted a maximum point of 3000 IU/g at 3 g/l total nitrogen content ($x_1=+0.33$) with CSL as the only N-component ($CSL/(NH_4)_2SO_4=1$; $x_2=+1$). This is in agreement with the experimental results measured in this point (see Table II) which verifies the validity of the response model. The final pH was around 7.5 in the maximum point with the highest xylanase yield. Addition of ammonium sulfate decreased the final fermentation pH which had a negative impact on xylanase production of *A. oryzae* NRRL 3485.

Table II. Experimental design and results of the 2 factors on 3*4 level design

Total N-content in wetting solution (g/l)	**$CSL/(NH_4)_2SO_4$ ratio**	**Xylanase activity (IU/g DM)**	
X_1	**X_2**	Flask I	Flask II
-1	-1	983	996
-1	0	1407	1407
-1	+1	1527	1532
-0.33	-1	489	687
-0.33	0	2593	2353
-0.33	+1	2749	2805
+0.33	-1	945	743
+0.33	0	2834	2720
+0.33	+1	3008	3184
+1	-1	673	702
+1	0	2339	2330
+1	+1	2595	2637

Table III. The regression coefficients, *t*- and *p* values of xylanase production 2 factors on 3*4 level design design

Factor	*Regression coefficient*	*t(18)-Value*	*p-Value*
Mean/Interc.	2559.438	19.9761	0.000000
(1) Total nitrogen (L)	300.750	3.73712	0.001509
Total nitrogen (Q)	-560.813	-4.15530	0.000594
(2) CSL/$(NH_4)_2SO_4$ (L)	863.678	11.75652	0.000000
CSL/$(NH_4)_2SO_4$ (q)	-606.938	-4.76986	0.000153
1L by 2L	317.137	3.21761	0.004774

R^2=0.918; Adj:.89572; MS Residual=86352

Moisture Content

The moisture content of the medium/substrate in SSF is of importance to the growth of microorganisms, production and expression of enzymes. The optimum moisture levels have to be determined for each system and microorganism (*24,25*). The importance of moisture content in SSF was demonstrated during cultivation of *A. oryzae* (*15*) and *A. sulphureus* (*2*). To examine the effect of moisture content on xylanase production of *A. oryzae* NRRL 3485 in SSF, moisture levels of 67, 75, 80, 83 and 86 % were examined (Fig.6). Although at 67% initial moisture content the growth of fungus was rather poor (by visual observation), the production of xylanase was relatively high. The highest xylanase production (3004 IU/g DM) however was observed at a 83% initial moisture content.

Initial pH of Substrate

The initial pH of the substrate had a strong effect on the xylanase production of *A. oryzae* NRRL 3485 in SSF. A maximum xylanase yield was observed at initial pH of 8.3 using both the optimized and non-optimized medium (Fig.7). However, the majority of *Aspergillus* strains prefer an initial pH of medium in the range of pH 4 to 7 (*23,27*). For instance, for a *A. fumigatus* strain, the optimal fermentation pH was found to be 3.0 because use of higher pHs enhanced production of proteases that partially degraded xylanase (*12*). On the other hand, an alkaline pH tolerant strain of *A. nidulans* produced maximum xylanase activities at a fermentation pH of 10 *(28)*.

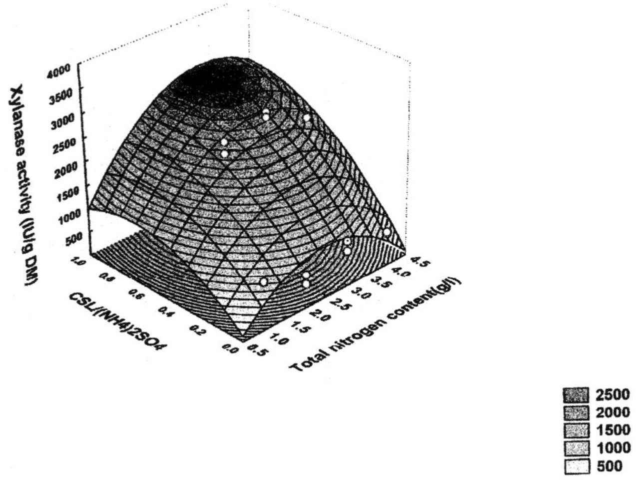

Figure 5. *Fitted surface of xylanase production by Aspergillus oryzae NRRL 3485 in SSF on eucalyptus pulp in terms of nitrogen content and composition (corn steep liquor and $(NH_4)_2SO_4$) in the wetting salt solution (T, 30°C; initial moisture content, 83%;, initial pH, 8.3, fermentation time, 4 d).*

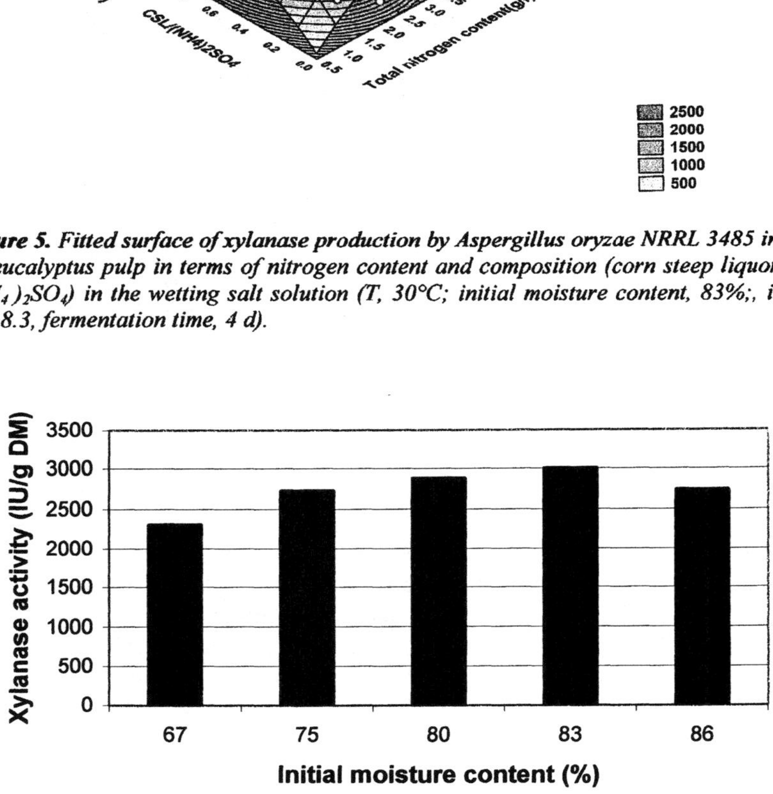

Figure 6. *Impact of initial moisture content of eucalyptus pulp on xylanase production by Aspergillus oryzae NRRL 3485 in SSF (T, 30°C; initial moisture content, 83%;, initial pH, 8.3, fermentation time, 4 d).*

Time-course of Xylanase Production

The time-course of xylanase production was studied on both non-optimized and optimized media at 83 % moisture content and initial pH of 8.3 (Fig.8). The peak in xylanase production on the optimized medium was on day 4. Through medium optimization, a 1.75-fold increase in xylanase yield was attained at a shorter fermentation time (Fig.8a). A simultaneous production of β-xylosidase and laccase enzymes was also detected (Fig.8b). The pH profiles of fermentation on the two media were different (Fig.8c) despite the fact that the initial pH was the same for both media (pH 8.3). For an *A. oryzae* strain, four days was reported as the optimum fermentation time in SSF (*15*). When *A. niger* was cultivated in SSF on wheat bran, 42 h was reported as the most favorable fermentation time for the production of mixtures of hydrolytic enzymes including xylanase (*29*).

Properties of *Aspergillus oryzae* Xylanase

Figs. 9 and 10 depict the effect of temperature and pH on hydrolysis of xylan by the SSF xylanase of *A. oryzae* NRRL 3485. The enzyme exhibited optimal activity at 65°C and pH 6.5. The xylanase was active over a broad range of pH (4.0-8.0) at 50°C. Previously, pH of 5.0 and 6.0 were reported for the xylanase of *A. oryzae* (*10,11,30*) while temperature optima of 55 and 70°C were observed for endoxylanase and β–xylosidase of this fungus, respectively (*11*).

The thermostability of the SSF enzyme extract (supernatant) strongly depended on the presence of substrate. Without substrate, the enzyme was stable only below 50°C (Fig.11). However, in the presence of pulp, the enzyme retained about 85 % of its activity even after 60 min (Fig.11). The thermostability curve at 60°C suggests the presence of a thermostable xylanase component (Fig.11). Two endo-xylanases (EXI, EXII) of *A. oryzae* were purified and characterized by Bailey *et al* (*10*). It was reported that while EX I was almost fully stable during a 4 h incubation at 55°C, EX II was completely inactive at this temperature. However, the endo-xylanase and β-xylosidase of an *A. oryzae* isolate were unstable over 50°C as described by other authors (*11*).

Biobleaching with *Aspergillus oryzae* Xylanase

As result of the increase of xylanase yields following media optimization, the biobleaching efficiency of the *A. oryzae* xylanase was also enhanced. The brightness gain over control increased by another 0.3 points as compared to the use of the SSF enzyme produced on the non-optimized medium. This increase in brightness was possible when the enzyme was used at a charge corresponding to

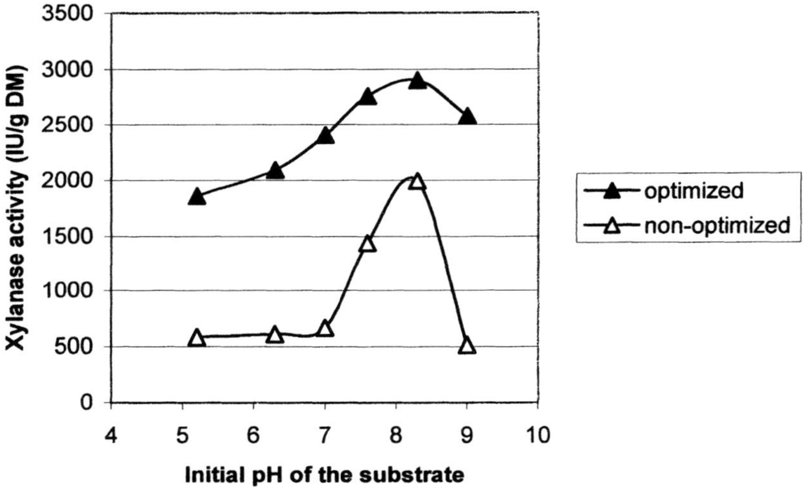

Figure 7. *Effect of initial pH of the substrate on xylanase production by Aspergillus oryzae NRRL 3485. Non-optimized wetting solution consisted of (g/l): NH_4NO_3, 5; corn steep liquor (50 % dry weight), 2; KH_2PO_4, 5; NaCl, 1; $MgSO_4$, 1; trace element solution I and II, 1 ml/l. Optimized wetting solution consisted of (g/l): corn steep liquor (50 % dry weight), 75; KH_2PO_4, 5; NaCl, 1; $MgSO_4$, 1; trace element solution I and II, 1 ml/l (T, 30°C; initial moisture content, 83%;, initial pH, 8.3, fermentation time, 4 d).*

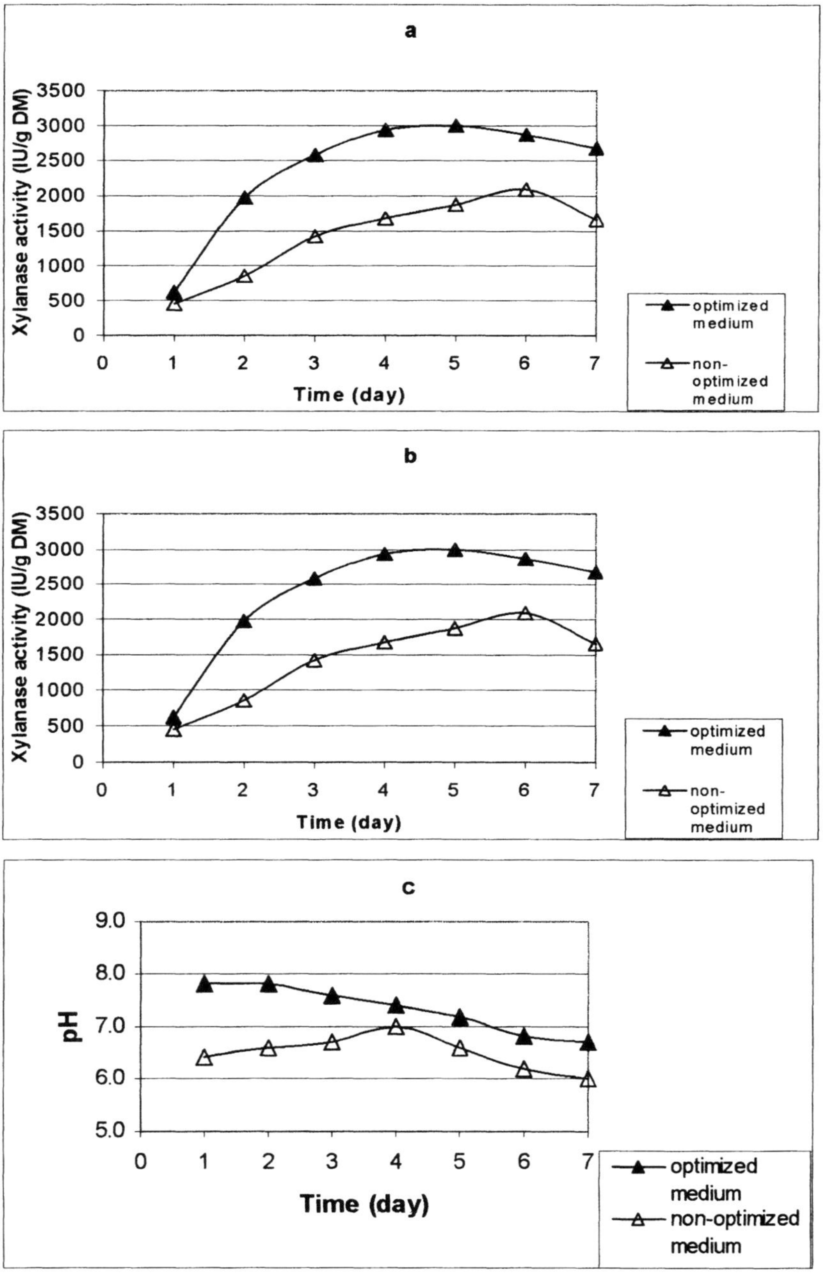

Figure 8. *Time course of xylanase, β–xylosidase and laccase production by Aspergillus oryzae NRRL 3485 in SSF on eucalyptus pulp (a: xylanase; b: β–xylosidase and laccase, c: pH).*

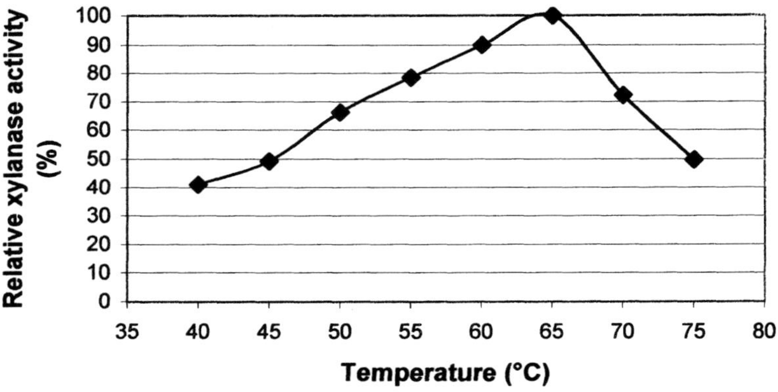

Figure 9. *Effect of temperature on hydrolysis of xylan by xylanase of Aspergillus oryzae NRRL 3485. Reaction was performed at pH 6.0 for 5 min.*

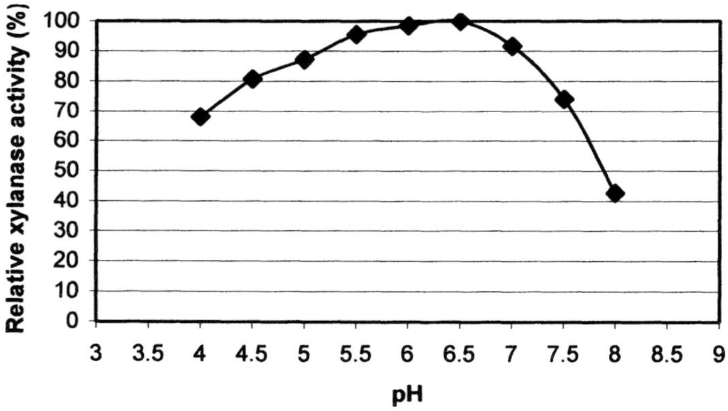

Figure 10. *Effect of pH on hydrolysis of xylan by xylanase of Aspergillus oryzae NRRL 3485. Reaction was performed at 50°C for 5 min.*

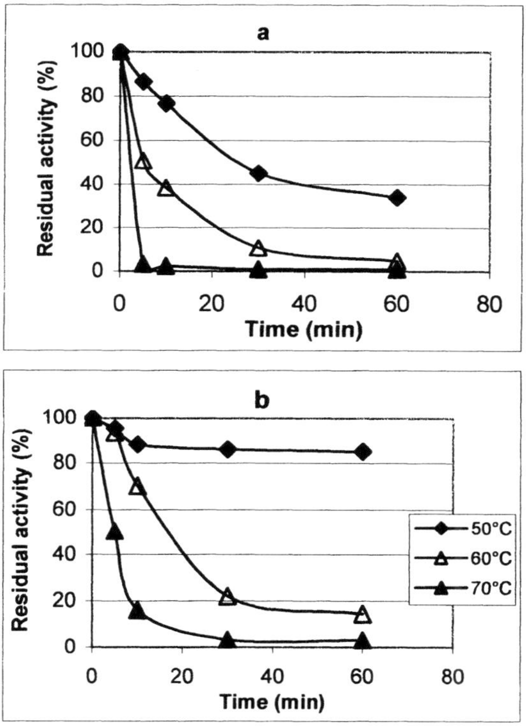

Figure 11. *Thermal stability of xylanase of Aspergillus oryzae NRRL 3485 without substrate (a) and in the presence of SSF solids (b) (pH 6.5; 0.05 M citrate-phosphate buffer).*

USD1/t pulp (Fig. 2a). The biobleaching effect could be translated into 28% savings of chlorine dioxide in comparison to the 24% reductions induced by the xylanase before media optimization (data not shown). Thus, at equal enzyme costs, the SSF xylanase produced under optimized conditions outperformed the commercial enzyme by 60%. At equal enzyme charges (5 IU/g pulp), the *A. oryzae* enzyme was 17% more efficient than the commercial product (Fig. 2b).

Conclusions

Solid state fermentation was used as a production method for xylanase by *Aspergillus oryzae* strains. Eucalyptus soda-aq pulp served as a fermentation substrate for the enzyme and thereafter as an enzyme carrier in pulp bleaching. This approach of enzyme production does not require a downstream processing of the enzyme prior to its utilization which has the advantage of lower enzyme production costs compared to the classical submerged fermentation of enzymes. A factorial experimental design, which was employed to optimize xylanase production in terms of nitrogen concentration and composition of the solid state fermentation medium, led to a 75% increase in the xylanase yields. The observed pH and temperature optima and thermostability exhibited by the xylanase of isolate NRRL 3485 is uncommon among xylanases of *A. oryzae*. The enzyme complex produced *in situ* on eucalyptus soda-aq pulp by solid state fermentation was able to bleach the same pulp effectively. When used at equal enzyme costs and charges, the *A. oryzae* enzyme outperformed the commercial xylanase control in its efficiency to improve pulp brightness. Because this particular isolate is a food-grade fungus that has been used for production of miso in Taiwan, the whole SSF material (*in situ* enzyme) may be used directly in pulp biobleaching without any health hazard.

Acknowledgements

This research was partly supported by Hungarian Science Foundation grant OTKA T 032690 and by the South Africa – Hungary Intergovernmental Science and Technology grants DAK-2/98 (1999-2001) and DAK-10/02 (2003-2005). The authors wish to thank the Sappi Management Services and the National Research Foundation (THRIP programme) for financial support and Sappi Management Services for granting permission to publish this work.

References

1. Tolan, J.S.; Olson, D; Dines, R.E. *Enzymes for Pulp and Paper Processing*; Jeffries, T.W.; Viikari, L., Eds.; ACS Symposium Series 655, American Chemical Society: Washington, DC, **1996**, pp. 25-35
2. Viikari, L.; Rauna, M.; Kantelinen, A.; Linko M.; Sunquist, J. *Proceedings of the 3rd International Conference on Biotechnology in the Pulp and Paper Industry,* **1986**, Stockholm, Sweden, pp. 67-69
3. Szendefy, J.; Szakacs, G.; Kemeny, S.; Christov, L. *Applications of Enzymes to Lignocellulosics;* Mansfield, S.D.; Saddler J.N., Eds.; ACS Symposium Series 855, American Chemical Society: Washington, DC, **2003**, pp. 255-284.
4. Pandey, A.; Selvakumar, P.; Soccol, C.R.; Nigam, P. *Curr.Sci.* **1999,** *77*, 149-162.
5. Tengerdy, R.P. *J. Sci. Ind. Res.* **1996**, *55,* 313-316
6. Tengerdy, R.P.; Szakacs, G.; *Biochem.Eng.J.* **2003**, *13,* 169-179.
7. Szakacs G.; Urbanszki, K.; Tengerdy, R.P. *Glycosyl Hydrolases for Biomass Conversion;* Himmel, M. E.; Baker, J. O.; Saddler, J. N., Eds., ACS Symposium Series 769, American Chemical Society: Washington, DC, **2001**, pp.190-203
8. Haltrich, D.; Nidetzky, B.; Kulbe, K.D.; Steiner, W.; Zupancic, S. *Biores. Technol.* **1996**, *58,* 137-161.
9. Bailey M. J.; Poutanen K. *Appl. Microbiol. Biotechnol,* **1989**, *1,* 5-10.
10. Bailey M. J.; Puls J.; Potanen K.; *Biotechnol. Appl. Biochem.* **1991**, *13,* 380-389.
11. Golubev, A.M.; Ibatullin, F.M.; Kilimnik A.Y.; Rodionova N.,A.; Neustroev K.N. *Biokhimiya* **1993,** *58,* 845-851.
12. Bailey M. J.; Viikari L. *World J. Microbiol. Biotechnol.* **1993,** *9,* 80-84.
13. Tenkanen M.; Buchert J.; Viikari L. *Enzyme Microb. Technol.* **1995,** *17,* 499-505.
14. Christov, L.P.; Szakacs, G.; Rele, M.V.; Balakrishnan, H. *Biotechnol.Techn.* **1999,** *13,* 313-316.
15. Yamane Y.; Fujita J.; Shimizu R.; Hiyoshi A.; Fukuda H.; Kizaki Y.; Wakbayashi S. *J. Biosci. Bioeng.* **2002**, *93*, 9-14.
16. Bailey, J.M.; Biely, P.; Poutanen, K. *J. Biotechnol.* **1992**, *23,* 257-270.
17. Miller, G.L. *Anal.Chem.* **1959**, *31,* 426-428.
18. Ghose, T.K. *Pure & Appl. Chem.* **1987**, *59,* 257-268.
19. Herr, D.; Baumer, F.; Dellweg, H. *European J. Appl. Microbiol. Biotechnol.* **1978**, *5*, 29-36.
20. Wolfenden, B S.; Wilson, R L. *J.Chem.Soc.Perkin Trans.* **1982**, *11,* 805-812.
21. Vries, R. P.; Kester, H.C.M.; Poulsen, C.H.; Bene, J.A.E.; Visser J. *Carbohyd. Res.* **2000,** *327,* 401-410.

22. Souza, D.F.; Souza, C.G.M.; Peralta R.M. *Process Biochem.* **2001**, *36*, 835-838.
23. Ghanem, N.B.; Yusef, H.H.; Mahrouse, H.K. *Biores. Technol.* **2000**, 73, 113-121.
24. Doelle, H.W.; Mitchell, D.A.; Rolz, C.E. *Solid Substrate Cultivation*; Elsevier Appl. Sci., London, **1992**.
25. Pandey, A.; Soccol, C.R.; Rodriguez-Leon, J.A.; Nigam, P. *Solid-State Fermentation in Biotechnology: Fundamentals and Applications;* Asiatech Publishers, Inc., New Delhi, **2001**.
26. Lu, W.; Li, D.; Wu, Y. *Enzyme Microb. Technol.* **2003**, *32*, 305-311.
27. Abdel-Sater, M. A.; El-Said, A.H.M. *Int. Biodeterior. & Biodegr.* **2001**, *47*, 15-21.
28. Taneja, K.; Gupta, S.; Kuhad, R. C. *Biores. Technol.* **2002**, *85*, 39-42.
29. Couri, S.; Costa Terzi, S.; Pinto, G.A. S.; Freitas S.P.; Costa A.C.A. *Process Biochem.* **2000**, *36*, 255-261.
30. Christov, L.P.; Szakacs, G.; Balakrishnan, H. *Process Biochem.* **1999**, *34*, 511-517.
31. Tenkanen, M. *Biotechnol. Appl. Biochem.* **1998**, *27*, 19-24.

Chapter 20

Laccase: A Harbinger to Kraft Pulping

T. J. Dyer and A. J. Ragauskas

Institute of Paper Science and Technology, Georgia Institute of Technology, 500 10th Street N.W., Atlanta GA 30332

Southern pine wood chips were subjected to a series of laccase-mediator (LMS) bio-pretreatments for kraft pulping employing 2,2'-azino-bis(3-ethylbenzthiazoline-6-sulfonate) (ABTS), 1-hydroxybenzotriazole (HBT), and violuric acid (VA) as mediators. These studies were the first of their kind in demonstrating the potential of LMS as a pretreatment to improve kraft pulping properties. The HBT bio-pretreatments had the greatest impact on kraft pulping, enhancing delignification while concomitantly increasing pulping yield. ^{13}C NMR analysis of black liquor lignins indicated increases in the ratio of substituted to unsubstituted aromatic carbons, suggesting that lignin from these bio-pretreatments may be more condensed than that obtained from a normal kraft pulp.

This paper is dedicated to
Loretta H. Dyer and Dr. Barry W. Crouse

Introduction

Kraft pulping is the cornerstone of modern chemical pulp production and a key operation in determining final pulp properties. Despite the predominance of this technology, the kraft pulping process will continue to adapt as challenges are directed toward the pulp and paper industry. Over the past several years, growing concerns over the use of chlorine and chlorine-containing chemicals have forced the pulp and paper industry to investigate not only alternative bleaching processes (*1-6*) but have also promoted renewed interest in enhancing delignification and increased digester productivity (*7-27*). The benefits of enhancing digester productivity or increasing pulp yields include reduced pulp manufacturing costs, improved wood utilization practices, reduced operating costs, and improved profits. In addition, the advantages of enhancing delignification include diminished environmental impacts due to decreased total reduced sulfur (TRS) emissions, reduced chlorinated organics in the bleach plant effluents, and reduced chemicals necessary for pulp bleaching.

One appealing approach for improving pulp yields consists of halting the kraft cook prior to reaching the terminal phase. In the terminal phase of kraft pulping, the selectivity between lignin and carbohydrates is significantly reduced, resulting in significant degradation of pulp carbohydrates. Halting the kraft cook prior to reaching the terminal phase reduces pulp carbohydrate losses but produces a pulp with a high lignin content (kappa number 40-50). The residual lignin in these high kappa kraft pulps must be removed through bleaching processes, typically an oxygen delignification stage. Several research groups have reported that pulp yields can be increased in the range of 2-6% by employing this approach (*4,8,10-12*).

Biotechnology provides an alternative approach for adapting to challenges faced by the pulp and paper industry (*28*). For example, lipase treatments have shown promise in degrading pitch (*29-34*) while concomitantly improving paper strength properties due to the formation of fatty acid groups (*31*). Xylanases have been used to enhance the bleachability of pulps by improving the diffusion of entrapped lignin from the cell wall (*35-52*). Meanwhile, pectinases are used for improved bark removal and to improve the cationic demand of pulps (*53*).

Several research groups have utilized fungal pretreatments in an attempt to improve delignification and digester productivity in kraft pulping (*54-66*). Wall et al. demonstrated that pretreating hardwood chips with a colorless strain of *Ophiostoma piliferum* improved kraft pulping efficiency by improving the penetration of pulping chemicals into the wood chips (*67*). They found a kappa number reduction of up to 29%, correlating to a 20% reduction in required active alkali. Furthermore, it was demonstrated that these fungal pretreatments provided an increased degree of polymerization and constant screened yield at a constant kappa number. Meanwhile, Wolfaardt et al. pretreated *Pinus patula* wood chips with a strain of *Stereum hirsutum* (*62*). The results of these studies suggested that this fungal strain can provide a substantial reduction in pulp lignin content or decrease the pulping time. However, the reduction in lignin content came at the expense of reduced pulp yield and increased alkali consumption. Hatakka et al. screened almost 300 wood rotting fungi and applied the most promising fungi as pretreatments to spruce wood chips (*68*). They showed that fungi having both good growth characteristics and selectivity for lignin are difficult to find. They also observed decreased lignin contents and pulp yields due to fungal pretreatment. As a result of these promising laboratory investigations, industrial studies are currently underway.

Jacobs et al. attempted to mimic a selective fungal pretreatment system by applying an enzyme mixture of hemicellulases, pectinases, and cellulases to sycamore wood chips (*69*). It was demonstrated that enzyme pretreatment of wood chips prior to kraft pulping enhanced delignification while maintaining similar viscosities and pulp yields. In addition, the pretreatments resulted in pulps with higher brightness and enhanced bleachability compared to the control kraft pulp. Jacobs et al. subsequently studied the impact of enzyme pretreatments on southern pine wood chips (*70*). They demonstrated that southern pine wood chips were delignified more efficiently after acetone extraction, producing pulps with lower kappa numbers and rejects while having higher pulp viscosities and yields.

Recent studies have shown the efficiency of employing lignin degrading enzymes, more specifically, laccase-mediator systems (LMS), to delignify high kraft pulps (*1,71-74*). Although laccase alone is ineffective in delignifying pulp fibers, it has been shown to effectively oxidize phenolic compounds and phenolic lignin model compounds (*75-79*). In the laccase-mediator concept, diffusion limitations due to the limited pore size of pulp fibers, as well as by the molecular size of extractable molecules, are circumvented by adding a low molecular weight compound known as a mediator (*80*). Bourbonnais et al. first demonstrated the laccase-mediator concept when they reported that laccase in the presence of 2,2'-azino-bis(3-ethylbenzthiazoline-6-sulfonate) (ABTS) could effectively delignify kraft pulps (*81*). Under the conditions employed for the LMS, laccase oxidizes a mediator in the presence of oxygen. The enzyme oxidized mediator then acts directly on lignin and results in efficient delignification. The reduced ABTS mediator is subsequently reoxidized by laccase, resulting in a reduction-oxidation cycle. A host of alternative mediators have been discovered, but the most effective mediators in delignification usually contain N-OH functional groups, such as 1-hydroxybenzotriazole (HBT), violuric acid (VA), and *N*-acetyl-*N*-phenylhydroxylamine (NHA) (*82,83*). Despite their efficacy in delignification, these alternative mediators are not as catalytic in nature as the ABTS mediator during the laccase-mediator reaction. The redox potential of mediators has a great influence on its efficiency and reactivity. Xu et al. examined the redox potential of N-hydroxy based mediators and found that for a given pH, the redox potential was greatest in HBT, followed by violuric acid and NHA, respectively (*84*). Earlier redox potential studies by Bourbonnais et al. indicated that the redox potential of ABTS was lower than violuric acid but greater than NHA (*85*). Although all the redox potential measurements fall within a very narrow range, the redox potential difference between the mediator and laccase is an important factor in determining the overall system reactivity and efficiency. Laccase-mediator systems have typically been utilized for investigations on lignin model compounds, low-lignin kraft, and sulfite pulps (*86-98*).

Chakar et al. have recently demonstrated that an LMS treatment, using HBT, VA, or NHA as mediators, can remove significant amounts of lignin from high-kappa kraft pulps (*1,71-74*). NMR analysis of the high kappa kraft residual lignin in the pulp after an LMS treatment indicated that the biotreatment extensively oxidizes C-5 noncondensed phenolic lignin structures, whereas C-5 condensed phenolic lignin structures were resistant to oxidation. The primary product detected from these oxidative treatments has been the formation of carboxylic acid groups. The efficiency of laccase mediator systems on high kappa kraft pulps has been well established (*99*).

It was envisaged that by applying laccase mediator systems to wood chips, we could enhance delignification during kraft pulping. The purpose of this study was to examine the feasibility of employing LMS as a bio-pretreatment for kraft pulping. In addition, this study investigates the impact of a LMS treatment on kraft pulp delignification, yield, and chromophore content as well as on the structural changes in LMS effluent samples and black liquor lignin.

Experimental

Chemicals

All chemicals were purchased from Aldrich Co., Milwaukee, WI, and used as received, except for *p*-dioxane and laccase. *p*-Dioxane was freshly distilled over $NaBH_4$ prior to using it for lignin isolation experiments. The laccase was NS51002 isolated from *Trametes villosa* and was donated by Novozymes Biotech, Franklinton, NC.

Wood

The softwood chips employed in this study originated from two 25-year-old *Pinus taeda* trees that were donated by Bowater Incorporated, Greenville, SC. The wood was debarked, split, and chipped at the Institute of Paper Science and Technology. After chipping, the chips were screened by thickness with a Rader screen. Accept wood chips were those that ranged from 2-8 mm in the thickness direction. The accept wood chips were subsequently sorted by size with a chip class screen according to SCAN-CM 40:94. The accept wood chips were those that passed through an oversize screen comprised of 45 mm round holes and an overthick screen having 8 mm slots but remained on the accepts screen which is composed of 7 mm holes.

Laccase assay

Laccase activity was measured by monitoring the rate of oxidation of syringaldazine. Laccase activity was measured according to Sealey (*87,88*). Briefly, one unit of activity (U) was defined as the change in absorbance at 530 nm of 0.001 per minute per milliliter of enzyme solution, in a 100 mM potassium phosphate buffer (2.2 ml) and 0.216 mM syringaldazine in methanol (0.3 ml, pH 6.7). The procedure was carried out at 23°C. The activity of the laccase was 4.27E + 07 U/ml of enzyme solution.

Laccase-mediator pretreatment procedure

A 1000-ml capacity stainless steel autoclave equipped with a pressure gauge and thermocouple was charged with 100 g of o.d. wood chips. The consistency was adjusted to 12% with distilled water. ABTS, HBT, or VA was charged in varying amounts to the vessel with subsequent mixing. The pH of the solution was adjusted to 4.5 with glacial acetic acid. Laccase (0.2 mL of enzyme solution/ g of o.d. wood chips) was added and the reactor was sealed and pressurized with oxygen to 150 psig. The autoclave was placed into an electrically-heated, rotating, multi-unit digester where the slurry was brought to a temperature of 45°C. The temperature was maintained throughout the incubation period. After mixing for 2 hours, the wood chips were removed and washed thoroughly with distilled water (5 L per 10 g of o.d wood chips) until the filtrate was pH neutral and colorless. The pretreated wood chips were then subjected to kraft pulping under varying conditions of alkali and sulfidity.

Kraft pulping

Conventional kraft pulping conditions were simulated in an electrically-heated, rotating, multi-unit digester. An exact amount of wood (100 o.d. grams) was charged to six individual 1 L stainless steel autoclaves. A mixture of sodium hydroxide and sodium sulfide (white liquor) was also charged to the autoclaves with additional make-up water to reach a constant liquor to wood ratio of 4:1. The exact amount of sodium hydroxide and sodium sulfide was adjusted to meet specific pulping conditions; more specifically, one pulping condition had a high effective alkali (EA) and low sulfidity (Condition A) and another had a low effective alkali and high sulfidity (Condition B). The individual vessels were then placed in a rotating, multi-unit digester. The temperature in each autoclave was increased as a ramp function from 23°C to a maximum cooking temperature of 170°C over 90 minutes. The kraft cook was interrupted at the appropriate H-factor and each vessel was cooled immediately in a cold water bath. The cooked wood chips were disintegrated in an industrial blender and screened and washed in a Valley screen. The pulping conditions employed to manufacture the pulps in this study are summarized in Table I.

Table I . Summary of Kraft Pulping Conditions

Condition	*% EA*	*% Sulfidity*	*H-factor*
A	21.4	23.2	863
B	14.6	56.8	1151

UV/vis Diffuse Reflectance Spectroscopy

The UV/vis spectra were recorded on a Perkin-Elmer Lambda 900 UV/vis spectrometer equipped with a diffuse reflectance and transmittance accessory (PELA-1000). The accessory is essentially an optical bench that includes double-beam transfer optics and a six-inch integrating sphere. Background corrections were recorded using a Labsphere SRS-99-020 standard. The reflectance data from were converted to *k/s* values by using the Kubelka-Munk theory (1931). The Kubelka-Munk equation describes the infinite reflectance as a function of absorption and scattering:

$$R_{\infty} = 1 + \frac{k}{s} - \sqrt{2\left(\frac{k}{s}\right) + \left(\frac{k}{s}\right)^2} \quad (1)$$

$$\frac{k}{s} = \left(\frac{q}{s}\right)c \qquad (2)$$

where R_∞ is the infinite reflectance, k is the absorption coefficient, s is the scattering coefficient, q is a proportionality constant and c is the chromophore concentration. The total visible reflectance was taken as the integral or area under the k/s vs. wavelength curve from 400 nm to 700 nm. Based on the Kubelka-Munk theory, we can assume that the total reflectance value is directly proportional to the total chromophore content. In addition, we can assume that changes in k/s ($\Delta k/s$) are directly proportional to changes in the chromophore content.

Isolation of black liquor lignins

The isolation of black liquor or kraft lignins was carried out following standard literature methods (*100,101*). In summary, the black liquors extracted at the end of each cook were isolated by precipitating the lignin from solution by acidification. The black liquors were first filtered through filter paper. Approximately 0.5 grams EDTA-2Na$^+$ was then added for every 100 mL black liquor. The pH of the liquors was subsequently adjusted to ~6.0 with 2 M H_2SO_4. The solutions were stirred vigorously for 1 hour, further acidified to a pH of 2.3, and frozen at -20°C. After thawing the solutions, the precipitates were washed thoroughly with cold water, collected, and air-dried. The kraft lignin precipitates were further purified by dissolving them in a 9:1 dioxane:water solution. The aqueous solutions were filtered with celite and the *p*-dioxane was removed under reduced pressure using a rotary evaporator. After acidification to pH 2.3, the solutions were washed thoroughly with distilled water and then freeze-dried. Finally, the kraft lignin samples were collected and Soxhlet extracted with pentane for 8 hours to remove sulfur and other impurities. The purified kraft lignins were dried under vacuum and used for subsequent NMR analysis.

^{13}C NMR Spectroscopy

Nuclear magnetic resonance (NMR) spectra were recorded on a Bruker DMX 400 MHz spectrometer. ^{13}C NMR spectra were recorded under quantitative conditions using an inverse gated pulse sequence with a 90° pulse, 14-s delay, and a sweep width of 222 ppm (*102*). All experiments employed 300-400 mg of lignin/mL of (dimethyl sulfoxide)-d_6 and were recorded at 50°C.

The free induction decays were Fourier transform processed with 10 Hz line broadening and analyzed according to reported chemical shifts for lignin functional groups (*103,104*). Lignin spectra were integrated and the area for the aromatic signals was assigned a value of 6 (*102*) for the six carbons in an aromatic group; other integration areas were integrated relative to this value. Hence, the integration regions were quantified as the number of carbon atoms per aromatic unit. This method may overestimate the aromatic contribution of lignin since some stilbenes and enol ethers are present in kraft and residual lignins. Studies by Gellerstedt and Lindfors (*105*) have detected the formation of enol ethers during kraft pulping, and these types of structures could potentially contribute to the apparent signal attributed to aromatic carbons, although their contribution would be less than 10%.

Typical accuracy values reported in the literature for integrating the various functional groups by ^{13}C NMR range from ±3-5% (*102*). In these experiments, the error analysis was based on reproducing the LMS biopretreatment utilizing VA as the mediator three separate times. The black liquor lignin from each was isolated and analyzed. The standard error calculated from these three treatments for the ^{13}C NMR experiments was ±3%.

Results and Discussion

To date, research efforts into biological wood chip pretreatments have focused on fungi or mixtures of enzymes to improve the kraft pulping process (*54-70,106-108*). Our research interests lie in applying laccase mediator systems as bio-pretreatments to kraft pulping. Chakar et al. have already shown that laccase mediator systems effectively delignify high-lignin kraft pulps (*1,71-74*). Utilizing such a technology may have positive impacts on wood utilization practices, since LMS have exhibited a high selectivity toward lignin and not toward carbohydrates. The purpose of this study was to examine the feasibility of employing LMS as a pretreatment to kraft pulping. This was accomplished by applying three different mediators, ABTS, HBT, and VA, during separate pretreatments. Since the conditions of each pretreatment were the same, we could relate changes in pulping responses to different mediators. These changes in kraft pulping were assessed by determining the structural differences in black liquor lignins as well as by kappa, yield, and total reflectance of the pulps.

Preliminary Studies

A series of preliminary experiments were employed to find the feasibility of using laccase as a precursor to kraft pulping. The resulting kraft pulps were analyzed in terms of the screened pulp yield and pulp kappa number.

Impact on Kappa Number

The kappa number responses of the control and LMS bio-pretreatments are found in Table II. The pretreated control samples labeled "control" employed conditions similar to the LMS with the exception that laccase and a mediator were not present. The pretreated mediator control experiments of ABTS, HBT, and VA were performed in the absence of laccase. In order to compare the delignification responses, the mediators were charged at the same molar equivalence in each experiment (i.e., all other experimental conditions such as enzyme dosage, temperature, time, O_2 pressure and pH were held constant). For these experiments, the molar dosage of mediator was 44 mmol/100 g o.d. wood chips. The pretreated laccase control experiments were performed in the absence of a mediator and at a dosage of 0.2 mL of enzyme solution/g of o.d. wood chips.

Table II. Delignification Responses to Bio-Pretreatment Conditions for Preliminary Studies[a]

Treatment	*Pulping Condition A*	*Pulping Condition B*
Normal kraft	29.8	30.9
Control	29.7	30.8
ABTS	29.7	30.7
HBT	31.3	31.4
VA	31.8	31.0
Laccase	29.9	31.2
Laccase + ABTS	30.8	32.4
Laccase + HBT	27.2	27.8
Laccase + VA	35.5	32.6

[a]Pulping Condition A had a confidence interval of ±1.5 kappa number units while Pulping Condition B had a confidence interval of ±1.4 kappa number units.

Each treatment condition consisted of two pulping conditions; more specifically, one pulping condition had a high EA (21.4%) and low sulfidity (23.2%) and another had a low EA (14.6%) and high sulfidity (56.8%). The high EA, low sulfidity condition, or Condition A, had an H-factor of 863 and a confidence interval of ±1.5 kappa number units while the low EA, high sulfidity condition (Condition B) had an H-factor of 1151 and a confidence level of ±1.4 kappa number units. The kappa number data in Table II suggests there is no significant difference among the delignification responses in the control experiments. Early investigations on LMS have demonstrated that the delignification response of a laccase treatment in the absence of a mediator is insignificant. (*87,109*) In addition, both the laccase and mediator must be present in order to achieve the greatest extent of delignification. The kappa number results from the LMS bio-pretreatments clearly suggest that the treatment having the most promising results employed HBT as the mediator. This pretreatment lead to nearly an additional 10% delignification compared to the normal kraft pulp and was independent of the pulping conditions employed. However, using ABTS and VA in the presence of laccase actually increased the kappa number. This observation indicates possible grafting of the mediator onto the pulp fiber (*86*). This was also observed by Lund and Felby when treating high-lignin content kraft pulp fibers with laccase and a mediator. They proposed that the wet tensile strength improved, likely as a result of grafting the mediator, thereby creating a cross-linked matrix around the fiber (*110*).

Impact on Pulp Yield

The screened pulp yield was also measured to further characterize the impact of these LMS bio-pretreatments on kraft pulping. Again, all treatment conditions involved two pulping conditions, Condition A and Condition B. A confidence interval of ±1.1% and ±0.9% was obtained for Condition A and Condition B, respectively. One noticeable difference between the Condition A and Condition B is obtained from the yield data in Table III. It was expected that the lower effective alkali and higher sulfidity conditions employed in Condition B would result in higher screened yields than Condition A. These milder conditions are more likely to lead to greater selectivity for lignin and higher retention of hemicelluloses (*111*). The pulp yield data in Table III are consistent with the kappa number data in Table II; that is, there is no significant difference among the yield responses for the control experiments. However, the data from the LMS bio-pretreatments clearly indicate an increase in pulp yield. The LMS treatments employing ABTS and VA had significant increases up to 3.6% for Condition A and 2.5% for Condition B. Based on the kappa number responses of these treatments, these increases further validate the conclusion that the mediator is likely grafting onto the pulp fiber. The LMS bio-pretreatments employing HBT, however, also experience significant increases in pulp yield. The yield increases are 1.6% and 2.5% for Condition A and B, respectively. Therefore, this pretreatment was able to achieve both enhanced delignification and increased yield under the conditions employed in this study.

Table III. Pulp Yield Responses (%) to Bio-Pretreatment Conditions for Preliminary Studies[a]

Treatment	*Pulping Condition A*	*Pulping Condition B*
Normal kraft	41.8	44.1
Control	42.7	43.4
ABTS	42.4	44.1
HBT	42.7	44.5
VA	42.7	44.7
Laccase	42.3	42.9
Laccase + ABTS	43.4	44.6
Laccase + HBT	43.4	46.7
Laccase + VA	45.4	46.7

[a]Pulping Condition A had a confidence interval of ±1.5 % yield while Pulping Condition B had a confidence interval of ±1.4 % yield.

Mediator Dosage Studies

The LMS bio-pretreatments employed in the feasibility portion of this study were effective in enhancing delignification with a concomitant increase in pulp yield when using HBT as the mediator. The other mediators, ABTS and VA, were not efficient in terms of enhancing delignification. Despite these results, we wanted to further investigate the impact of mediator dosage on the performance of these bio-pretreatments.

Impact on Lignin Content and Yield

The LMS bio-pretreatments for this portion of the study were performed in a similar manner to those previously described in the experimental section. The exception is that the mediator concentration was varied by employing dosages of 11, 22, and 44 mmol/100 g o.d. wood chips. In addition, this portion of the study employed only one pulping condition since we were interested in the efficacy of the mediator and not the impact of specific pulping conditions. The

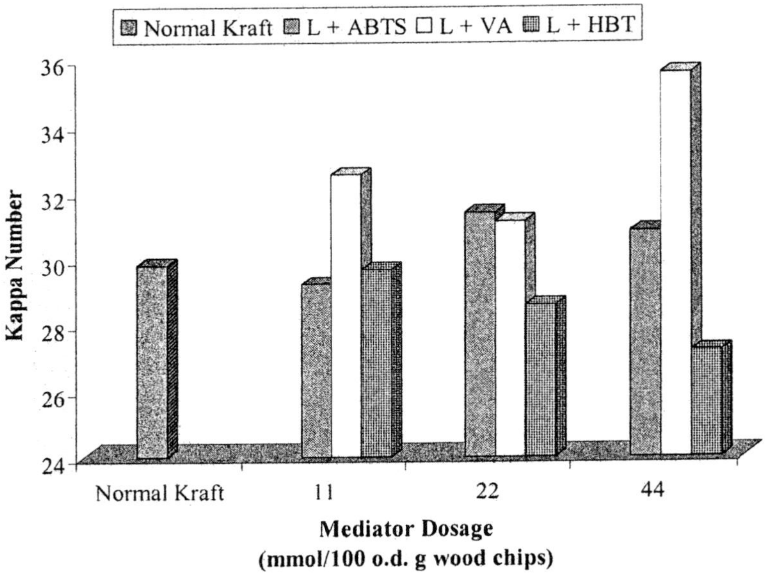

Figure 1. Mediator dosage study results of kappa number measurements of LMS bio-pretreatments using ABTS, VA, and HBT as mediators. The confidence level associated with this pulping condition is ±1.5 kappa number units.

pulping condition employed for these experiments involved 21.4% effective alkali and 23.2% sulfidity in the pulping liquor. The confidence interval associated with this pulping condition is ±1.5 kappa number units. Each treatment condition was kraft pulped to an equivalent H-factor of 863. Delignification responses of the LMS bio-pretreatments are shown in Figure 1.

The kappa number of the LMS bio-pretreatments employing ABTS as the mediator remained equal to or greater than the normal kraft pulp. However, the kappa number did decrease with decreasing dosage of ABTS. The treatments employing VA as the mediator responded differently. The minimum kappa number for these treatments occurred at the 22 mmol/100 o.d. g wood chips concentration. Meanwhile, the kappa number of treatments performed with HBT increased with decreasing mediator dosage. Further experimentation may give rise to the optimal mediator dosage in each case. However, the enhanced delignification obtained with HBT compared to ABTS and VA suggests that using HBT as the mediator is the most promising mediator technology of those employed under the conditions of these experiments. Chakar et al. demonstrated that the delignification performance of an LMS employing VA as the mediator was superior to that of HBT on a high-kappa kraft pulp (*72*). The results achieved in LMS bio-pretreatments are likely different since we are employing this technology on wood chips. There are several potential differences in applying LMS to wood chips compared to pulp. For example, wood chips contain a higher proportion of extractives, hemicellulose, and lignin compared to pulp. Therefore, the changes in kappa number could be a result of oxidative chemistry with extractives. Wood chips also have less exposed surface area than pulp and are even more diffusion-limited in terms of the reactions that can take place than pulp. The smaller exposed surface area and diffusion limitations in wood chips could potentially lead to decreased delignification responses and different mediator efficiencies compared to those demonstrated for high- and low-kappa kraft pulp.

To further characterize the impact of varying mediator dosages for these LMS bio-pretreatments, the pulp yield was also measured. Figure 2 summarizes the results of these measurements. The confidence interval associated with the pulp yield is ±1.1% for the high effective alkali (21.4%) and low sulfidity (23.2%) pulping condition. The pulp yield for treatments employing ABTS followed that of the kappa number; that is, as the kappa number decreased, the pulp yield also increased. The treatments involving VA responded with decreasing pulp yields as the mediator dosage decreased. Pulp yields for treatments involving HBT also increased with increasing mediator concentration with the greatest change occurring between 22 and 44 mmol/100 o.d. g wood chips.

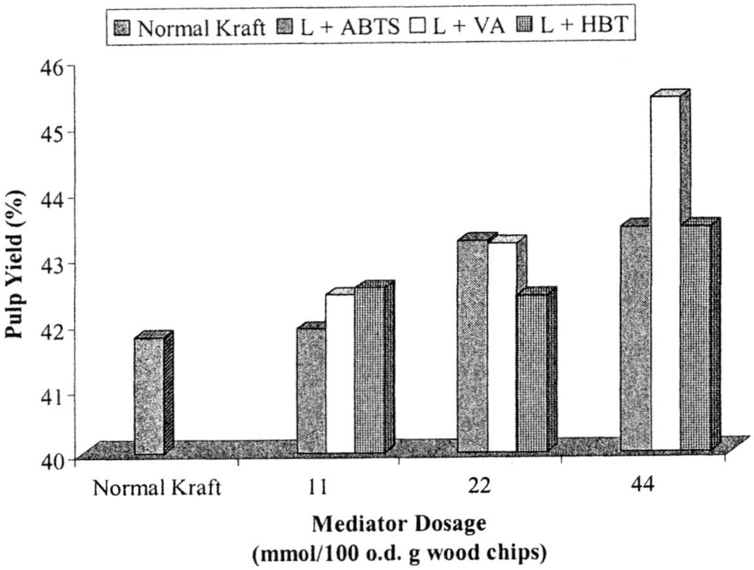

Figure 2. Mediator Dosage study results of pulp yield measurements on LMS bio-pretreatments employing ABTS, VA, and HBT as mediators. The confidence interval associated with the pulp yield is ±1.1%.

These yield results, together with kappa number results, reinforce that employing LMS bio-pretreatments are feasible for kraft pulping. Further experimentation may be required to find the most suitable mediator, mediator dosage, etc., to further enhance the delignification and yield properties of these pulps.

Impact on Diffuse Reflectance

Having characterized the bio-pretreated pulps for lignin content and pulp yield, we proceeded further with our study by examining the total visible reflectance of the pulp samples. Solid-state UV/vis diffuse reflectance spectroscopy is a powerful technique for studying changes in chromophores and in chromophore content of lignin-containing materials. By transforming reflectance data to *k/s* data via the Kubelka-Munk theory (Equations 1 and 2), we generated *k/s* vs. wavelength spectra. Subsequently, the total visible reflectance was taken as the area under the *k/s* vs. wavelength curve from 400 nm to 700 nm.

The resulting total reflectance values are directly proportional to the chromophore content of each pulp. Each total reflectance value is the average of five measurements. Figure 3 summarizes the total visible reflectance results from the mediator dosage study. The confidence level associated with the total visible reflectance is ±2.3 units for the pulping condition utilizing a high effective alkali (21.4%), low sulfidity (23.2%) and H-factor of 863.

Bio-pretreatments employing ABTS as the mediator had similar total reflectance to an untreated, normal kraft pulp. These results appear to be a direct reflection of the kappa number, which decreased with decreasing mediator concentration. The treatments involving VA as the mediator had consistently higher total reflectance than the normal kraft pulp. In addition, the mediator dosage of these bio-pretreatments had little impact on the total reflectance. Such is also the case with the LMS bio-pretreatments employing HBT as the mediator. These treatments had approximately the same total reflectance despite significant decreases in the mediator dosage. Also, the reflectance values of pretreatments employing HBT were much smaller than normal kraft pulp. These results suggest that the LMS bio-pretreatments removed leucochromophores, or potential chromophores, prior to kraft pulping.

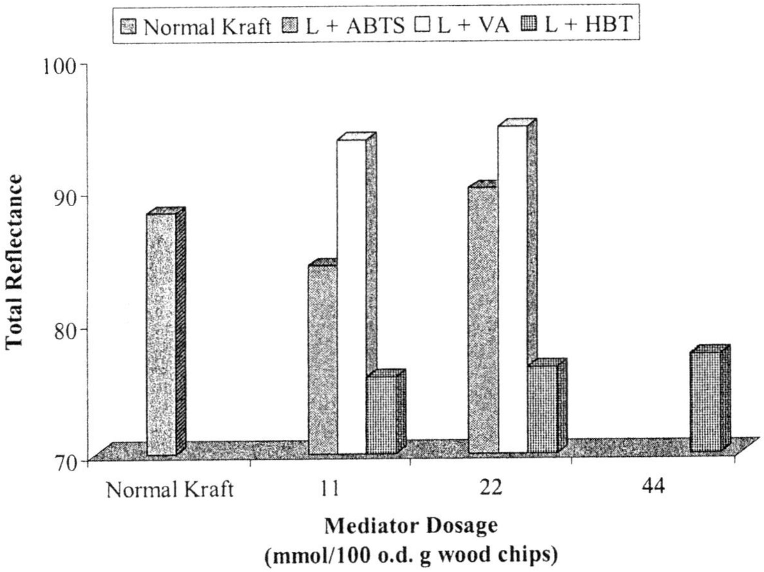

Figure 3. Results of the total reflectance measurements on LMS bio-pretreatments employing ABTS, VA, and HBT as mediators. The confidence interval associated with the total reflectance is ±2.3 units.

Interestingly, the laccase-mediator pretreated wood chips exhibited similar color characteristics to that of the respective mediators while those wood chips pretreated with only the mediator did not exhibit these characteristics. However, after kraft pulping, the observed color differences in the laccase-mediator pretreated wood chips were no longer apparent in the resulting pulps. Instead, the pulps produced from laccase + HBT pretreated wood chips had lower absorbances for all visible wavelength ranges while laccase + ABTS pretreated wood chips had lower absorbances at all visible wavelength ranges than the laccase + VA pretreated wood chips.

Analysis of Black Liquor Lignins

The impacts of the LMS bio-pretreatments in this study were further analyzed by examining structural changes in the black liquor lignin. The black liquor lignins from the control and bio-pretreatments were isolated and characterized via quantitative ^{13}C NMR. This effective technique facilitated the characterization of several important lignin functional groups including carboxylic acids, β-O-4 aryl ether structures, aromatics, and methoxyl groups. Changes in the relative proportions of these functional groups provide critical information on the nature of reactions involved in modern pulping reactions as well as these LMS bio-pretreatments (*112*). Tables IV and V illustrate the results of these investigations. The pulping conditions utilized for the isolated black liquor lignins were 21.4% EA and 23.2% sulfidity in the white liquor and an H-factor of 863. Meanwhile, the mediator concentration of the LMS bio-pretreatments was 22 mmol/100 o.d. g wood chips.

The results shown in Table IV indicate that relative to normal kraft pulp, the methoxyl content did not significantly change in any of the LMS bio-pretreatments. The confidence level associated with the methoxyl content was ±0.03 methoxyl units/aromatic ring. There have been several studies on both low- and high-lignin content kraft pulps which indicated a general decrease in methoxyl content with an LMS treatment (*81,88,99,109*). However, the amount of lignin removed during kraft pulping is significantly greater than that removed during previous LMS bio-bleaching investigations. Therefore, the change in methoxyl content in these black liquor lignins may not be apparent simply due to the abundance of lignin removed during the kraft pulping. In addition, these results may indicate that methoxyl structures in wood chips are resistant to the LMS bio-pretreatments whereas they are not as resistant to LMS bio-bleaching treatments. The confidence level for the β-O-4 aryl ether content was ±0.01 β-

O-4 aryl ether units/aromatic ring. The β-O-4 aryl ether structures also did not significantly change in the LMS bio-pretreatments relative to the normal kraft pulp. As shown in previous studies, the β-O4 aryl ethers also tend to be resistant toward LMS treatments on low- and high-lignin content pulps (*87,99*). Although LMS treatments on lignin model compounds have suggested that such structures are reactive toward LMS (*79,81*), the lignin macromolecule in wood and pulp is much more complex than a dimeric model compound and a different selectivity may be occurring in pulp and wood. This phenomenon may be due to the LMS having a higher affinity toward lignin functional groups other than β-O-4 aryl ether structures (*99*).

Table IV. Black Liquor Lignin Analysis of Bio-Pretreatment Conditions

Treatment	*Methoxyl Content (per aromatic ring)*[a]	*β-O-4 Aryl Ether Content (per aromatic ring)*[b]
Normal kraft	1.1	0.33
Laccase	1.1	0.34
Laccase + ABTS	1.2	0.33
Laccase + VA	1.0	0.29
Laccase + HBT	1.1	0.29

[a]The confidence level is ±0.03 methoxyl units/aromatic ring.
[b]The confidence level is ±0.03 β-O-4 aryl ether units/aromatic ring.

The trend in carboxylic acid functional groups in Table V indicates no relative change in this functional group relative to the normal kraft pulp. The exception here is the LMS bio-pretreatment employing VA as the mediator. Previous studies on high- and low-lignin content kraft pulps suggested a general increase in such moieties (*88,99*). The carboxylic acid content had a confidence interval of ±0.01 carboxylic acid units per aromatic ring. The results in Table V also demonstrate that relative to the normal kraft pulp, the ratio of the substituted to unsubstituted aromatics is enriched with the LMS bio-pretreatments. This suggests that the lignin from these treatments may be more condensed than the normal kraft pulp. These results reinforce the results of previous studies that indicated LMS treatments are more selective toward noncondensed phenolic structures (*88,89,99*). Overall, it may be important to postulate that the eclectic behavior of LMS bio-treatments toward lignin functional groups is influenced by the content and type of lignin present in the wood or pulp.

Table V. ^{13}C Black Liquor Lignin Analysis of Bio-Pretreated Pulps

Treatment	*Carboxylic Acid Content (per aromatic ring)[a]*	*Ratio of substituted to unsubstituted aromatic C*
Normal kraft	0.27	1.2
Laccase	0.26	1.3
Laccase + ABTS	0.33	1.1
Laccase + VA	0.40	1.8
Laccase + HBT	0.23	1.9

[a]The confidence interval is ±0.01 carboxylic acid units/aromatic ring.

Conclusions

In summary, the enhanced delignification response clearly indicates that employing laccase-mediator systems as a bio-pretreatment for kraft pulping is feasible on wood chips. In addition, of the mediators employed in these studies, HBT was the most effective in enhancing delignification while concomitantly increasing pulp yield. The efficacy of the mediators could be enhanced by adjusting parameters such as mediator dosage, treatment consistency and the like. In addition to improving delignification performance and increasing pulp yield, the total reflectance of the LMS bio-pretreatments was decreased. Spectral analysis of black liquor lignins isolated after kraft pulping revealed an enrichment of condensed structures which suggests that the mediators preferentially react with phenolic noncondensed structures while the phenolic condensed structures remain intact. Nonetheless, our investigations into LMS bio-pretreatments for kraft pulping suggest that differences exist in the delignification chemistry and selectivity of HBT, VA, and ABTS. This is despite the fact that VA and HBT operate by way of nitroxyl radicals (*92*). As in the case with laccase mediator biobleaching applications, the practical viability of this novel discovery will require additional breakthroughs in the development of more effective laccase mediator systems.

Acknowledgments

The authors would like to thank Drs. Elder, Lucia, McDonough, and Runge for their guidance as well as the Institute of Paper Science and Technology and its member companies for their support of these ongoing studies. The authors would also like to express their gratitude to T. Kolterman and M. Dye at Bowater Incorporated for providing the investigators with wood supplies and Novozymes Biotech, Franklinton, NC, for furnishing the laccase needed for these studies. Portions of this work were used by T.J.D. as partial fulfillment of the requirements for the Ph.D. degree at the Institute of Paper Science and Technology.

References

1. Chakar, F.S., Ragauskas, A.J., and Nilvebrant, N.-O.; *TAPPI Pulping Conference Proceedings*. Montreal. 1998, 109-118.
2. Lucia, L.A., Ragauskas, A.J., and Chakar, F.S.; *Industrial & Engineering Chemistry Research*. **2002,** *41,* 5171-5180.
3. Lucia, L.A., Goodell, M.M., Chakar, F.S., and Ragauskas, A.J.; In *Oxidative Delignification Chemistry,* D.S. Argyropoulos, Editor ACS Symposium Series. p. 92-107 (2001).
4. Chakar, F.S., Lucia, L.A., and Ragauskas, A.J.; *International Pulp Bleaching Conference Proceedings*. Halifax, NS. 2000, 123-130.
5. Dyer, T.J., Lucia, L.A., and Ragauskas, A.J.; *TAPPI Pulping Conference Proceedings*. Boston, MA. 2000, 9 pp.
6. Dyer, T.J. and Ragauskas, A.J.; *TAPPI Pulping Conference Proceedings*. Seattle, WA. 2001, 17 pp.
7. McDonough, T.J., Burton, J.T., and Yethon, A.E.; *TAPPI Pulping Conference Proceedings*. Orlando, FL. 1999, 171-182.
8. Steffes, F., Bokstrom, M., and Norden, S.; *TAPPI Breaking the Pulp Yield Barrier Symposium Proceedings*. Atlanta, GA. 1998, 183-195.
9. Bokstrom, M. and Norden, S.; *Appita Annual General Conference Proceedings*. 1998, 327-334.
10. Magnotta, V., Kirkman, A., Jameel, H., and Gratzl, J.; *TAPPI Breaking the Pulp Yield Barrier Proceedings*. Atlanta, GA. 1998, 165-182.
11. Parthasarathy, V.; *TAPPI Pulping Conference Proceedings*. San Francisco. 1997, 739-757.
12. Parsad, B., Gratzl, J., Kirkman, A., Jameel, H., Rost, T., and Magnotta, V.; *Tappi Journal*. **1994,** *77,* 135-147.
13. Rutkowski, J.; *Przeglad Papierniczy*. **2001,** *57,* 85-90.
14. Rutkowski, J. and Perlinska-Sipa, K.; *Przeglad Papierniczy*. **2000,** *56,* 644-647.
15. George, J., Lachenal, D., and Robert, D.; *Journal of Pulp and Paper Science*. **2000,** *26,* 271-274.
16. Pu, Y., Zhan, H., Hue, B., and He, W.; *Huanan Ligong Daxue Xuebao, Ziran Kexueban*. **2000,** *28,* 30-35.
17. Huang, G., Zhang, Z., Han, K., Wu, S., and Zheng, Z.; *Huanan Ligong Daxue Xuebao, Ziran Kexueban*. **2000,** *28,* 66-73.
18. Liu, J., Huang, G., and Zhang, Z.; *Zhongguo Zaozhi*. **2000,** *19,* 1-6.

19. Huang, G., Zhang, Z., Han, K., Zhan, K., and Zheng, Z.; *Zhongguo Zaozhi.* **1999,** *18,* 9-15.
20. Li, J., MacLeod, J.M., Berry, R.M., and McPhee, F.J.; *Pulp & Paper Canada.* **1999,** *100,* 48-52.
21. Chivukula, S., Chang, H., Jameel, H., and Wizani, W.; *TAPPI Pulping Conference Proceedings.* Montreal. 1998, 419-429.
22. Li, J., MacLeod, M., and Berry, R.; *TAPPI Pulping Conference Proceedings.* Montreal. 1998, 407-418.
23. Li, J., Macleod, J.M., McPhee, F.J., Berry, R.M., O'Hagan, T.J., Thomas, B.C., Tristram, R.W., and Kowalska, R.E.; *TAPPI Pulping Conference Proceedings.* San Francisco. 1997, 577-586.
24. George, J., Lachenal, D., and Perez, C.; *TAPPI Pulping Conference Proceedings.* San Francisco. 1997, 889-895.
25. Parthasarathy, V.R.; *TAPPI Pulping Conference Proceedings.* San Francisco. 1997, 739-757.
26. Courchene, C.E.; *Breaking the Pulp Yield Barrier Symposium Proceedings.* Atlanta, GA. 1998, 11-20.
27. Minja, R.J.A., Christensen, P.K., and Moe, S.T.; *TAPPI Pulping Conference Proceedings.* Nashville, TN. 1996, 319-324.
28. Ragauskas, A.J.; In *Progress in Biotechnology,* L. Viikari and R. Lantto, Editors. p. 7-12 (2002).
29. Rundloef, M., Eriksson, M., Stroem, H., and Wagberg, L.; *Cellulose.* **2002,** *9,* 127-137.
30. Wang, L., Zhang, P., Zhang, Y., Li, Z., Meng, X., and Liu, C.; *Dalian Qinggongye Xueyuan Xuebao.* **2001,** *20,* 177-179.
31. Mustranta, A., Buchert, J., Spetz, P., and Holmbom, B.; *Nordic Pulp & Paper Research Journal.* **2001,** *16,* 125-129.
32. Zhang, X., Stebbing, D., Saddler, J.N., Beatson, R.P., and Kruus, K.; *Journal of Wood Chemistry and Technology.* **2000,** *20,* 321-335.
33. Chen, J., Huang, R., Chen, J., Lu, S., and Chen, B.; *Fujian Shifan Daxue Xuebao.* **2000,** *16,* 94-96.
34. Zhang, X.; *Pulp & Paper Canada.* **2000,** *101,* 59-62.
35. Damiano, V.B., Bocchini, D.A., Gomes, E., and Da Silva, R.; *World Journal of Microbiology & Biotechnology.* **2003,** *19,* 139-144.
36. Liu, L., Liu, H., Li, H., Li, C., Pi, S., and Li, Y.; *Zhongguo Zaozhi Xuebao.* **2002,** *17,* 37-40.
37. Roncero, M.B., Torres, A.L., Colom, J.F., and Vidal, T.; *Bioresource Technology.* **2003,** *87,* 315-323.
38. Stafford, M., Genco, J.M., and Ghanem, A.; *TAPPI Pulping/Process and Product Quality Conference Proceedings.* Boston, MA. 2000, 415-422.
39. Wong, K.K.Y., Allison, R.W., and Spehr, S.; *Journal of Pulp and Paper Science.* **2001,** *27,* 229-234.

40. Yamaguchi, H. and Iwasaki, M. WO2000/328477, 2000.
41. Wong, K.K.Y., James, C.S., and Campion, S.H.; *Journal of Pulp and Paper Science*. **2000,** *26,* 377-383.
42. Kim, D.-H. and Paik, K.-H.; *Journal of Industrial and Engineering Chemistry (Seoul)*. **2000,** *6,* 194-200.
43. Shah, A.K., Cooper, D., Adolphson, R., and Eriksson, K.E.L.; *Journal of Pulp and Paper Science*. **2000,** *26,* 8-11.
44. Mansfield, S.D., Wong, K.K.Y., and Dickson, A.R.; *Wood and Fiber Science*. **2000,** *32,* 105-115.
45. Jeffries, T.W., Davis, M., Rosin, B., and Landucci, L.L.; *7th International Conference on Biotechnology in the Pulp and Paper Industry Proceedings*. Vancouver, B.C. 1998, C41-C43.
46. Buchert, J., Oksanen, T., Pere, J., Siika-Aho, M., Suurnakki, A., and Viikari, L.; *Trichoderma and Gliocladium*. **1998,** *2,* 343-363.
47. Nelson, P.J., Chin, C.W.J., Grover, S.G., and Ryynanen, H.; *International Symposium on Wood and Pulping Chemistry Proceedings*. Helsinki, Finland. 1995, 331-336.
48. Viikari, L., Suurnakki, A., Buchert, J., Hejnesson, A., Carlsson, G., Strom, G., and Westermark, U.; *International Symposium on Wood and Pulping Chemistry Proceedings*. Helsinki, Finland. 1995, 255-262.
49. De Jong, E., Wong, K.K.Y., and Saddler, J.N.; *Holzforschung*. **1997,** *51,* 19-26.
50. Suurnakki, A., Clark, T.A., Allison, R.W., Viikari, L., and Buchert, J.; *Tappi Journal*. **1996,** *79,* 111-117.
51. Farrell, R.L., Viikari, L., and Senior, D.; *Pulp Bleaching*. **1996,** 363-377.
52. Buchert, J., Tenkanen, M., Kantelinen, A., and Viikari, L.; *Bioresource Technology*. **1994,** *50,* 65-72.
53. Itoh, K., Uegaki, Y., Okazaki, K., and Tachibana, S.; *Sen'i Gakkaishi*. **1998,** *54,* 253-260.
54. Ander, P.; *Polpu, Chongi Gisul*. **1993,** *25,* 70-6.
55. Takano, M., Nishida, A., and Nakamura, M.; *Journal of Wood Science*. **2001,** *47,* 63-68.
56. Qin, W., Yu, H., Ai, Y., and Huang, X.; *Xianweisu Kexue Yu Jishu*. **1999,** *7,* 23-28.
57. Parra, C., Oses, R., Ruiz, J., Urtubia, I., Baeza, J., Freer, J., and Rodriguez, J.; *7th International Conference on Biotechnology in the Pulp and Paper Industry Proceedings*. Vancouver, B.C. 1998, A161-A164.
58. Wolfaardt, J.F., Bosman, J.L., Jacobs, A., Male, J.R., and Rabie, C.J.; *6th International Conference on Biotechnology in the Pulp and Paper Industry Proceedings*. Vienna. 1996, 211-216.
59. Lee, S.H., Murata, S., Kondo, R., Sakai, K., Nishida, T., and Takahara, Y.; *Mokuzai Gakkaishi*. **1993,** *39,* 1089-94.

60. Hunt, K.; *Pulp & Paper Canada.* **1978,** *79,* 65-8.
61. Copur, Y., Kiemle, D., Stipanovic, A., Koskinen, J., and Makkonen, H.; *Paperi ja Puu.* **2003,** *85,* 158-162.
62. Wolfaardt, F., Rabie, C.J., Taljaard, L., Jacobs, A., Male, J.R., and Singels, A.; *TAPPI Pulping/Process and Product Quality Conference Proceedings.* Boston, MA. 2000, 624-632.
63. Bajpai, P., Bajpai, P.K., Akhtar, M., and Jauhari, M.B.; *Journal of Pulp and Paper Science.* **2001,** *27,* 235-239.
64. Akhtar, M., Scott, G.M., Swaney, R.E., and Kirk, T.K.; *ACS Symposium Series Proceedings.* 1998, 15-26.
65. Ahmed, A., Scott, G.M., Akhtar, M., and Myers, G.C.; *North American Nonwood Fiber Symposium Proceedings.* Atlanta, GA. 1998, 231-238.
66. Ferraz, A., Guerra, A., Souza-Cruz, P.B., and Mendonca, R.; In *Progress in Biotechnology,* L. Viikari and R. Lantto, Editors. p. 73-80 (2002).
67. Wall, M.B., Stafford, G., Noel, Y., Fritz, A., Iverson, S., and Farrell, R.L.; *International Conference on Biotechnology in the Pulp and Paper Industry Proceedings.* Vienna, Austria. 1996, 205-210.
68. Hatakka, A., Maijala, P., Mettala, A., Hakala, T., Hauhio, L., and Ellmen, J.; In *Progress in Biotechnology,* L. Viikari and R. Lantto, Editors. p. 81-88 (2002).
69. Jacobs, C.J., Venditti, R.A., and Joyce, T.W.; *Tappi Journal.* **1998,** *81,* 143-147.
70. Jacobs-Young, C.J., Gustafson, R.R., and Heitmann, J.A.; *Paperi ja Puu.* **2000,** *82,* 114-119.
71. Chakar, F.S., Allison, L., Kim, D.-H., Ragauskas, A.J., and Elder, T.; *TAPPI Pulping/Process and Product Quality Conference Proceedings.* Boston, MA. 2000, 938-951.
72. Chakar, F.S. and Ragauskas, A.J.; In *Oxidative Delignification Chemistry,* D.S. Argyropoulos, Editor ACS Symposium Series. p. 444-455 (2001).
73. Chakar, F.S. and Ragauskas, A.J.; *Holzforschung.* **2000,** *54,* 647-653.
74. Chakar, F.S. and Ragauskas, A.J.; *Journal of Wood Chemistry and Technology.* **2000,** *20,* 169-184.
75. Xu, F.; *Biochemistry.* **1996,** *35,* 7608-7614.
76. Yaropolov, A.I., Skorobogat'ko, O.V., Vartanov, S.S., and Varfolomeyev, S.D.; *Applied Biochemistry and Biotechnology.* **1994,** *49,* 257-280.
77. Higuchi, T.; *Wood Science and Technology.* **1990,** *24,* 23-63.
78. Kawai, S., Umezawa, T., Shimada, M., and Higuchi, T.; *FEBS LETTERS.* **1988,** *236,* 309-311.
79. Kawai, S., Umezawa, T., and Higuchi, T.; *Archives of Biochemistry and Biophysics.* **1988,** *262,* 99-110.
80. Bourbonnais, R. and Paice, M.G.; *FEBS LETTERS.* **1992,** *267,* 99-102.

81. Bourbonnais, R., Paice, M.G., Reid, I.D., Lanthier, P., and Yaguchi, M.; *Applied and Environmental Microbiology*. **1995,** *61,* 1876-1880.
82. Call, H.-P. WO/9429425, 1994.
83. Amann, M.; *9th International Symposium on Wood and Pulping Chemistry Proceedings*. Montreal, Canada. 1997, F4-1thru F4-5.
84. Xu, F., Kulys, J.J., Duke, K., Li, K., Krikstopaitis, K., Deussen, H.-J.W., Abbate, E., Galinyte, V., and Schneider, P.; *Applied and Environmental Microbiology*. **2000,** *66,* 2052-2056.
85. Bourbonnais, R., Leech, D., and Paice, M.G.; *Biochimica et Biophysica Acta*. **1998,** *1379,* 381-390.
86. Sealey, J., Ragauskas, A.J., and Elder, T.; *Holzforschung*. **1999,** *53,* 498-502.
87. Sealey, J. and Ragauskas, A.J.; *Journal of Wood Chemistry and Technology*. **1998,** *18,* 403-416.
88. Sealey, J. and Ragauskas, A.J.; *Enzyme and Microbial Technology*. **1998,** *23,* 422-426.
89. Poppius-Levlin, K., Wang, W., Tamminen, T., Hortling, B., Viikari, L., and Niku-Paavola, M.-L.; *Journal of Pulp and Paper Science*. **1999,** *25,* 90-94.
90. Poppius-Levlin, K., Wang, W., Ranua, M., Niku-Paavola, M.-L., and Viikari, L.; *TAPPI Pulping Conference Proceedings*. San Francisco, CA. 1997, 327-333.
91. Potthast, A., Rosenau, T., Koch, H., and Fischer, K.; *Holzforschung*. **1999,** *53,* 175-180.
92. Potthast, A., Koch, H., and Fischer, K.; *9th International Symposium on Wood and Pulping Chemistry Proceedings*. Montreal, Canada. 1997, F-2-1 - F-2-4.
93. Balakshin, M.Y., Chen, C.-L., Gratzl, J.S., Kirkman, A.G., and Jakob, H.; *Holzforschung*. **2000,** *54,* 171-175.
94. Balakshin, M.Y., Chen, C.L., Gratzl, J., Kirkman, A., and Jakob, H.; *Holzforschung*. **2000,** *54,* 165-170.
95. Crestini, C. and Argyropoulos, D.S.; *Bioorganic & Medicinal Chemistry*. **1998,** *6,* 2161-2169.
96. Call, H.P. and Mucke, I.; *Journal of Biotechnology*. **1997,** *53,* 163-202.
97. Bourbonnais, R. and Paice, M.G.; *Tappi Journal*. **1996,** *79,* 199-204.
98. Paice, M.G., Bourbonnais, R., and Reid, I.D.; *Tappi Journal*. **1995,** *78,* 161-169.
99. Chakar, F.S. *Institute of Paper Science and Technology*. Dissertation. Atlanta. 2001.
100. Froass, P.M., Ragauskas, A.J., and Jiang, J.E.; *Holzforschung*. **1998,** *52,* 385-390.
101. Gellerstedt, G. and Lindfors, E.-L.; *International Pulp Bleaching Conference Proceedings*. Stockholm, Sweden. 1991, 73-88.

102. Robert, D.; In *Methods in Lignin Chemistry,* S.Y. Lin and C.W. Dence, Editors. Berlin, Heidelberg: Springer-Verlag. p. 251-273 (1992).
103. Nimz, H.M., Nemr, M., Schmidt, P., Margot, C., Schaub, B., and Schlosser, M.; *Journal of Wood Chemistry and Technology.* **1982,** *2,* 371-382.
104. Kringstad, K. and Morck, R.; *Holzforschung.* **1983,** *37,* 237-244.
105. Gellerstedt, G. and Lindfors, E.-L.; *Nordic Pulp & Paper Research Journal.* **1987,** *2,* 71-75.
106. Jacobs-Young, C., Heitmann, J.A., and Venditti, R.A.; *AIChE Symposium Series Proceedings.* 1998, 1-15.
107. Jacobs-Young, C.J., Venditti, R.A., and Joyce, T.W.; *ACS Symposium Series Proceedings.* 1998, 55-65.
108. Jacobs-Young, C.J., Venditti, R.A., and Joyce, T.W.; *Tappi Journal.* **1998,** *81,* 261-266.
109. Paice, M.G., Bourbonnais, R., Reid, I.D., Archibald, F.S., and Jurasek, L.; *Journal of Pulp and Paper Science.* **1995,** *21,* J280-J284.
110. Lund, M. and Felby, C.; *Enzyme and Microbial Technology.* **2001,** *28,* 760-765.
111. Kocurek, M.J.; In *Pulp and Paper Manufacture.* ed. T.M. Grace and E.W. Malcolm. Joint Textbook Committee of the Pulp and Paper Industry: 1989; Vol. 5. pp
112. Gellerstedt, G. and Gustafsson, K.; *Journal of Wood Chemistry and Technology.* **1987,** *16,* 347-365.

Chapter 21

Biocatalytic Conversion of Renewable Feedstocks to Industrial Chemicals

Manoj Kumar, Jeff Pucci, Gopal Chotani, and Karl Sanford

Process Science Department, Genencor International Inc., 925 Page Mill Road, Palo Alto, CA 94304

Biocatalytic systems to economically produce biochemicals, using less expensive feedstocks like cellulose are an attractive alternative to petroleum based chemical synthesis. This is due, in part, to significant advances in our understanding and application of genetic and metabolic engineering of host organisms and biomass hydrolyzing enzymes. This chapter explores an enzymatic process for conversion of lignocellulose-based feedstocks to glucose and the concomitant conversion of glucose by microorganisms and/or enzymes to desired chemical products. In this process, glucose conversion to products parallels its formation by enzymatic catalysis. This reduces enzyme inhibition and speeds the conversion of cellulose to products. Furthermore, such a process provides key glucose-controlled biocatalytic conditions, which are critical for many fermentative processes.

Introduction

Industrial fermentations are used for manufacturing of bioproducts[1a]. They predominantly use glucose as a feedstock for the production of proteins, enzymes and chemicals. These fermentations can be batch, fed-batch, or continuous, and operate under controlled-substrate feeding and conditions that form minimal byproducts. Substrate-controlled feeding, pH, dissolved oxygen, and medium composition are critical operating conditions that must be controlled during fermentation in order to optimize fermentation time, yield and efficiency. Currently used methods and feedstocks, however, have drawbacks that reduce the efficiency of the fermentation processes.

Glucose is useful as a starting substrate in a multitude of chemical and biological synthetic applications. However, the use of purified glucose for fermentation processes is disadvantageous for several reasons. Glucose syrups of purity levels greater than 90% are relatively expensive. In addition, the presence of even moderate glucose concentrations increases the susceptibility of the fermentation system to microbial contamination, thereby resulting in an adverse effect upon the production efficiency. Another disadvantage is that even the presence of low to moderate levels of glucose in the fermentation vat adversely affects its conversion to the desired end product due to enzymatic inhibition and/or catabolite repression[1b]. Various attempts have been made to reduce the costs of industrial fermentation by utilizing less expensive substrates than glucose[2]. However, despite numerous approaches, there remains a need to develop methods to more efficiently produce desired chemicals and biochemicals from economical, non-glucose substrates[3].

Renewable feedstocks[4] such as corn-stover, corn-fiber and rice-straw are comprised of three major components:[5] cellulose (35-45%) hemicellulose (25-40%) lignin (5-30%). These feedstocks are naturally recalcitrant to enzyme degradation and need pretreatment[6] to allow enzymatic digestibility. Two known pretreatment processes currently used are steam & dilute acid hydrolysis[7] and wet alkaline oxidation[7]. During the pretreatment process, part of the major components of biomass feedstocks gets converted to undesired byproducts that are in general inhibitory to the the growth of microorganisms and thus can't be used directly in fermentative settings[8]. For example, lignins generate phenolic monomers, cellulosic components get converted to furans, and hemicellulose forms carboxylic acids[8].

Enzymes responsible for the hydrolysis of cellulose fractions of lignocellulosics to glucose perform sub-optimally mainly due to product and substrate inhibition[9,10,11]. Therefore, keeping cellulose available to cellulases and product concentration to minimal levels can help maintain the stability and activity of cellulosic enzymes.

Genencor's proprietary concept[12] of continuous biocatalytic systems using sequential enzyme reactions for processing the cellulosic component of biomass to biochemicals minimizes these problems. This concept addresses issues related to *i)* substrate inhibition, *ii)* enzyme inactivation, *iii)* cofactor instability, *iv)* intermediate inhibition, and *v)* mass transfer limitations and thus overcomes key existing limitations for biomass conversion to industrial chemicals[13].

Advantages of Genencor's biocatalytic conversion of biomass[12] to value-added chemicals include, *a)* commercial viability, simplicity and economic feasibility; *b)* prevention of product inhibition of cellulosic enzymes by concurrent conversion to bioproducts; *c)* feasibility of quantitative conversion, *d)* elimination of byproducts; *e)* higher productivity and yield on carbon; *f)* production capacity enhancement. This biocatalytic conversion concept is novel because a multienzyme process for converting renewable biomass to value-added commercial ingredients has not yet been commercially demonstrated.

Various industrial chemicals such as organic acids (gluconic, citric, succinic, acetic, ascorbic, lactic, 3-hydroxy propionic acid), solvents (acetone, ethanol, glycerol, butanol, 1,3-propanediol), aminoacids (glutamate, lysine, methionine), antibiotics (penicillin), and industrial enzymes can be made from fermentable sugars derived from lignocellulosics using Genencor's biocatalytic systems. In this chapter, we will illustrate a case study of gluconic acid production from cellulose using a biocatalytic system and compare it with a current manufacturing process[14,15]. We will provide data to support the use of this method for conversion of biomass derived feedstocks to other industrially relevant biochemicals.

In-vitro Biocatalytic Conversion of Lignocellulosic Feed-stock to Gluconic Acid

Today, gluconic acid is one of the major industrial biochemicals produced using fermentation technology[14]. Both, sodium gluconate and gluconic acid are used in various applications such as dairy, food, pharmaceutical, cleaning, textiles, cement, and metallurgy. The current gluconate and gluconic acid markets are substantial (Table 1) and are growing rapidly. The key producers of gluconic acid at present include ADM, Fuzisawa, Akzo, Roquette, Glucona, and Jungbunzlauer.

Chemical	Volume (M T/yr)	Price ($$/Kg)	Market (MM$/yr)
Sodium Gluconate	60,000	1.50	90
Glucono-δ-lactone	15,000	3.50	53

Table 1: Gluconic acid and glucono-δ-lactone market data (Chemical Market Reporter, 1998, July 20)

Current Gluconic Acid Processes: Several processes for the production of gluconic acid by chemical and biochemical means are reported in literature[15,17]. Chemically, gluconic acid is produced by the oxidation of the aldehyde group in glucose and the subsequent reaction of the acid with an appropriate base to make gluconate. The current biochemical manufacturing process for gluconic acid is by *Aspergillus niger* fermentation of glucose[15]. The formation of gluconic acid by this process is described in the following equation.

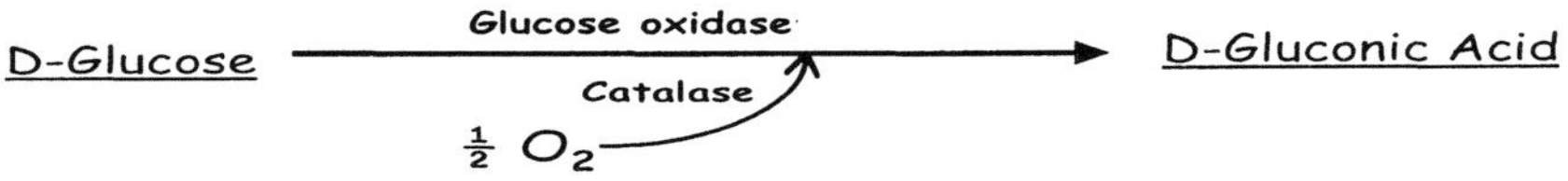

The *Aspergillus* process requires a high concentration of glucose (>25 wt%), high aeration rates and dissolved oxygen, the latter achieved by applying up to four bar of air pressure. This viable whole cell process can have a maximum of 30% solids in sugar content as further increase in sugar inhibits the cellular metabolism. An alternative to this whole cell process would be to consider a patented[16] but not practiced technology of using enzymes and glucose. However, this method is not competitive with the whole cell process due to its economic need for using 30-60% glucose, concentrations at which glucose is inhibitory to enzymes. Using a dilute glucose feed would yield a dilute gluconic acid product stream, which would be economically nonviable to recover solid gluconate. Thus, this patented[16] but not practiced enzymatic process not only requires high enzyme dosage but results in high viscosity of the reaction medium due to increased sugar concentration, thereby reducing the rate of oxygen transfer.

New Gluconic Acid Process: The biomass based gluconic acid production process[12] can successfully compete with the current *A. niger* gluconic acid technology for several reasons:

1) This process needs significantly lower dosage of enzymes,

2) The steady state concentration of glucose is low in the bioreactor, resulting in reduced substrate and/or product-based inhibition of enzymes,
3) The cost of raw materials is significantly less; biomass vs. D-glucose,
4) There is control over feed stock concentration, and thus no oxygen transfer limitations exist,
5) Better carbon efficiency as putting *A. niger* cell mass in place in current process uses 20% of total glucose,
6) Waste disposal issues do not exist and
7) The down-stream recovery process is simpler.

Key advantages of this simple and economical *in-vitro* biocatalytic process for biomass conversion to gluconate are: a) prevention of product inhibition of cellulosic enzymes by concurrent conversion to gluconate, b) feasibility of quantitative conversion c) elimination of byproduct formation, d) higher productivity, e) increased production capacity, and f) higher yield on carbon.

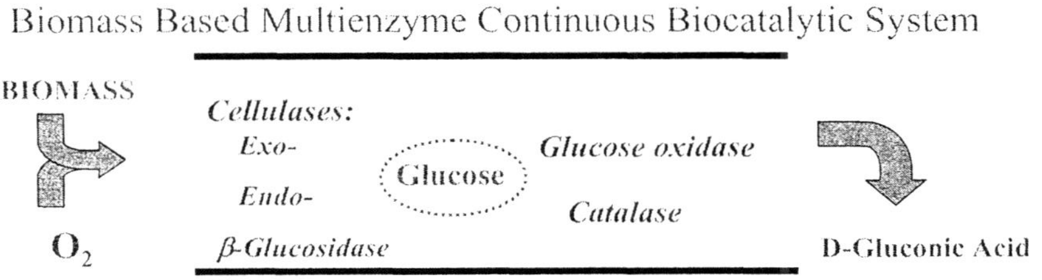

Figure 1: Genencor's concept for production of gluconic acid from biomass

Experimental Methods: Enzymes used in this work were assayed for activity using methods adopted form Bergmeyer HU et al (1983) in: Methods of enzymatic analysis, (Bergmeyer, HU eds.) 3rd edition. All products and substrates were measured in a reaction using HPLC and quantitation of products was referenced to a concentration based calibration plot that was generated. Glucose concentration was assayed using both HPLC and Monarch glucose analyzer. Before analysis samples for the enzymatic reaction were appropriately diluted (10-100 fold) to meet the linear range of the calibration curve generated. All the experiments were performed in 1-liter laboratory fermenters (Applicon) equipped with air supply system, stirring, pH control, temperature control, foam control, and dissolved oxygen probe. Unless otherwise mentioned, all the experiments were done using the following parameters: temperature 45^{0}C, pH 5 (buffered using 50 mM citrate), agitation between 600-900 rpm to maintain dissolved oxygen above 30% saturation, working volume 300 ml, pH control using 10% sodium hydroxide solution. In the calculation of the conversion, the dilution effect of adding sodium hydroxide solution was taken into account. Enzyme activities of glucose oxidase, catalase and cellulose are expressed in Titrimetric Units, Baker Units, and Buffered Filter Paper Units. One Titrimetric unit will oxidize 3.0 mg of glucose to gluconic acid in 15 minutes under assay conditions of 35^{0}C at pH5.1. One Baker unit of catalase decomposes 264 mg of hydrogen peroxide in one-hour under assay conditions of 25^{0}C at pH 7. One Buffered Filter Paper Unit (BFPU) of Spezyme CP liberates one micromole of reducing sugar (expressed as glucose equivalent) in one minute under assay conditions of 50^{0}C at pH 5.

Experimental Results: Cellulose (Avicel; 30 g 10 wt%) slurry was prepared in 270 g of 50 mM citrate buffer pH 5.0 and its conversion to glucose was started by adding 10 ml (1000 BFPU; dosed at 30 mgs of total protein per gram of cellulose) of Genencor enzyme Spezyme CP. The degree of hydrolysis was measured over the course of reaction (Figure 2). Preliminary experiments[12] illustrated excellent conversion of cellulose derived from biomass (such as Avicel and acid-pretreated corn-stover (NREL)) to gluconate using this enzymatic system. Glucose and cellobiose were produced and converted to the final product thus preventing their accumulation, which is inhibitory to the cellulolytic enzymes[18]. This method thus provides an attractive alternative for production of gluconate from cellulose when compared to direct conversion[16] of glucose to gluconic acid using glucose oxidase and catalase at the same enzyme dosage level.

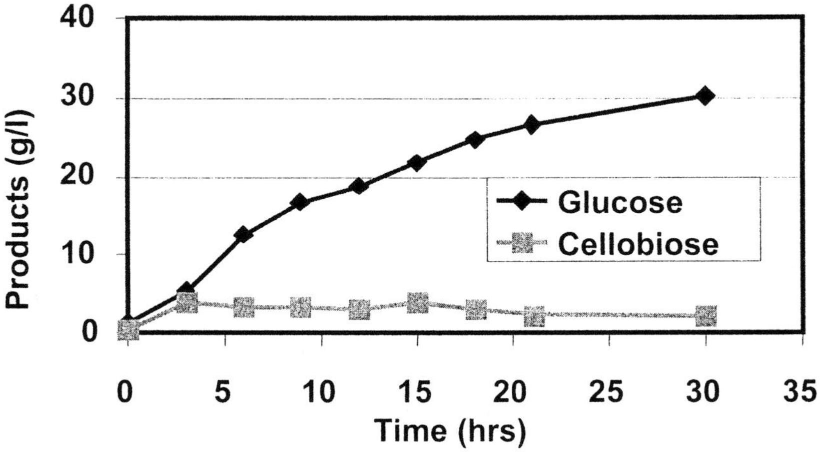

Figure 2: In vitro Biocatalytic Conversion of Cellulose (Avicel, Lattice 20) with Cellulase (Spezyme) to Glucose

In the next experiment, cellulose (Avicel; 30 g 10 wt%) slurry prepared in 270 g of 50 mM citrate buffer pH 5.0 was mixed with 1.5 ml (2250 Titr Units) of OxyGO 1500^{TM} (Genencor glucose oxidase) and 2 ml (2000 Baker units) of Fermcolase 1000^{TM} (Genencor catalase) along with 10 ml of Genencor Spezyme CP (1000 BFPU), which converted cellulose to gluconic acid (Figure 3) at an improved rate compared to the rate of the production of glucose from cellulose in the control experiment (Figure 2). The steady-state concentration of glucose in the reaction fell below detection limits. These results established that in the same period of time where 30g/l glucose was produced from Avicel in the control experiment (Figure 2), over 50 g/l gluconic acid was produced from Avicel using the enzyme blend of OxyGO/Fermcolase/Spezyme (Figures 3, 4). In 48 hrs, we were able to convert 60% of technical grade Avicel (Lattice 20) to gluconate. It was noticed that by keeping the glucose and cellobiose concentration to a minimum, we could keep the cellulose hydrolyzing enzymes stable during the reaction (Figure 4). Using acid pre-treated and washed corn-stover (NREL, 30 g, 10 wt% slurry) under identical conditions as in Figure 3, 70% of cellulosic content of corn-stover was converted to gluconate (Figure 5).

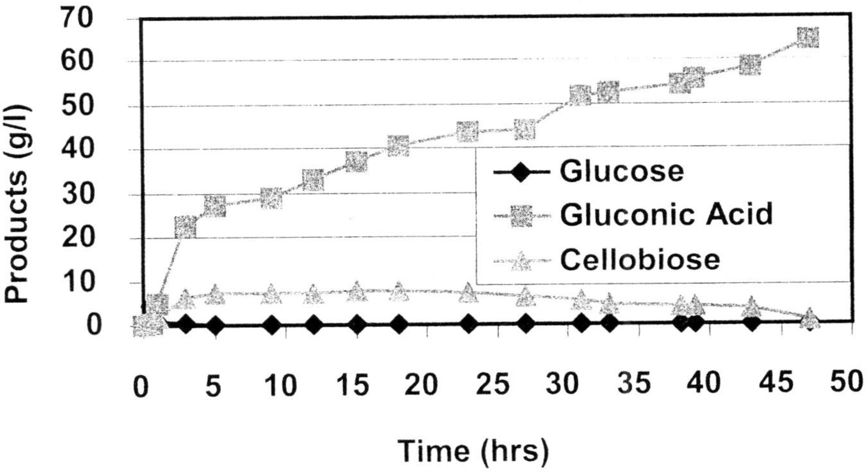

Figure 3: In vitro Biocatalytic Conversion of Cellulose (Avicel, Lattice 20) with Cellulase (Spezyme), Glucose Oxidase (OxyGO), and Catalase (Fermcolase) to Gluconic Acid

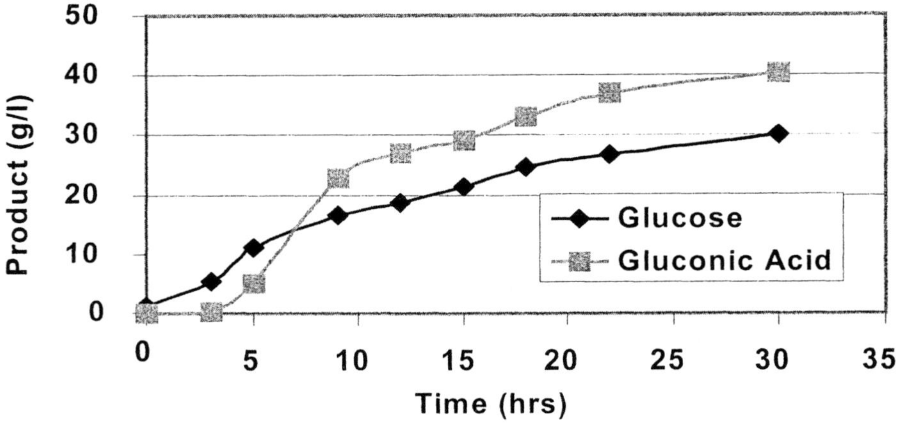

Figure 4: Comparative analysis of glucose and gluconic acid production from cellulose demonstrating inhibition of cellulolytic enzymes by cellulose hydrolysis products

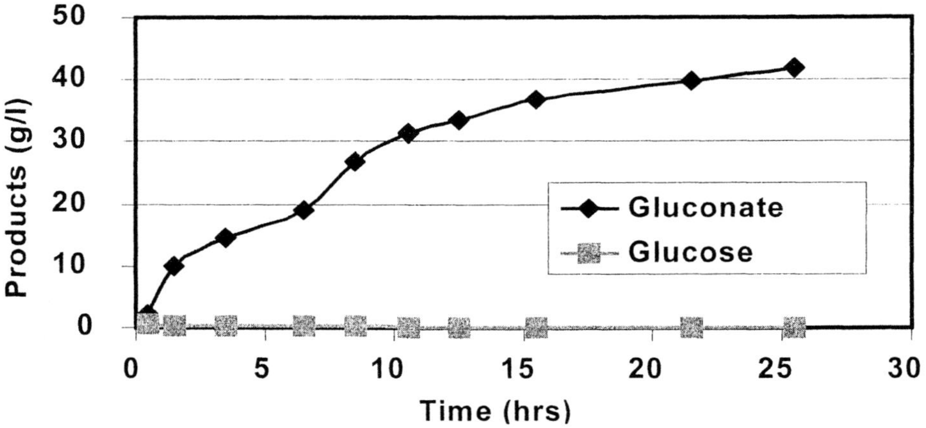

Figure 5: In vitro biocatalytic conversion of acid-pretreated corn-stover to gluconic acid with Spezyme, OxyGO and Fermcolase.

These results clearly demonstrate that the rate of gluconate production from cellulose was improved over that of glucose from cellulose when a biocatalytic system was used. The effective glucose concentration in the bioreactor remained near zero alleviating inhibition of the cellulolytic enzymes. All the enzymes remained stable during this biocatalytic process. These experiments showed at least 60% cellulose conversion to gluconate in a 48 hr period using pure cellulose and conversion of 70% available cellulose from corn-stover to gluconate in 24 hr. One of the key factors for successful implementation of this technology commercially is making sure all enzymes needed for cellulose conversion to gluconate remain stable. Hydrogen peroxide, produced as a byproduct in the conversion of glucose to gluconate, is an oxidative irreversible inactivator of biocatalysts. In the biocatalytic system, since the glucose concentration in the reaction system is effectively zero, hydrogen peroxide buildup does not occur. It is thus possible to take these bioconversion reactions to theoretically maximal yields by preventing inactivation of enzymes by H_2O_2.

Fermentation of Biomass to chemicals

A biocatalytic system for converting biomass to industrial chemicals is not only applicable to enzymatic conversions but also to fermentative conversion using cellulose[19]. We report here three examples of fermentative conversion of cellulose to chemicals namely 1,3 propanediol, lactic acid, and succinic acid.

Experimental Methods: Experiments to convert cellulose to 1,3-propanediol[20] and succinic acid were performed at $34^{0}C$ and pH 6.7, and for conversion to lactic acid were done at $34^{0}C$ and pH 6.4. Cellulose (technical grade, AVICEL® Lattice 20) was used in a slurry form (equivalent to 100 g/L glucose) following sterilization at 121°C for 30 min. Desired enzymes and requirements specific for 1,3-propanediol production (20 mg spectinomycin and 1 mg vitamin B12) were added as 0.2 micron filtered solutions in DI water. TM2 fermentation medium consists of following components: Potassium dihydrogen phosphate 13.6 g/l, dipotassium hydrogen phosphate 13.6 g/l, magnesium sulfate hexahydrate 2 g/l, citric acid monohydrate 2 g/l, ferric ammonium citrate 0.3 g/l, ammonium sulfate 3.2 g/l, and yeast extract 5 g/l.

Experimental Results: For control experiments, 20% cellulose slurry was made in a 2L flask by combining it with 200 mL of TM2 medium. Following addition of 13 ml (1300 BFPU) SPEZYME® CP (Genencor) and 20 mg spectinomycin and 1 mg vitamin B12, samples were taken from the vessel at varying reaction times, centrifuged, and the supernatants frozen to terminate enzyme action. The supernatants were subjected to HPLC analysis. This experiment monitored degradation of cellulose by measuring glucose formation. It was determined that 12.19 g/l glucose accumulated in 98.7 hours. Conversion of cellulose to glucose was demonstrated at a rate of 0.12 g/l-hour at 34°C and pH 6.7 (data not shown).

For measurement of 1, 3-propanediol production, cellulose slurry was prepared in a minimal fermentation medium in a 1L bioreactor. Following addition of 13 ml (1300 BFPU) SPEZYME® CP (Genencor) and 20 mg spectinomycin and 1 mg vitamin B12, 60 mls of OD 1.2 measuring (at 550nm) inoculum of 1,3-propanediol-producing *E. coli* strain TTaldABml/p109f1 WS#2 (Genencor International internal culture collection), taken from a frozen vial and prepared in soytone-yeast extract-glucose medium (Difco), was added to the bioreactor. Samples taken from the reaction vessel at varying times were centrifuged and the supernatants frozen to terminate the enzyme action. The supernatants were subjected to HPLC analysis. This experiment monitored fermentation of cellulose to 1,3-propanediol by measuring glucose formation and its conversion to glycerol (1,3-propanediol pathway intermediate) and then to 1,3-propanediol. In 24.4 hours, the accumulation of glycerol and 1,3-propanediol amounted to 1.02 and 4.73 g/l, respectively (Figure 6). The fermentative bioconversion of biomass to glycerol and 1,3-propanediol occurred at a rate of 0.24 g/l-hour at 34°C and pH 6.7. The rate of conversion of cellulose to 1,3-propanediol was twice that of conversion of cellulose to glucose under identical conditions. This further illustrates the issue of product inhibition and stability of cellulosic enzymes that is remedied by concomitant conversion of cellulose hydrolysis products to relevant chemicals.

To further assess the utility of the biocatalytic system, cellulose conversion to lactic acid[21] and succinic acid[22] was also examined.

Cellulose fermentation to lactic acid was monitored by measuring glucose formation from cellulose using SPEZYME® CP (Genencor) and its subsequent conversion to lactate using the lactate producing strain *Lactobacillus casei*. Cellulose in a slurry form was mixed in *Lactobacilli* MRS medium (Difco), and 22ml of the enzyme was added to it. 46ml of OD 24.2 measuring (at 550nm) inoculum of the lactate producing strain of *Lactobacillus casei* (ATCC 393), grown at 34°C with nitrogen sparge at 0.6 slpm was added to the bioreactor. Samples taken from the reaction vessel at various times were centrifuged, and supernatants were frozen to terminate the enzyme action. The supernatants were subjected to HPLC analysis. This experiment monitored conversion of cellulose to lactate by measuring glucose formation and its conversion to lactate. In 48 hours, accumulation of lactate amounted to 3.93 g/l (Figure 7).

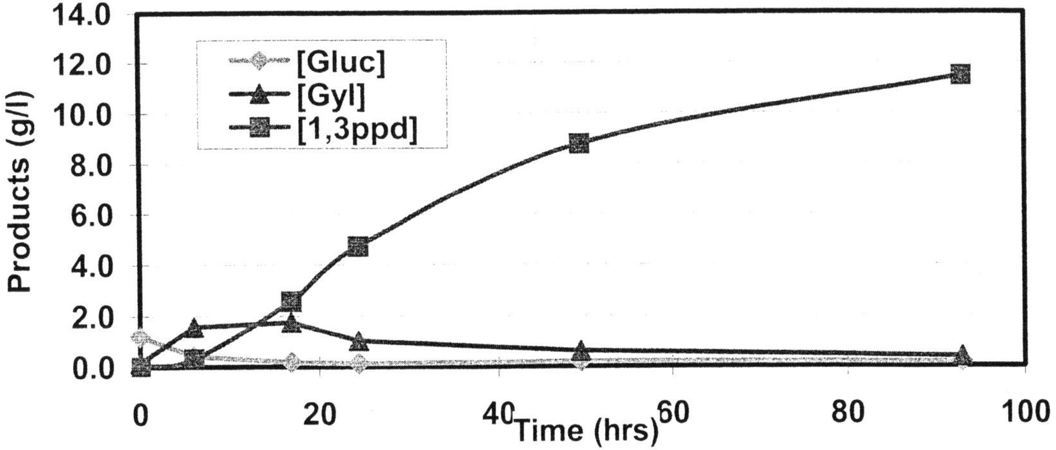

Figure 6: Biocatalytic conversion and concomitant fermentation of cellulose to 1, 3-Propanediol

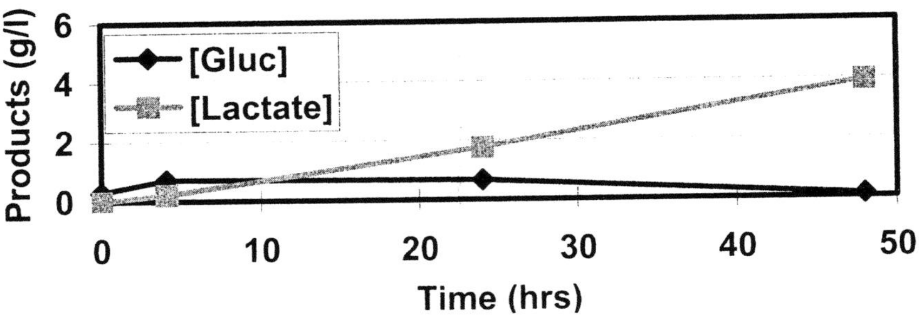

Figure 7: Biocatalytic conversion and concomitant fermentation of cellulose to lactic acid

Fermentative conversion of cellulose to succinic acid was carried out in a 1L bioreactor by measuring glucose formation from cellulose using Spezyme CP (Genencor) and its subsequent conversion to succinate, using a succinate producing *E. coli* strain, 36 1.6 ppc (Genencor International internal culture collection). Cellulose slurry was prepared in the minimal fermentation medium, and following enzyme addition (22ml), 60ml of OD 9.3 measuring inoculum of succinate-producing strain 36 1.6ppc *E. coli*, grown in TM2 + 10g/L glucose medium from a frozen vial under nitrogen sprage at 0.6 slpm was added to the bioreactor. Samples from the reaction vessel, taken at various times, were centrifuged and supernatants frozen. The supernatant was subjected to HPLC analysis. This experiment monitored fermentative conversion of cellulose to succinate by measuring glucose formation and its conversion to succinate (Figure 8). In 48 hours, accumulation of succinate amounted to 2.73 g/l.

Summary

Bioconversion of renewable feedstocks to industrial chemicals has been described. In this chapter, we have illustrated technology that Genencor International has developed to harness biomass as carbon feedstocks for conversion to industrial products and to make available its bioengineered enzymes to convert biomass into fermentable sugars. This research effort provides a means for the production of desired bioproducts by enzymatic conversion of biomass-based feedstock substrates. This concept of using cellulosic biomass for manufacturing industrial chemicals has several incentives that can be explored and implemented.

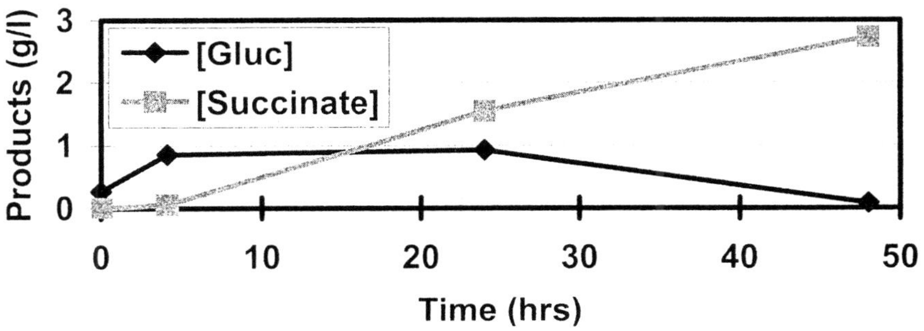

Figure 8: Biocatalytic conversion and fermentation of cellulose to succinic acid

Acknowledgements

Authors wish to acknowledge Dr. Roopa Ghirnikar for editorial help in preparing this manuscript and Dr. Kamrin MacKnight in intellectual property submission.

References:

1. a) Chotani G, Dodge T, Hsu A., Kumar M, LaDuca R, Trimbur D, Weyler W, Sanford K; Biochemica Biophysica Acta, **2000**, 1543, 434-455. b) Dien BS, Nicols NN, and Bothast RJ; J Ind. Microbiol. Biotechnol. **2002**, 29, 221-227
2. Linko P, Biotechnol Adv. , **1985**, 3, 39-63.
3. Olsen ES, Aulich TR, Sharma R, Timpe RC; Appl Biochem Biotechnol. **2003**, 105-108, 843-514.
4. CE Wyman; Biotechnology Prog. **2003**, 19, 254-262.
5. McMillan JD, *et al.* ACS Symposium Series 566, page 8.
6. JP Weil, K. Westgate, K. Kohlman, and MR Ladisch, Cellulose Enzyme and Microbial Technology **1994**, 16, 1002-1112.
7. Schell DJ, Farmer J, Neuman M; McMillan JD; Appl Biochem Biotechnol. **2003,** 105-108, 69-85.
8. Klinke HB, Ahring BK, Schmidt AS, Thomsen AB, Bioresour Technol. **2002,** 82, 15-26.
9. Zaldivar J, Martinez A, Ingram LO Biotechnol Bioeng. **2000**, 68, 524-30.
9. Wald S, Wilke CR, Blanch HW, Biotechnol Bioeng. **1984**, 26, 221-30
10. Desai SG, Converse AO, Biotechnol. Bioeng. **1997**, 56, 650-655.

11. Todorovic R, Grujic S, Matavulj M, Microbios Letters **1987**, 36, 113-119.
12. Methods for producing end Products form a carbon substrate; Kumar M, Pucci J, Chotani GK, Sanford KJ Serial Number of US and WTO Patent Application **2003**, US03/03532.
13. Schugerl K, Adv Biochem Eng Biotechnol. **2000**, 70, 41-76.
14. Fong WS, Stanford Research Institute, **1975,** Report No. 95.
15. Milsom PE and Meers JL, Comprehensive biotechnology **1985**, 3, 681-702.
16. Enzymatic production of gluconic acid or its salts Vroemen AJ and Beverini M US Patent 1999, US005897995.
17. Anastassiadis S, Aivasidis A, Wandrey C; Continuous gluconic acid production by isolated yeast-like mould strains of Aureobasidium pullulans. Appl Microbiol Biotechnol. **2003**, 61, 110-7 and references there in.
18. Valjamae P, Petterson G, and Johansson G, Eur. J. Biochem. **2001**, 268, 4520-4526.
19. a) Sanford K, Soucaille P, Gregg, Chotani GK; Current Opinion in Microbiology, **2002**, 5, 318-322; b) http://www.genencor.com
20. Zeng AP and Biebl H. Adv. Biochem. Eng. Biotechnol. **2002** 74, 239-262.
21. Lawford HG, Rousseau JD, Appl Biochem Biotechnol. **2002**, 98-100, 429-48.
22. Barnes SP, Keller J, Water Sci Technol. **2003**, 48, 155-6

Indexes

Author Index

Subject Index

D

G

H

N

O

P

R

T

U

V

W

X

Y

Z